# ELEMENTARY THEORY
## OF
## STRUCTURES

*Prentice-Hall International Series in Dynamics*
Y. C. FUNG, *editor*

# ELEMENTARY THEORY
# OF
# STRUCTURES

## Second Edition

### YUAN-YU HSIEH

Professor of Civil Engineering
National Taiwan University

PRENTICE-HALL, INC. *Englewood Cliffs, New Jersey* 07632

*Library of Congress Cataloging in Publication Data*

HSIEH, YUAN-YU, 1922–
　　Elementary theory of structures.

　　(Prentice-Hall international series in
dynamics)
　　Includes index.
　　1. Structures, Theory of.　I. Title.
II. Series.
TA645.H75　1982　　　624.1'7　　　81-11980
ISBN　0-13-261545-2　　　　　　AACR2

Editorial/production supervision
　and interior design by *Daniela Lodes*
Manufacturing buyer: *Joyce Levatino*

Printed in the United States of America
10　9　8　7　6　5　4　3

ISBN　0-13-261545-2

Prentice-Hall International, Inc., *London*
Prentice-Hall of Australia Pty. Limited, *Sydney*
Prentice-Hall of Canada, Ltd., *Toronto*
Prentice-Hall of India Private Limited, *New Delhi*
Prentice-Hall of Japan, Inc., *Tokyo*
Prentice-Hall of Southeast Asia Pte. Ltd., *Singapore*
Whitehall Books Limited, *Wellington, New Zealand*

# CONTENTS

# PREFACE

In preparing this new edition, the author has been guided by a number of criteria: the transition to the use of SI units; the growing role of matrix presentation in structural analysis; the usefulness of the material for teaching and study.

The arrangement of the material, for the most part, remains the same; classical methods of solution are accompanied by the modern ones. However, many sections have been revised and several chapters reorganized. These include the rewriting of the treatment of nonprismatic members and deleting of the matrix algebra for structural engineers. Over eighty interesting examples and problems have been added, most of which are expressed in SI units. There is a considerable amount of new material to reinforce the contents of matrix formulation of structural theories throughout the whole book.

The author wishes to express his appreciation to many university professors and students in various part of the world who have read, used, and supported this book during the past decade. Particular acknowledgement is due to Dr. Y. C. Fung, professor at U.C.S.D., for his constant encouragement throughout the writing of this volume; to Dr. Benjamin Koo, professor at University of Toledo, and Dr. Z. A. Lu of U. C. Berkeley for their kind interest in the development of the revision.

The author also wishes to take this opportunity to thank the National Taiwan University for a sabbatical leave. The leisure provided plus the Southern California's cozy sunshine made the writing of a book almost an enjoyable occupation.

Finally, the author is indebted to his wife, Nelly, who typed the manuscript.

Y. Y. Hsieh
Monterey Park, California

# ELEMENTARY THEORY
## OF
## STRUCTURES

# 1

# INTRODUCTION

## 1-1. ENGINEERING STRUCTURES

The word *structure* has various meanings. By an *engineering structure* we mean roughly something constructed or built. The principal structures of concern to civil engineers are bridges, buildings, walls, dams, towers, and shell structures. Structures as such are composed of one or more solid elements so arranged that the whole structures as well as their components are capable of holding themselves without appreciable geometric change during loading and unloading.

To design a structure involves many considerations among which are two major objectives that must be satisfied:

1. The structure must meet the performance requirement.
2. The structure must carry loads safely.

Consider, for example, the roof truss resting on columns shown in Fig. 1-1. The purposes of the roof truss and of the columns are, on the one hand, to hold in equilibrium their own weights, the load of roof covering, and the wind and snow (if any); and, on the other hand, to provide rooms for housing a family, for a manufacturing plant, or for other uses. In this book, we are concerned only with the load-carrying function of structures.

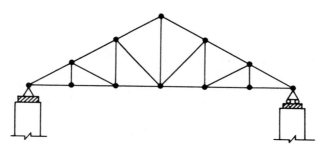

**Fig. 1-1**

## 1-2. THEORY OF STRUCTURES DEFINED

The complete design of a structure is outlined in the following stages:

1. *Developing a general layout.* The general layout of a structure is selected from many possible alternatives. The primary concern is the purpose for which the structure is to be built. This stage involves the choice of structural type, the selection of material, and a tentative estimation of cost based on a reasonable analysis of a preliminary structural design. It may also involve selecting the best location or adapting the structure to a site that has not been predetermined. Proper attention should also be given to the aesthetic value of the structure, since the requirement of good appearance can reasonably be met without much increase in cost. There are many other considerations, including the legal, financial, sociological, cultural, and environmental aspects. It is clear that this stage of design calls for an engineer with a high order of experience, skill, general knowledge, and imagination.

2. *Investigating the loads.* Before a refined structural analysis can be carried out, it is necessary to determine the loads for which a given structure should be designed. General information about the loads imposed on a structure is usually given in the specifications and codes. However, it is part of the designer's responsibility to specify the load conditions and to take care of exceptional cases.

The weight of the structure itself together with the material permanently attached to it is called *dead load* and is regarded as fixed in magnitude and location. Since the dead load must be approximated before the structure is designed, the preliminary data are only tentative. Revision must be made if the initial estimation is not satisfactory.

All loads other than dead load may be called *live loads.* Live loads are generally classified as movable loads and moving loads. *Movable loads* are loads that may be transported from one location to another on a structure without dynamic impact; for example, people, furniture, and goods on a building floor, or snow or ice on a roof. *Moving loads* are loads that move continuously over the structure, such as railway trains or tracks on a bridge, wind on a roof or wall, or hydrostatic pressure on an abutment. Moving loads may also be applied suddenly to the structure; for example, the centrifugal and longitudinal forces induced by the acceleration of vehicles and the dynamic forces generated by earthquakes.

In an ordinary structural design all loads are treated as static loads in order to simplify the analysis. In this way the impact due to a moving live load is expressed as a percentage of the live load, and the earthquake force is commonly considered to be a horizontal force equal to a fraction of the weight of a structure.

Other load considerations may include thermal effects and resistance to bomb blasting.

2

3. *Stress analysis.* Once the basic form of the structure and the external loads are defined, a structural analysis can be made to determine the internal forces in various members of the structure and the displacements at some controlling points. When live loads are involved, it is important to determine the maximum possible stresses in each member being considered. The principles governing this phase of design are usually discussed in the theory of structures.

4. *Selection of elements.* The selection of suitable sizes and shapes of members and their connections depends on the results of the stress analysis together with the design provisions of the specifications or codes. A trial-and-error approach may be used in the search for a proportioning of elements that will be both economical and adequate. A sound knowledge of strength of material and process of fabrication is also essential.

5. *Drawing and detailing.* Once the makeup of each part of the structure has been determined, the last stage of design can begin. This final stage includes the preparation of contract drawing, detailing, job specification, and final cost; this information is necessary for construction to proceed.

These five stages are interrelated and may be subdivided and modified. In many cases they must be carried out more or less simultaneously. The subject matter of the theory of structures is stress analysis with occasional reference to loadings. The emphasis of structural theory is usually on the fundamentals rather than on the details of design.

## 1-3. THEORIES OF STRUCTURES CLASSIFIED

Structural theories may be classified from various points of view. For convenience, we shall characterize them by the following aspects:

1. *Statics versus dynamics.* Ordinary structures are usually designed under static loads. Dead load and snow load are static loads that cause no dynamic effect on structures. Some live loads, such as trucks and locomotives moving on bridges, are also assumed as concentrated static load systems. They do cause impact on structures; however, the dynamic effects are treated as a fraction of the moving loads to simplify the design.

The particular, specialized branch that deals with the dynamic effects on structures of accelerated moving loads, earthquake loads, wind gusts, or bomb blasts is *structural dynamics.*

2. *Plane versus space.* No structure is really planar. However, structural analyses for beams, trussed bridges, or rigid frame buildings are usually treated as plane problems, although they are never two-dimensional in themselves. On the other hand, in some structures, such as towers and framing for domes, the stresses are interrelated between members not lying in a plane in such a way that the analysis cannot be simplified on the basis of component planar struc-

tures. Such structures must be considered as space frameworks under non-coplanar force system.

3. *Linear versus nonlinear structures.* Linear structure means that a linear relationship is assumed to exist between the applied loads and the resulting displacements in a structure. This assumption is based on the following conditions:

a. The material of the structure is elastic and obeys Hooke's law at all points and throughout the range of loading considered.

b. The changes in the geometry of the structure are so small that they can be neglected when the stresses are calculated.

Note that if the principle of superposition is to apply, a linear relationship must exist, or be assumed to exist, between loads and displacements. The principle of superposition states that the total effect at some point in a structure due to a number of causes (forces or diaplacements) acting simultaneously is equal to the sum of effects for the causes acting separately.

A nonlinear relationship between the applied actions and the resulting displacements exists under either of two conditions:

a. The material of the structure is inelastic.

b. The material is within the elastic range, but the geometry of the structure changes significantly during the application of loads.

The study of nonlinear behavior of structures includes *plastic analysis of structures* and *buckling of structures.*

4. *Statically determinate versus statically indeterminate structures.* The term *statically determinate structure* means that structural analysis can be carried out by statics alone. If this is not so, the structure is statically indeterminate.

A statically indeterminate structure is solved by the equations of statics together with the equations furnished by the geometry of the elastic curve of the structure in linear analysis. We note that the elastic deformations of the structure are not only associated with the applied loads on the structure but are also affected by the material properties (e.g., the modulus of elasticity $E$) and by the geometric properties of the member section (e.g., the cross-sectional area $A$ or the moment of inertia $I$). Thus, loads, material properties, and geometric properties are all involved in solving a statically indeterminate structure, while load factor alone dominates in a statically determinate problem.

5. *Force versus displacement.* The force and displacement are two categories of event occurring to a structure. The objective of a structural analysis is to determine forces and displacements pertaining to the structure, and their relationships specified by the geometric and material properties of structural elements. Structural analysis can then be divided into two categories: the force method and the displacement method. In the *force method*, we treat the forces as the basic unknowns and the displacements are expressed in terms of forces; whereas in the *displacement method*, we regard the displacements as the fundamental unknowns, and the forces are expressed in terms of displacements. In matrix analysis of linear structures, the force method is often referred to as the *flexibility method*, and the displacement method is called the *stiffness method.*

## 1-4. ACTUAL AND IDEAL STRUCTURES

All analyses are based on some assumptions not quite in accordance with the facts. It is impossible for an actual structure to correspond fully to the idealized structure on which the analysis is based. The materials of which the structure is built do not follow exactly the assumed properties, and the dimensions of the actual structure do not coincide with their theoretical values.

To illustrate, let us take a simple example. In designing a reinforced concrete beam of rectangular section, the values of $E$ and $I$ are usually assumed to be constant. However, the amount of reinforcing steel placed in the beam varies with the stresses; therefore, the values of $E$ and $I$ are not constant throughout the span. Besides, there is great uncertainty involved even in choosing a constant $E$ or $I$. Even without considering other factors, such as the supports, the connections, and the working dimensions of the structure, we find that the behavior of an actual structure often deviates from that of an idealized structure by a considerable amount. However, it does not follow from this that the results of analysis are not useful for practical purposes. We must set an idealized model in order to carry out practical analysis, and from practical analyses we make the idealization more and more consistent with actuality.

## 1-5. SCOPE OF THIS BOOK

Three major types of basic structures are thoroughly discussed in this book:

1. *Beam*
2. *Truss*
3. *Rigid frame*

A *beam*, in its narrow sense, is a straight member subjected only to transverse loads. A beam is completely analyzed when the values of bending moment and shear are determined.

A *truss* is composed of members connected by frictionless hinges or pins. The loads on a truss are assumed to be concentrated at the joints. Each member of a truss is considered as a two-force member subjected to axial forces only.

A *rigid frame* is built of members connected by rigid joints capable of resisting moment. Members of a rigid frame, in general, are subjected to bending moment, shear, and axial forces.

The present book is confined exclusively to the static, planar, and linear aspects of structure. The first seven chapters deal with the basic concepts of structure and an analysis of statically determinate structures, in which only forces are involved. The remaining chapters deal primarily with an analysis of statically indeterminate structures, with both forces and displacements involved.

5

# 2

## STABILITY AND DETERMINACY

## OF STRUCTURES

### 2-1. EQUATIONS OF EQUILIBRIUM
### FOR A COPLANAR FORCE SYSTEM

The first and major function of a structure is to carry loads. Beams, trusses, and rigid frames all have this element in common: Each sustains the burden of certain loads without showing appreciable distortions. In structural statics all force systems are assumed to act on rigid bodies. Actually, there are always some small deformations that may cause some small change of dimension in structure and a shifting of the action lines of the forces. However, such deviations are neglected in stress analysis.

A structure is said to be *in equilibrium* if, under the action of external forces, it remains at rest relative to the earth. Also, each part of the structure, if taken as a free body isolated from the whole, must be at rest relative to the earth under the action of the internal forces at the cut sections and of the external forces thereabout. If such is the case, the force system is balanced, or in equilibrium, which implies that the resultant of the force system (either a resultant force or a resultant couple) imposed on the structure, or segment thereof, must be zero.

Since this book is confined to planar structures, all the force systems are coplanar. The generally balanced coplanar force system must then satisfy the following three simultaneous equations:

$$\sum F_x = 0 \qquad \sum F_y = 0 \qquad \sum M_a = 0 \qquad (2\text{-}1)$$

where $\sum F_x$ = summation of $x$ component of each force in the system

$\qquad \sum F_y$ = summation of $y$ component of each force in the system

The subscripts $x$ and $y$ indicate two mutually perpendicular directions in the Cartesian coordinate system;

$\qquad \sum M_a$ = summation of moment about any point $a$ in the plane due to each force in the system

Note that $\sum F_x$ also represents the $x$ component of the resultant of the force system, $\sum F_y$ the $y$ component of the resultant of the force system, and $\sum M_a$ the moment about $a$ of the resultant of the force system.

The alternative to Eq. 2-1 may be given by

$$\sum F_y = 0 \qquad \sum M_a = 0 \qquad \sum M_b = 0 \qquad (2\text{-}2)$$

provided that the line through points $a$ and $b$ is not perpendicular to the $y$ axis, $a$ and $b$ being two arbitrarily chosen points and the $y$ axis being an arbitrarily chosen axis in the plane. Or

$$\sum M_a = 0 \qquad \sum M_b = 0 \qquad \sum M_c = 0 \qquad (2\text{-}3)$$

provided that points $a$, $b$, and $c$ are not collinear, $a$, $b$, and $c$ being three arbitrarily chosen points in the plane.

Explanation of Eq. 2-2 is as follows:

1. Let $R$ denote the resultant of the force system. Assume that $R \neq 0$. Since $\sum M_a = 0$ and $\sum M_b = 0$, the resultant $R$ cannot be a couple. It must be a force through $a$ and $b$ and by assumption is not perpendicular to the $y$ axis.

2. By $\sum F_y = 0$, we mean that the resultant has no $y$-axis component and must therefore be perpendicular to the $y$ axis.

The foregoing contradictory statements lead to the conclusion that the force is also zero. Therefore, Eq. 2-2 is the condition for $R = 0$.

A similar explanation is given here for Eq. 2-3:

1. Assume that $R \neq 0$. Since $\sum M_a = 0$, $\sum M_b = 0$, and $\sum M_c = 0$, the resultant $R$ cannot be a couple. It must be a force through $a$, $b$, and $c$.

2. But by assumption $a$, $b$, and $c$ are not collinear.

These statements lead us to conclude that the force is also zero. Therefore, Eq. 2-3 is the condition for $R = 0$.

Two special cases of the coplanar force system in equilibrium are worth noting:

1. *Concurrent forces.* If a system of coplanar, concurrent forces is in equilibrium, then the forces of the system must satisfy the following equations:

$$\sum F_x = 0 \qquad \sum F_y = 0 \qquad (2\text{-}4)$$

Another set of independent equations necessary and sufficient for the equilibrium of the forces of a coplanar, concurrent force system is

$$\sum F_y = 0 \qquad \sum M_a = 0 \qquad (2\text{-}5)$$

provided that point $a$ is not on the line through the concurrent point of forces and perpendicular to $y$ axis.

A third set of equations of equilibrium for a coplanar, concurrent force system is

$$\sum M_a = 0 \qquad \sum M_b = 0 \tag{2-6}$$

where $a$ and $b$ are any two points in the plane of the forces, provided that the line through $a$ and $b$ does not pass through the concurrent point of forces.

2. *Parallel forces.* If a coplanar, parallel force system is in equilibrium, the forces of the system must satisfy the equations

$$\sum F_y = 0 \qquad \sum M_a = 0 \tag{2-7}$$

where the $y$ axis is in the direction of the force system and $a$ is any point in the plane.

Another set of independent equations of equilibrium for a system of coplanar, parallel forces may be given as

$$\sum M_a = 0 \qquad \sum M_b = 0 \tag{2-8}$$

where $a$ and $b$ are any two points in the plane, provided that the line through $a$ and $b$ is not parallel to the forces of system.

There are two simple, special cases of equilibrium that deserve explicit mention:

1. *Two-force member.* In Fig. 2-1 is shown a body subjected to two external forces applied at $a$ and $b$. If the body is in equilibrium, then the two forces cannot be in random orientation, as shown in Fig. 2-1(a), but must be directed along $ab$, as shown in Fig. 2-1(b). Furthermore, they must be equal in magnitude and opposite in sense. This can be proved by using first the equations $\sum M_a = 0$ and $\sum M_b = 0$. In order for the moment $a$ to vanish, the force $F_b$ must pass through $a$. Similarly, the force $F_a$ must pass through $b$. Next, since $\sum F = 0$, it is readily seen that $F_a = -F_b$.

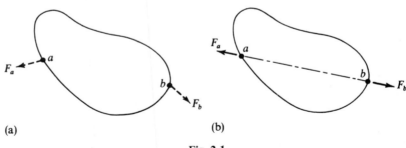

(a)                                              (b)

**Fig. 2-1**

2. *Three-force member.* Figure 2-2 shows a body subjected to the action of three external forces applied at $a$, $b$, and $c$. If the body is in equilibrium, then the three cannot be in random orientation, as shown in Fig. 2-2(a). They must be concurrent at a common point $O$, as shown in Fig. 2-2(b); otherwise,

the total moment about the intersection of any two forces could not vanish. A limiting case occurs when point $O$ moves off at infinite distance from $a$, $b$, and $c$, in which case the forces $F_a$, $F_b$, and $F_c$ are parallel.

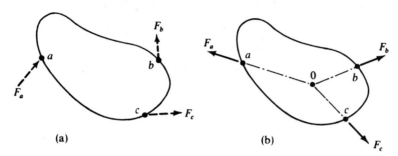

(a)                                                          (b)

**Fig. 2-2**

## 2-2. SUPPORT REACTIONS

Structures are either partially or completely restrained so that they cannot move freely in space. Such restraints are provided by supports that connect the structure to some stationary body, such as the ground or another structure. The first step in structural analysis is to take the structure without the supports and calculate the forces, known as *reactions*, exerted on the structure by the supports. The reactions are considered part of the external forces other than the loads on the structure and are to balance the other external loads in a state of equilibrium.

Certain symbols used to designate supports must first be described. There are generally three different types of support: the *hinge*, the *roller*, and the *fixed support*. The distribution of the reactive forces of a support may be very complicated, but in an idealized state the resultant of the forces may be represented by a single force completely specified by three elements—the *point of application*, the *direction*, and the *magnitude*. It may be noted that, in analysis, the direction simply means the slope of the action line, while the magnitude of force may be positive or negative, thus indicating not only its numerical size but also the sense of the action line.

*Hinge support.* A hinge support is represented by the symbol 〼 or 〼 . It can resist a general force $P$ in any direction but cannot resist the moment of the force about the connecting point, as illustrated in Fig. 2-3.

The reaction of a hinge support is assumed to be through the center of the connecting pin; its magnitude and slope of action line are yet to be determined. It is therefore a reaction with two unknown elements which could equivalently be represented by the unknown magnitudes of its horizontal and vertical com-

**Fig. 2-3**

ponents both acting through the center of the hinge pin. This is justified by the following equations from statics:

$$|R| = \sqrt{R_x^2 + R_y^2} \qquad \theta_x = \tan^{-1}\frac{R_y}{R_x} \qquad (2\text{-}9)$$

where $|R|$ = magnitude of the reaction $R$

$R_x$ = $x$ component of $R$

$R_y$ = $y$ component of $R$

$\theta_x$ = angle that $R$ makes with $x$ direction

The magnitude and direction of $R$ can be determined if the unknown magnitudes of $R_x$ and $R_y$ are found.

From this, a hinge support can also be replaced by two links along the horizontal and vertical directions through the center of the connecting pin as shown in Fig. 2-4(a). Each link is a two-force member, the axial force of which represents an element of reaction ($R_x$ or $R_y$). In general, a hinge support is equivalent to two supporting links provided in any two different directions, not necessarily an orthogonal set, through the connecting point as shown in Fig. 2-4(b), where $R_1$ and $R_2$ indicate the axial forces in two links. The reaction $R$ at the pin can always be determined if the magnitudes of $R_1$ and $R_2$ are obtained.

(a)                                              (b)

**Fig. 2-4**

*Roller support.* A roller support is represented by either the symbol ![symbol] or ![symbol] . The support mechanism used is such that the reaction acts normal

to the supporting surface through the center of the connecting pin, as shown
in Fig. 2-5(a)–(c). The reaction may be either away from or toward the supporting surface. As such, the roller support is incapable of resisting moment and
lateral force along the surface of support.

A roller support supplies a reactive force, fixed at a known point and in a
known direction, the magnitude of which is unknown. It is therefore a reaction
with one unknown element.

A link support, shown in Fig. 2-5(d), is also of this type since the link is a
two-force member and the reaction must be along the link.

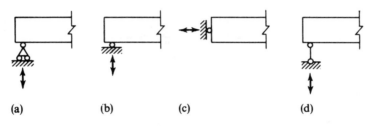

(a)          (b)          (c)          (d)

**Fig. 2-5**

*Fixed support.* A fixed support is designated by the symbol ⫣⊢ . It is
capable of resisting force in any direction and moment of force about the
connecting end, thus preventing the end of the member from both translation
and rotation. The reaction supplied by a fixed support may be represented by
the unknown magnitudes of a moment called $M_o$, a horizontal force $R_x$, and a
vertical force $R_y$ acting through the centroid of the end cross section $O$, as
shown in Fig. 2-6(a). These three unknown elements can be expressed as equal
to a single force $R$ with its three elements—the magnitude, direction, and
point of application—yet to be determined, as shown in Fig. 2-6(b). Now the
magnitude and direction of $R$ can be related to its components $R_x$ and $R_y$ by
Eq. 2-9,

$$|R| = \sqrt{R_x^2 + R_y^2} \qquad \theta_x = \tan^{-1}\frac{R_y}{R_x}$$

and the point of application $O'$ can be located by the distance $d$ from $O$, which,

(a)                          (b)

**Fig. 2-6**

in turn, is related to $M_o$ by

$$d = \frac{M_o}{R_x}$$

Since fixed support provides moment resistance, it is one step beyond the hinge support in rigidity.

Two devices equivalent to the fixed support are shown in Fig. 2-7. Each is composed of a hinge and a roller and represents three elements of reaction capable of resisting both force and moment.

**Fig. 2-7**

### 2-3. INTERNAL FORCES AT A CUT SECTION OF A STRUCTURE

A truss structure is composed of pin-connected members and is assumed to be pin-loaded, as is shown in Fig. 2-8(a). Now if any one of the members is taken from its connecting pins as a free body, the forces exerted on the member must be concentrated at the two ends of the member through the centers of pins. Furthermore, these two systems of concurrent forces can be combined into two resultant forces that must be equal, opposite, and collinear, as indicated in Fig. 2-8(b). In other words, each member of a truss is a *two-force* member.

(a)                              (b)                              (c)

**Fig. 2-8**

Hence, the internal forces existing in any cut section of a truss member (assumed straight and uniform) must be a pair of equal and opposite axial forces to balance the axial forces exerted on the ends, as shown in Fig. 2-8(c).

The fact that each member of a truss represents an unknown element of internal force enables us to obtain the total number of unknown elements of internal force by counting the total number of members of which the truss is composed.

Members of structures, such as beam and rigid frame, are acted on by more than two forces. Let us investigate the elements of internal force in any cut section $A$–$A$ of the beam in Fig. 2-9(a) or of the rigid frame in Fig. 2-9(b).

We begin by taking the free bodies of the portions to the left and right of the section $A$–$A$, as shown in Fig. 2-9(c) and (d). It is obvious that forces of internal constraint must exist between these two portions in order to hold them together. Such internal forces, of course, always occur in pairs of equal and opposite forces. The actual distribution of these internal forces cannot be easily discovered. To maintain the equilibrium of the free body, however, the internal forces must be statically equal and opposite to the system of forces acting externally on the portion considered, and the internal forces can always be represented by a force applied at the centroid $O$ of the cross section together with a couple of moment $M$. Furthermore, the force can, in turn, be resolved into a normal component $N$ and a tangential component $V$. Thus, in Fig. 2-9(c) and (d) we represent the stress resultant on any section $A$–$A$ by the three unknown magnitudes of $N$, $V$, $M$, called, respectively, the *normal force*, the *shearing force*, and the *resisting moment* at that section.

From the foregoing discussion, we remember that to take a free body from a beam or rigid frame we must assume three unknown elements of internal force generally existing in the cut section.

**Fig. 2-9**

## 2-4. EQUATIONS OF CONDITION OR CONSTRUCTION

Structures such as truss, beam, and rigid frame may sometimes be considered to be one rigid body sustained in space by a number of supports. Out of several such rigid bodies a compound form of structure may be built by means of connecting devices, such as hinges, links, or rollers, and mounted on a number of supports. In either the simple or the compound type of structure the external force system of the entire structure, consisting of the loads on the structure and the support reactions, must satisfy the equations of equilibrium if the structure is to remain at rest. However, in the compound type of structure the connecting devices enforce further restrictions on the force system acting on the structure, thus providing additional equations of statics to supplement the equations of equilibrium. Equations supplied by the method of special construction (other than external supports) are called *equations of condition or construction*. We discuss these further in Sec. 2-6.

## 2-5. STABILITY AND DETERMINACY OF A STRUCTURE WITH RESPECT TO SUPPORTS WHEN THE STRUCTURE IS CONSIDERED A MONOLITHIC RIGID BODY

When one considers the design of a structure, careful thought must be given to the number and arrangement of the supports directly related to the statical stability and determinacy of the structure. In the following discussions we shall treat the structures as one monolithic rigid body mounted on a number of supports. Thus, there will be no internal condition involved, and the stability and determinacy of the structure will be judged solely by the stability and determinacy of supports.

1. Two elements of reaction supplied by supports, such as two forces each with a definite point of application and direction, are not sufficient to ensure the stability of a rigid body, because the two are either collinear, parallel, or concurrent. In each of these cases, the condition of equilibrium is violated, not because of the lack of strength of supports, but because of the insufficient number of support elements. This is referred to as *statical instability*.

If two reactive forces are collinear [see Fig. 2-10(a)], they cannot resist an external load that has a component normal to the line of reactions. If they are parallel [see Fig. 2-10(b)], they cannot prevent the body from lateral sliding. If they are concurrent [see Fig. 2-10(c) or (d)], they cannot resist the moment about the concurrent point $O$ due to any force not through $O$.

Algebraically, in each of the cases above, one equilibrium condition is not satisfied. For instance, in Fig. 2-10(a) or (b) the condition $\sum F_x = 0$ is violated ($x$ indicates the direction normal to the line of reaction); whereas in Fig. 2-10(c) or (d) the condition $\sum M_o = 0$ is not fulfilled. The body is, therefore, not in equilibrium. It is unstable.

Only under some very special conditions of loading can the body be stable, such as those shown in Fig. 2-11. In Fig. 2-11(a) the applied loads acting on the body are themselves in equilibrium; therefore, no reaction is required. In Fig. 2-11(b) the applied load is in the same direction as the reactions so

(a)

(b)

(c)

(d)

**Fig. 2-10**

(a)

(b)

(c)

(d)

**Fig. 2-11**

that equilibrium can be maintained for the parallel force system; and in Fig. 2-11(c) or (d) the applied load is through the concurrent point $O$; therefore, equilibrium can also be established.

Structures stable under special conditions of loading but unstable under general conditions of loading are said to be in a state of *unstable equilibrium* and are classified as unstable structures.

2. At least three elements of reaction are necessary to restrain a body in stable equilibrium. Consider each of the cases shown in Fig. 2-12. The rigid body is subjected to restraints by three elements of reaction, and the restraints can be solved by the three available equilibrium equations. The satisfaction of all the three equilibrium equations, $\sum F_x = 0$, $\sum F_y = 0$, and $\sum M = 0$, for loads and reactions acting on the body guarantees, respectively, that the body will neither move horizontally or vertically nor rotate. The system is said to be *statically stable and determinate*.

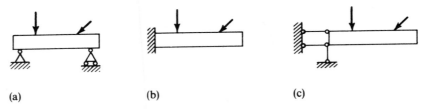

(a)                          (b)                          (c)

**Fig. 2-12**

3. If there are more than three elements of reaction, as in each of the cases shown in Fig. 2-13, the body is necessarily more stable, because of the additional restraints. Since the number of unknown elements of reaction is more than the number of equations for static equilibrium, the system is said to be *statically indeterminate* with regard to the reactions of support.

(a)                          (b)                          (c)

**Fig. 2-13**

4. That the number of elements of reaction should be at least three is a necessary but not a sufficient condition for an externally stable structure. There are many cases that are obviously not stable with respect to the support system even though three or more than three elements of reaction are supplied. When, for example, the lines of reaction are all parallel, as in Fig. 2-14(a), the body

is unstable, because it is vulnerable to lateral sliding. Another case is shown in Fig. 2-14(b), where lines of the three reaction elements are originally concurrent at point $O$. The system is also unstable because, even though complete collapse probably will not result, a small initial rotation about $O$ because of the moment caused by any force not through $O$ will certainly occur until the three reaction lines form the triangle indicated by the crosshatched lines.

The above-mentioned instability, which results from the inadequacy of arrangement of supports, is referred to as *external geometric instability*.

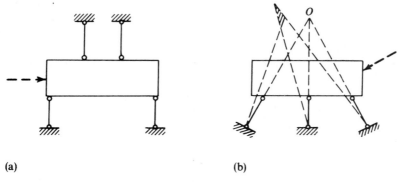

(a)                                         (b)

**Fig. 2-14**

5. A monolithic rigid body is rigid by definition; hence, it will have no problem of internal instability. Furthermore, at any cut section of a monolithic rigid body, the elements of internal force, which are no more than three in number, can always be determined by the equations of equilibrium, once the reactions are completely defined. Therefore, the stability and determinacy of the entire system, mentioned in this section, are solely determined by the stability and determinacy of supports and reactions.

Let us sum up the main points of foregoing discussions as follows:

1. If the number of unknown elements of reaction is fewer than three, the equations of equilibrium are generally not satisfied, and the system is said to be unstable.

2. If the number of unknown elements of reaction is equal to three and if no external geometric instability is involved, then the system is statically stable and determinate.

3. If the number of unknown elements of reaction is more than three, then the system is statically indeterminate; it is stable provided that no external geometric instability is involved. The excess number $n$ of unknown elements designates the $n$th degree of indeterminacy. For example, in each case of Fig. 2-13 there are five unknown elements of reaction. Thus, $5 - 3 = 2$, which indicates an indeterminacy of second degree.

## 2-6. GENERAL STABILITY AND DETERMINACY OF STRUCTURES

Structural stability and determinacy must be judged by the number and arrangement of the supports as well as the number and arrangement of the members and the connections of the structure. They are determined by inspection or by formula. For convenience, we shall deal with the general stability and determinacy of beams, trusses, and rigid frames in separate sections.

*2-6a. General Stability and Determinacy of Beams.* If a beam is built up without any internal connections (internal hinge, roller, or link), the entire beam may be considered as one piece of monolithic rigid body placed on a number of supports, and the question of the stability and of the determinacy of the beam is settled solely by the number and arrangement of supports, as discussed in Sec. 2-5.

Now let us proceed to investigate what will happen if a certain connecting device is inserted in a beam. Let us suppose that a hinge is introduced into the statically stable and determinate beam of Fig. 2-15(a) or (b). The beam in each case will obviously become unstable under general loading as the result of a relative rotation between the left and the right portions of the beam at the internal hinge, as indicated in Fig. 2-15(c) or (d). That the hinge has no capacity to resist moment constitutes a restriction on the external forces acting on the structure; that is,

$$M = 0$$

about the hinge. In other words, the moment about the hinge calculated from the external forces on either side of the hinge must be zero in order to guarantee that these portions will not rotate about the hinge.

Referring to Fig. 2-15(c) or (d), we see that in each case there are three elements of reaction supplied by supports, whereas there are four conditions of

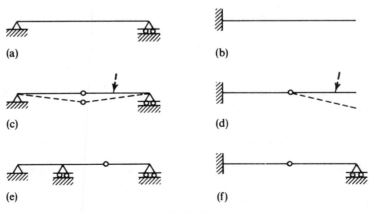

(a)  (b)  (c)  (d)  (e)  (f)

**Fig. 2-15**

statics to restrict the external forces—three from equilibrium plus one from construction. This means that the number of unknown elements of reaction is one fewer than the independent equations of statics available for their solution. Therefore, the equations of statics for the force system are generally not satisfied. The beam is unstable, unless we provide at least one additional element of reaction, such as the additional roller support shown in Fig. 2-15(e) or (f), which makes the total number of unknown elements of reaction equal to the total number of independent equations of statics needed to determine the elements. If this is done, the beam will be restored to a statically stable and determinate state.

Next, let us suppose that a link (or a roller) is introduced into a section of the statically stable and determinate beam of Fig. 2-15(a) or (b). We expect that this beam will be less stable than that of a hinged connection because the link (roller) cannot resist both moments about the link pin and forces normal to the link. The beam will collapse under general types of loading as a result of the relative rotation and the lateral translation of the left and right portions of the beam at the link, as indicated in Fig. 2-16(a) or (b).

(a)                                    (b)

(c)                                    (d)

**Fig. 2-16**

That a link (or roller) has no capacity to resist lateral forces and moments constitutes two restrictions to the external forces acting on the structure, namely,

$$H = 0 \quad \text{and} \quad M = 0$$

about the link (either of two pins). $H$ is the sum of forces on either side of the link in the direction normal to the link. The satisfaction of condition $H = 0$ for the portion of the structure on either side of the link prevents movement in the direction normal to the link of one portion of the structure relative to the other. Satisfying condition $M = 0$ for the portion of the structure on either side of the link ensures that these portions will not rotate about the pins of the link.

Referring to Fig. 2-16(a) or (b), we find that in each case there are three elements of reaction supplied by the support system, while there are five conditions of statics to restrict them—three from equilibrium and two from construction. Since the number of elements in reaction is two fewer than the number of statical equations to determine them, the beam is, therefore, quite unstable unless we supply at least two more elements of reaction, such as the hinged support shown in Fig. 2-16(c) or (d), to balance the situation. This done, the beam will be restored to its statically stable and determinate state.

There are beams for which the number of reaction elements is greater than the total number of independent equations of statics available. The beams are then classified as *statically indeterminate*; and the excess number of unknown elements indicates the degree of indeterminacy.

Geometric instability is most likely to occur whenever internal connections are introduced into an originally stable structure. Consider, for example, Fig. 2-17(a). The beam is statically indeterminate to the first degree. Now if a hinge is inserted into the beam, as shown in Fig. 2-17(b), it seems to be statically determinate. However, when a load is applied, a small initial displacement will result and will not be resisted elastically by the structure. In such a case, the beam is unstable not because of the inadequacy of the supports but because of the inadequacy of the arrangement of members. This is referred to as *internal geometric instability*. Very often when this occurs, the structure will collapse. In the present case collapse will not occur; the beam will come to rest in a position such as that marked by the dashed lines shown in Fig. 2-17(b).

(a)                                                      (b)

**Fig. 2-17**

From the foregoing discussions, a criterion may be established for the stability and determinacy of beams. Let $r$ denote the number of reaction elements and $c$ the number of equations of condition ($c = 1$ for a hinge; $c = 2$ for a roller; $c = 0$ for a beam without internal connection).

1. If $r < c + 3$, the beam is unstable.
2. If $r = c + 3$, the beam is statically determinate provided that no geometric instability (internal and external) is involved.
3. If $r > c + 3$, the beam is statically indeterminate.

Further illustrations are given in Table 2-1.

***2-6b. General Stability and Determinacy of Trusses.*** A truss is composed of a number of bars connected at their ends by a number of pinned joints so as to form a network, usually a series of triangles, and mounted on a number of supports, such as the one shown in Fig. 2-18(a). Each bar of a truss is a two-

TABLE 2-1

| Beam | $r$ | $c$ | $r \lessgtr c + 3$ | Classification |
|---|---|---|---|---|
| | 5 | 2 | $5 = 5$ | Stable and determinate |
| | 6 | 2 | $6 > 5$ | Stable and indeterminate to the first degree |
| | 5 | 2 | $5 = 5$ | Unstable* |
| | 4 | 3 | $4 < 6$ | Unstable |
| | 6 | 3 | $6 = 6$ | Stable and determinate |
| | 7 | 2 | $7 > 5$ | Unstable* |

*Internal geometric instability; a possible form of displacement is indicated by the dotted lines.

**(a)**           **(b)**

**Fig. 2-18**

force member; hence, each represents one unknown element of internal force (see Sec. 2-3). The total number of unknown elements for the entire system is counted by the number of bars (internal) plus the number of independent reaction elements (external). Thus, if we let $b$ denote the number of bars and $r$ the number of reaction components, the total number of unknown elements of the entire system is $b + r$. Now if the truss is in equilibrium, every isolated portion must likewise be in equilibrium. For a truss having $j$ joints, the entire system may be separated into $j$ free bodies, as illustrated in Fig. 2-18(b), in which each joint yields two equilibrium equations, $\sum F_x = 0$ and $\sum F_y = 0$, for the concurrent force system acting on it. From this a total of $2j$ independent equations, involving $(b + r)$ unknowns, is obtained. We may thus establish a criterion for the stability and the determinacy of truss by counting the total unknowns and the total equations.

   1. If $b + r < 2j$, the system is unstable.
   2. If $b + r = 2j$, the system is statically determinate provided that it is also stable.
   3. If $b + r > 2j$, the system is statically indeterminate.

The satisfaction of condition $b + r \gtrless 2j$ does not ensure a stable truss. For the truss to be stable requires fulfillment of further conditions. First, the value of $r$ must be equal to or greater than the three required for statical stability of supports. Next, there must be no inadequacy in the arrangement of supports and bars so as to avoid both external and internal geometric instability.

Basically, a stable truss can usually be obtained by starting with three bars pinned together at their ends in the form of a triangle and then by extending from it by adding two new bars for each new joint, as shown in Fig. 2-18(a). Since this truss satisfies $b + r = 2j(b = 13, r = 3, j = 8)$, it is therefore statically determinate.

Suppose that this truss form is changed, as shown in Fig. 2-19. The number of bars and joints remains the same; the criterion equation is still satisfied. But it is geometrically unstable, since there is no bar to carry the vertical force (shear) in the panel where the diagonal is omitted. Other examples are given in Table 2-2.

**Fig. 2-19**

Figure 2-20 shows a long-span trussed bridge, which we may consider to be composed of three rigid trusses connected by a hinge $A$ and a link $BC$ and mounted on a number of supports. These connections are not completely rigid, so that certain equations of condition are introduced to restrict the external forces acting on the structure. In this case the hinge at $A$ provides one condition equation $M_A = 0$, which means that the moment about $A$ of the forces on either side of $A$ must be zero. The hanger $BC$ provides two condition equations, $M_B = 0$ (or $M_C = 0$) and $H = 0$, which means that the moment about $B$ (or $C$) of the forces on either side of $B$ (or $C$) must be zero and also that the sum of

TABLE 2-2

| Truss | $b$ | $r$ | $j$ | $b + r \gtrless 2j$ | Classification |
|-------|-----|-----|-----|---------------------|----------------|
| | 7 | 3 | 5 | $10 = 10$ | Stable and determinate |
| | 7 | 3 | 5 | $10 = 10$ | Unstable* |
| | 7 | 3 | 5 | $10 = 10$ | Unstable** |
| | 6 | 3 | 5 | $9 < 10$ | Unstable |
| | 6 | 4 | 5 | $10 = 10$ | Stable and determinate |
| | 8 | 4 | 5 | $12 > 10$ | Stable and indeterminate to the second degree |
| | 6 | 4 | 5 | $10 = 10$ | Unstable*** |

*Internal geometric instability due to three pins $a$, $b$, $c$ on a line; possible displacement as indicated by dotted lines.

**External geometric instability due to parallel lines of reaction.

***Internal geometric instability due to lack of lateral resistance in panel $abcd$.

the horizontal forces on either side of the hanger must be zero since the vertical hanger is incapable of resisting horizontal forces.

The stability and determinacy of the truss may be investigated by first counting the number of bars, joints, and reaction elements. It is found that the equation $b + r = 2j$ is satisfied by the truss since $b = 40$, $r = 6$, and $j = 23$. Thus, the necessary condition for the system to be statically determinate is

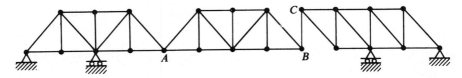

**Fig. 2-20**

fulfilled. Next, there is no obvious instability either in the formation of the truss or in the supports. Both the portions to the left of *A* and to the right of *BC* are rigidly formed and adequately supported. The portion in the center span is also rigidly formed. Its connection to the side portions by a hinge and a hanger constitutes three elements of support. As regards reactions, there are a total of six elements which can just be determined by six statical equations, three from equilibrium and three from construction. Thus, the entire system is stable and statically determinate; furthermore, it is stable and statically determinate as regards support reactions.

There are certain cases in which the stability or instability of a truss is not obvious. One way of determining stability is to attempt a stress analysis and to discover whether the results are consistent or not. An inconsistent result indicates that the answer is not unique but infinite and indeterminate. If such is the case, then the truss is said to be *unstable*. We discuss this further in Secs. 4-4 and 4-5.

**2-6c. General Stability and Determinacy of Rigid Frames.** A rigid frame is built of beams and columns connected rigidly, such as the one shown in Fig. 2-21(a). The stability and determinacy of a rigid frame may also be investigated by comparing the number of unknowns (internal unknowns and reaction unknowns) with the number of equations of statics available for their solution. Like a truss, a rigid frame may be separated into a number of free bodies of

(a)                              (b)

**Fig. 2-21**

24

joints, as shown in Fig. 2-21(b), which requires that every member of the frame be taken apart. As discussed in Sec. 2-3, there are usually three unknown magnitudes ($N$, $V$, $M$) existing in a cut section of a member. However, if these quantities are known at one section of a member, similar quantities for any other section of the same member can be determined. Hence, there are only three independent, internal, unknown elements for each member in a frame. If we let $b$ denote the total number of members and $r$ the reaction elements, then the total number of independent unknowns in a rigid frame is ($3b + r$).

Next, a rigid joint isolated as a free body will generally be acted upon by a system of forces and couples, as indicated in Fig. 2-21(b), since a rigid joint is capable of resisting moments. For equilibrium of such a joint, this system, therefore, must satisfy three equilibrium equations $\sum F_x = 0$ $\sum F_y = 0$, and $\sum M = 0$. Thus, if the total number of rigid joints is $j$, then $3j$ independent equilibrium equations may be written for the entire system.

It may happen that hinges or other devices of construction are introduced into the structure so as to provide additional equations of statics, say a total of $c$. Then the total number of equations of statics available for the solution of the ($3b + r$) unknowns is ($3j + c$). The criteria for the stability and the determinacy of the rigid frame are thus established by comparing the number of unknowns ($3b + r$) with the number of independent equations ($3j + c$):

1. If $3b + r < 3j + c$, the frame is unstable.
2. If $3b + r = 3j + c$, the frame is statically determinate provided that it is also stable.
3. If $3b + r > 3j + c$, the frame is statically indeterminate.

It should be recalled from the similar discussion dealing with the criterion for trusses that satisfaction of the condition $3b + r \gtrless 3j + c$ does not warrant a stable frame unless $r \gtrless 3$ and, also, that no geometric instability is involved in the system.

Consider the frame in Fig. 2-21(a). There are six joints (including those at supports), six members, and six reaction elements, but no condition of construction. Thus, $3b + r = 18 + 6 > 3j + c = 18 + 0$. The excess number six in unknowns indicates that the frame is statically indeterminate to the sixth degree. Further examples for classifying frame stability and determinacy are given in Table 2-3.

Criteria such as the above are general and useful; but many problems, which may be investigated by a formula, can readily be settled by inspection through cutting frame members and reducing the structure to several simple parts.

Suppose that we wish to analyze the degree of indeterminacy of the frame shown in Fig. 2-22(a). The best approach is to cut members as indicated in Fig. 2-22(b) so that the structure is separated into three statically determinate and stable parts. The number of restraints removed to accomplish this result gives the degree of indeterminacy of the frame. Since each cut involves three internal

TABLE 2-3

| Frame | $b$ | $r$ | $j$ | $c$ | $3b + r \lessgtr 3j + c$ | Classification |
|---|---|---|---|---|---|---|
| | 10 | 9 | 9 | 0 | $39 > 27$ | Indeterminate to the 12th degree |
| | 10 | 9 | 9 | 4 | $39 > 31$ | Indeterminate to the eighth degree |
| | 10 | 9 | 9 | 1 | $39 > 28$ | Indeterminate to the 11th degree |
| | 10 | 9 | 9 | 3* | $39 > 30$ | Indeterminate to the ninth degree |
| | 10** | 6 | 9 | 0 | $36 > 27$ | Indeterminate to the ninth degree |

*If a pin is inserted in a rigid frame, generally, $c$ = the number of members meeting at the pin minus one. In this case $c = 4 - 1 = 3$.

**The overhanging portions, such as $ab$ and $cd$ on the right side of the frame, should not be counted in the number of members.

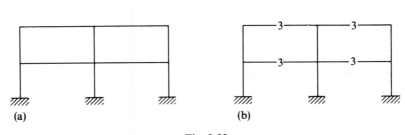

(a)         (b)

**Fig. 2-22**

unknown elements, the total number of restraints removed by four cuts is $(4)(3) = 12$; the frame is statically indeterminate to the 12th degree.

The advantage of this approach over counting the number of bars and joints and reaction elements will easily be seen when we come to determine the degree of indeterminacy of the frame of a tall building, such as the one shown in Fig. 2-23. Since the building can be separated into 12 stable and determinate parts by 77 cuts in the beams, it is statically indeterminate to the 231st degree.

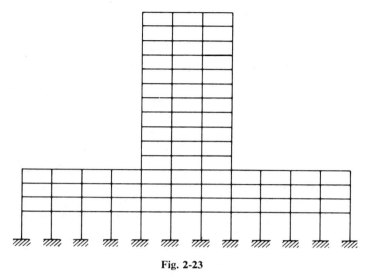

Fig. 2-23

# PROBLEMS

**2-1.** Discuss the stability and determinacy of the beams shown in Fig. 2-24.

(a)

(b)

(c)

(d)

Fig. 2-24

**2-2.** Discuss the stability and determinacy of the trusses shown in Fig. 2-25.

(a)

(b)

Fig. 2-25

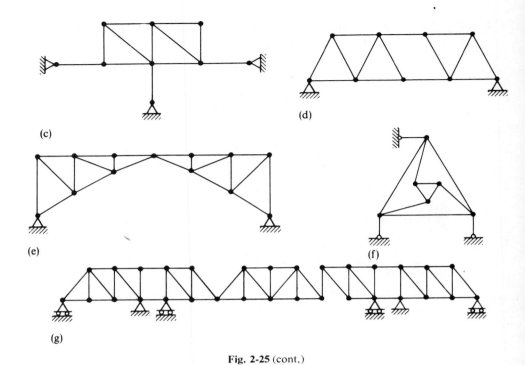

Fig. 2-25 (cont.)

2-3. Discuss the stability and determinacy of the rigid frames shown in Fig. 2-26.

Fig. 2-26

28

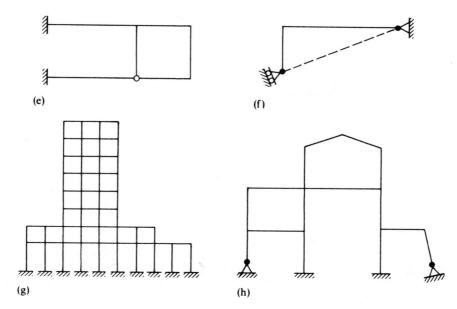

(e)

(f)

(g)

(h)

**Fig. 2-26** (cont.)

# 3

# STATICALLY DETERMINATE BEAMS

## 3-1. GENERAL

A *beam* is defined as a structural member predominantly subjected to bending moment. In this chapter and later chapters we limit our discussions to beams of symmetrical section, the centroidal axis of such beams being a straight line. Furthermore, we assume that the beam is acted on by only transverse loading and moment loading and that all the loads and reactions lie in the plane of symmetry. Form these, it follows that such a beam will be subjected to bending and shear in the plane of loading without axial stretching and twisting.

Statically determinate beams may be classified as:

1. Simple beam. A beam that is supported at its two ends with a hinge and a roller [see Fig. 3-1(a)], or their equivalents [see Fig. 3-1(b)], is termed a *simply supported beam* or *simple beam*.

2. Cantilever beam. Figure 3-1(c) shows a *cantilever beam*, which is fixed or built-in at one end and free at the other end. Another form of cantilever is shown in Fig. 3-1(d), in which the three elements of reaction of fixed support are provided by a hinge and a roller closely placed at one end segment.

3. Simple beam with overhang. A beam may be simply supported at two points and have its end portion (or portions) extend beyond the support, as in Fig. 3-1(e) or (f); it is then called a *simple beam with overhang*.

4. Compound beam. The beams indicated above may be connected by internal hinges or rollers to form a *compound beam*, such as those shown in Fig. 3-1(g) and (h). Care must be used in providing the connections so that instability is not produced.

## 3-2. ANALYSIS OF STATICALLY DETERMINATE
### BEAMS

To illustrate the general procedure in analyzing a beam, let us consider the loaded beam in Fig. 3-2(a).

The first step in the analysis is to find the reactions at ends *a* and *b* denoted

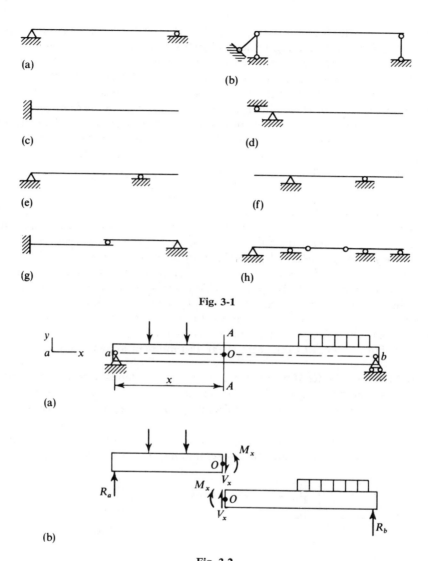

Fig. 3-1

Fig. 3-2

by $R_a$ and $R_b$, respectively. This can readily be accomplished by applying the equilibrium equations:

$$\sum M_a = 0 \qquad \sum M_b = 0$$
or
$$\sum M_b = 0 \qquad \sum F_y = 0$$

Next we investigate the shear force and bending moment at each transverse cross section of the beam. The *shear force* at any transverse cross section of the beam, say section $A$–$A$, at a distance $x$ from the left end [see Fig. 3-2(a)], is the algebraic sum of the external forces (including those of reaction) applied to the portion of the beam on either side of $A$–$A$. The *bending moment* at section $A$–$A$ of the beam is the algebraic sum of the moments taken about an axis through $O$ (the centroid of section $A$–$A$) and normal to the plane of loading of all the external forces applied to the portion of the beam on either side of $A$–$A$. By considering either the left or the right portion as the free body, as shown in Fig. 3-2(b), we readily see that the *shear resisting force* at the section $V_x$ is equal and opposite to the shear force for that section just defined; and the *resisting moment* at the section $M_x$ is equal and opposite to the bending moment for the section just defined. The values of $V_x$ and $M_x$ can be found from the two equations of equilibrium

$$\sum F_y = 0 \quad \text{and} \quad \sum M_o = 0$$

for the portion considered.

Since the shear and bending moment in a transversely loaded beam will, in general, vary with the distance $x$ defining the location of the cross section on which they occur, both are therefore the functions of $x$. It is advisable to plot curves or diagrams from which the value of functions ($V_x$ and $M_x$) at any cross section may readily be obtained. To do this, we let one axis, the $x$ axis, coincide with the centroidal axis of beam, indicating the position of beam section, and the other axis, the $y$ axis, indicate the value of function $V_x$ or $M_x$. The graphic representation is called *shear* or *moment curve*.

Our sign conventions for beam shear and moment are as follows:

1. Shear is considered positive at a section when it tends to rotate the portion of the beam in the clockwise direction about an axis through a point inside the free body and normal to the plane of loading; otherwise, it is negative [see Fig. 3-3(a)].

2. Bending moment is considered positive at a section when it tends to bend the member concave upward; otherwise, it is negative [see Fig. 3-3(b)].

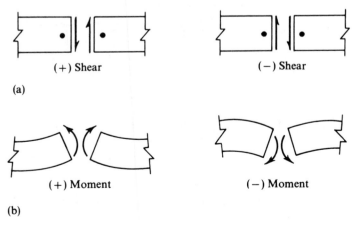

(+) Shear        (−) Shear

(a)

(+) Moment        (−) Moment

(b)

**Fig. 3-3**

(a)

$R_A = \dfrac{wl}{2}$      $R_B = \dfrac{wl}{2}$

w per unit length

$V_x = \dfrac{wl}{2} - wx$

(b)

$M_x = \dfrac{wl}{2}x - \dfrac{wx^2}{2}$

$\dfrac{wl^2}{8}$

(c)

Fig. 3-5

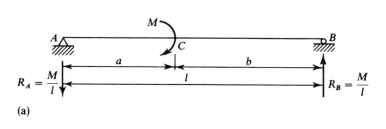

(a)

$R_A = \dfrac{M}{l}$      $R_B = \dfrac{M}{l}$

(b)

$V_x = -\dfrac{M}{l}$

$\dfrac{Mb}{l}$    $M_x = M(l - x)/l$

$M_x = -Mx/l$    $\dfrac{Ma}{l}$

(c)

Fig. 3-6

The process of moment changes at $C(x = a)$, the total change being $M$ (from $-Ma/l$ to $Mb/l$). The moment diagram is shown in Fig. 3-6(c), and the point of inflection (zero moment) is at $C$ where the moment curve passes the $x$ axis.

If the beam is subjected to several loads (or several groups of loads), the shear or the moment diagram may be plotted separately for each load (or each group of loads) and then combined into one diagram by the principle of super-position.

Table 3-1 shows that application of this principle in drawing qualitative moment diagram for a simple beam (or a general member) restrained by end moments.

TABLE 3-1

| Case | Separate Moment Diagram | Combined Moment Diagram* |
|------|------------------------|--------------------------|

*With large end moment (or moments) the combined moment diagrams for Cases (3), (4), (5), or (6) could be all negative without point of inflection.

From Table 3-1, we see that if the beam carries no load but end moments, the moment curve is a straight line with one point of inflection or no point of inflection. If the beam carries a concentrated load or a uniform load together with the end moment on one end or both ends, the moment curve may pass zero at one, or two, or no point on the beam axis.

It may be worth mentioning here that if the beam is made of elastic material, the beam will be deformed under load; the elastic deformations of beams are primarily caused by bending. With reference to the moment diagram and points of zero moment, we can easily sketch the deflected elastic curve.

**Example 3-4.** In Fig. 3-7(a) is shown a cantilever carrying a distributed load the intensity of which varies linearly from $w$ per unit length at the fixed end to zero at free end.

At any section distance $x$ from the free end $a$,

$$V_x = -\left(\frac{wx}{l}\right)\left(\frac{x}{2}\right) = -\frac{wx^2}{2l}$$

$$M_x = -\left(\frac{wx^2}{2l}\right)\left(\frac{x}{3}\right) = -\frac{wx^3}{6l}$$

These are plotted in Fig. 3-7(b) and (c), respectively.

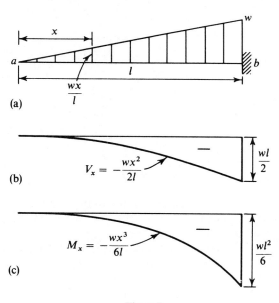

**Fig. 3-7**

**Example 3-5.** In Fig. 3-8(a) is shown a simple beam with an overhang. We wish to plot the shear and moment diagrams.

It may be convenient to analyze the beam by separating it into two parts,

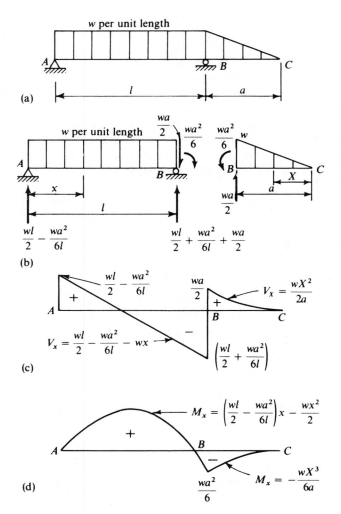

**Fig. 3-8**

a simple beam and the overhanging portion and then by superposing the results. To do this, we cut the beam into two free bodies by passing a section at the immediate right side of support $B$, as shown in Fig. 3-8(b). The analysis of the overhanging portion, similar to that of a cantilever, illustrated in Example 3-4, is first laid out. At any section distance $X$ from end $C$ (origin is taken at $C$), we have

$$V_X = \frac{wX^2}{2a} \qquad M_X = -\frac{wX^3}{6a}$$

The shear and moment at the cut section just to the right of support $B$ are thus found to be

$$V_B = \frac{wa}{2} \text{ (up)} \qquad M_B = \frac{wa^2}{6} \text{ (counterclockwise)}$$

38

Next, consider the portion $AB$, which is subjected to a uniform load over the entire span together with an end shear of $wa/2$ (down) and an end moment of $wa^2/6$ (clockwise) at $B$ exerted on by the portion $BC$. The value of $wa/2$ will go to support $B$ and cause no bending and shear on $AB$. At any section distance $x$ from $A$ (origin is taken at $A$),

$$V_x = \left(\frac{wl}{2} - \frac{wa^2}{6l}\right) - wx$$

which is of linear variation; and

$$M_x = \left(\frac{wl}{2} - \frac{wa^2}{6l}\right)x - \frac{wx^2}{2}$$

which represents a parabola. Note that the construction of the moment curve for the portion of the simple beam may be done by superposing the positive moment and the negative moment, as suggested in Table 3-1.

The shear and moment diagrams for the entire beam are plotted in Fig. 3-8(c) and (d).

**Example 3-6.** Analyze the loaded compound beam of Fig. 3-9(a), which is composed of a simple beam and a cantilever connected by an internal hinge.

We note that the number of unknown reaction elements, including the horizontal reaction of support $e$, is four. These are to be solved by four inde-

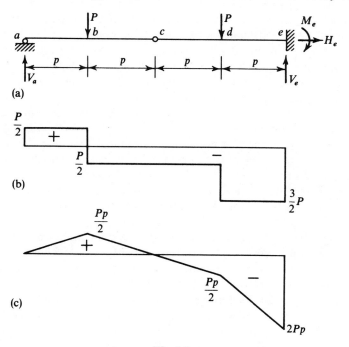

**Fig. 3-9**

pendent equations of statics (i.e., three equilibrium equations and one condition equation):

$$\sum F_x = H_e = 0$$
$$\sum F_y = V_a + V_e - 2P = 0$$
$$\sum M_e = (V_a)(4p) - P(3p + p) + M_e = 0$$
$$M_c = (V_a)(2p) - Pp = 0$$

From the above, we obtain $V_a = P/2$, $V_e = 3P/2$, $H_e = 0$, and $M_e = 2Pp$.

After the reactions are determined, it is not difficult to construct the shear and moment diagrams, shown in Fig. 3-9(b) and (c), respectively. Note that the moment curve must pass zero at $c$ as required.


### 3-3.  PRINCIPLE OF VIRTUAL WORK
### AND BEAM ANALYSIS

The principle of virtual work as applied to statically determinate plane structures may be stated as follows: *If a small virtual displacement consistent with the constraints is given to an ideal system in equilibrium under the action of a balanced force system, the total work done by all active forces is equal to zero.*

By an *ideal system* we mean a system of rigid bodies connected and supported smoothly and rigidly so that no energy is stored or dissipated during a small displacement. Note that if such a system is in equilibrium under the action of a balanced force system, no displacement can actually occur, and no work can actually be done by the forces. However, an arbitrary, fictitious displacement, called *virtual displacement*, for the system may be assumed to occur. This has nothing to do with the real displacement of the body but only serves as a technique in solving equilibrium problems. Virtual displacement has no finite value; otherwise, it might cause some shifting of the lines of action of the forces so that the system would be no longer in equilibrium. For this reason, the virtual displacement is generally assumed to be vanishingly small. It is customary to use $\delta s$ to denote the linear virtual displacement and $\delta\theta$ the angular virtual displacement, as distinguished from real infinitesimal displacements $ds$ and $d\theta$. The corresponding work done by the force system is called *virtual work*, and is usually denoted by $\delta W$.

Also note that virtual displacements for a system should be so chosen that they are compatible with the constraints on the system. Thus for the bar in Fig. 3-10(a) possible rotation about the supporting hinge at $A$ may occur so that bar $AB$ will turn through an angle $\delta\theta$ at $A$ as indicated. For the beam in Fig. 3-10(b) horizontal displacement may take place as a result of the lack of lateral resistance of the roller supports so that end $A$ will move to $A'$ and $B$ to $B'$; and $AA' = BB' = \delta x$. Likewise, for the beam shown in Fig. 3-10(c) end $A$ may move to the left from $A$ to $A'$, $AA' = \delta x$, and end $B$ to $B'$ downward,

Fig. 3-10

$BB' = \delta y$. In all cases, the structure must be in an unstable form; otherwise, the displacement is impossible.

*Active force* is the opposite of *workless force*. The work done by a constantly applied force $P$ during a displacement $s$ is defined as the product of the magnitude of the force, the magnitude of the displacement, and the cosine of the angle $\theta$ between the force and the displacement:

$$W = Ps \cos \theta \qquad (3-1)$$

And the work done by a couple of constant moment $M$ is defined as the product of the magnitude of the moment and of the magnitude of the angular displacement $\phi$ of the couple:

$$W = M\phi \qquad (3-2)$$

A force produces positive work when the displacement and the projection of the force along the displacement are in the same direction; a couple does positive work when the angular displacement and the couple have the same sence. From this we see that such forces as the following are termed *workless*:

1. The force that is normal to the displacement
2. The force that does not move or the couple that does not turn
3. The internal forces or moments that exist in pairs with equal magnitude but opposite direction so that they do an equal amount of positive work and negative work for any displacement assigned to the system

Note that the internal forces do produce work in an elastic body known as *strain energy*, as will be discussed in Sec. 8-3.

To prove the principle of virtual work, we begin with the case of single particle $A$ under a balanced coplanar concurrent force system $P_1, P_2, \ldots, P_n$ as shown in Fig. 3-11. Imagine that $A$ is given a small arbitrary displacement $\delta s$ as shown. Then the total virtual work done by this force system is

$$\delta W = (P_1 \cos \theta_1 + P_2 \cos \theta_2 + \cdots + P_n \cos \theta_n) \, \delta s$$

The expression in parenthesis is the projection of resultant force in the direction of the displacement, and it must be zero since the particle is in equilibrium. Therefore,

$$\delta W = 0 \qquad (3-3)$$

41

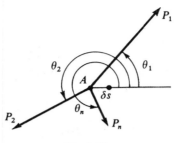

**Fig. 3-11**

meaning that the virtual work done on a particle during any virtual displacement is zero.

Next, let us consider a rigid body in equilibrium. A rigid body may be regarded as a collection of particles, say $n$ particles, and the system is in equilibrium if every one of its $n$ constituent particles is in equilibrium. Therefore, for an isolated particle $i$, the total work done by the forces acting on it due to any virtual displacement given to the body is zero according to Eq. 3-3:

$$\delta W_i = 0$$

For the system considered as the summation of the total particles we may write the expression of virtual work as the sum of the amount of virtual work done on all the particles of the body:

$$\delta W = \delta W_1 + \delta W_2 + \cdots + \delta W_n = \sum_{i=1}^{n} \delta W_i = 0 \qquad (3\text{-}4)$$

Now the forces involved in Eq. 3-4 are of two kinds, external and internal. The internal forces occur in self-canceling pairs; only the active part of external forces contributes to the virtual work. Thus, we narrow down to the statement that the total virtual work done by all active forces acting on a rigid body during any virtual displacement is zero.

Finally, let us extend the principle to a balanced ideal system that consists of a number of rigid bodies, say $m$ bodies, joined together and supported smoothly and rigidly. For any isolated rigid body $j$ in the system, the total work done by the forces acting on it because of any small compatible virtual displacement assigned to the system is zero according to Eq. 3-4:

$$\delta W_j = 0$$

Since there is no energy loss or gain in each ideal connection or means of support, we can write an expression of virtual work for the system based on the summation of virtual works done on all of the individual rigid bodies:

$$\delta W = \delta W_1 + \delta W_2 + \cdots + \delta W_m = \sum_{j=1}^{m} \delta W_j = 0 \qquad (3\text{-}5)$$

However, of the forces involved in the foregoing expressions, some are internal forces exerted by connections that produce positive virtual work on one body and the same amount of negative virtual work on the other of each connected pair; others are workless support reactions, and only the active part of the external forces on the system contributes to the virtual work. This completes our proof for the statement.

In solving certain problems of statics, we find that methods of solution based on the principle of virtual work have a decided advantage over the equilibrium equations based on the procedure of taking a free body. The principle is of general use. However, here we shall limit its application to the analysis of statically determinate beams, as illustrated in the following examples.

**Example 3-7.** Figure 3-12(a) shows a simple beam $AB$ subjected to a single concentrated load $P$ acting at point $C$. We wish to find the reaction at end $A$ by the method of virtual work.

To do this, we start by removing the constraint at $A$ and substituting for it a vertical force $R_A$, as shown in Fig. 3-12(b). Note that in applying the principle of virtual work, we must remove certain constraints in a stable structure since otherwise no displacement is possible.

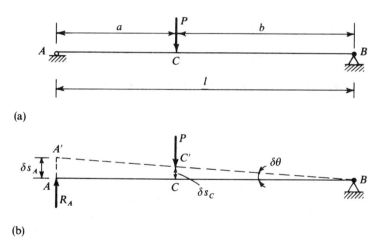

(a)

(b)

**Fig. 3-12**

Now if an infinitesimal virtual angular displacement $\delta\theta$ of rigid beam $AB$ about hinge at $B$ is assumed to take place, the beam will deflect as shown by the dashed line $A'C'B$ in Fig. 3-12(b). The corresponding virtual displacements at $A$ and $C$, denoted by $\delta s_A$ and $\delta s_C$ respectively, are found to be

$$\delta s_A = l\,\delta\theta$$

$$\delta s_C = b\,\delta\theta$$

Both $\delta s_A$ and $\delta s_C$ can be considered to be vertical, since the arcs can be taken as coincident with their own vertical tangents for a small rotation. Thus, the equation of virtual work can be written as

$$(R_A)(l\,\delta\theta) - (P)(b\,\delta\theta) = 0$$

from which

$$R_A = \frac{Pb}{l}$$

**Example 3-8.** In Fig. 3-13(a) is shown a simple beam subjected to a concentrated load $P$ applied at midspan. Let us find the shear and moment at section $C$ by the method of virtual work.

(a)

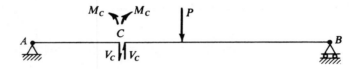

(b)

(c)

(d)

**Fig. 3-13**

Before starting, we should note that the internal forces of shear and moment at any section exist in pairs equal in magnitude but opposite in direction. They are workless when both acquire the same displacement; they produce work only when the section is cut and a relative displacement occurs between the two sides of the cut section.

To analyze the internal forces at section $C$ of Fig. 3-13(a), we first let the beam be cut at section $C$ and replaced by shearing forces $V_C$ and resisting moments $M_C$ in pairs, as shown in Fig. 3-13(b), thus placing the beam in a state of unstable equilibrium. The virtual displacement is so chosen that each time we introduce only one active unknown element into the system, to avoid solving simultaneous equations. Thus, to find $V_C$ we let portion $AC$ turn about $A$ through a virtual angular displacement $\delta\theta$ and the portion $BC$ rotate about $B$ with same virtual angular displacement $\delta\theta$. This leaves the internal moment forces workless, since their contributions ($M_C \, \delta\theta - M_C \, \delta\theta$) cancel each other. In other words, we allow a small, relative transverse sliding between the two portions $AC$ and $BC$ at $C$ and at the same time prevent them from relative rotation. The displacement diagram for the beam consistent with the assumed angular displacements is shown by the dashed lines $AC_1C_2B$ in Fig. 3-13(c). Based on the load diagram in Fig. 3-13(b) and the displacement diagram in Fig. 3-13(c), the equation of virtual work can be wirtten

$$V_C(a \, \delta\theta + b \, \delta\theta) - P\left(\frac{l}{2} \, \delta\theta\right) = 0$$

or

$$V_C l \, \delta\theta - P\frac{l}{2} \, \delta\theta = 0$$

from which

$$V_C = \frac{P}{2}$$

Similarly, to find $M_C$ we let both portions $AC$ and $BC$ move transversely the same virtual displacement $\delta s$ at $C$, thus leaving the shearing forces inactive in work. In other words, we allow relative rotation between the two portions $AC$ and $BC$ at $C$ but at the same time prevent them from relative transverse sliding. The displacement diagram for the beam, consistent with the assumed $\delta s$, is given in Fig. 3-13(d) by dashed line $AC'B$. Based on the load diagram in Fig. 3-13(b) and the displacement diagram in Fig. 3-13(d), we may write the equation of virtual work as

$$M_C\left(\frac{\delta s}{a} + \frac{\delta s}{b}\right) - P\left(\frac{l \, \delta s}{2b}\right) = 0$$

or

$$M_C\left(\frac{l \, \delta s}{ab}\right) - P\left(\frac{l \, \delta s}{2b}\right) = 0$$

from which

$$M_C = \frac{Pa}{2}$$

**Example 3-9.** The technique demonstrated in Example 3-8 can be used to solve more complicated cases, like the compound beam loaded as shown in Fig. 3-14(a). Let us determine the reactions at supports $A$ and $E$.

(a)

(b)

(c)

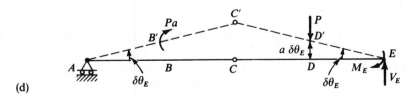

(d)

**Fig. 3-14**

To find the reaction at $A$, called $R_A$, we remove the support and introduce $R_A$ at $A$. Furthermore, we let end $A$ move vertically a small virtual displacement $\delta s_A$. (Note that, as explained before, this is equivalent to rotating the portion $AC$ through a small angle about the hinge at $C$.) The whole beam undergoes deflections as shown in Fig. 3-14(b) by $A'B'CDE$. Applying the principle of virtual work, we have

$$(R_A)(\delta s_A) + (Pa)\left(\frac{\delta s_A}{2a}\right) = 0$$

from which
$$R_A = -\frac{P}{2}$$

Next, to determine the reaction at $E$, we remove the constraint at $E$ and introduce in its place the reaction components $V_E$ and $M_E$. Solving $V_E$, we let point $E$ move a small vertical distance $\delta s_E$ without rotating the beam axis at $E$; the beam will deflect as shown by the dashed line $AB'C'D'E'$ in Fig. 3-14(c). Applying the virtual work equation gives

$$(V_E)(\delta s_E) - (P)(\delta s_E) - (Pa)\left(\frac{\delta s_E}{2a}\right) = 0$$

from which
$$V_E = \frac{3P}{2}$$

Note that in this case $M_E$ is inactive in work.

Solving $M_E$, we let $CE$ rotate about the point $E$ through a small angular displacement $\delta\theta_E$ without producing linear movement at $E$. The deflected beam consistent with this is shown by the dashed line $AB'C'D'E$ in Fig. 3-14(d). Thus,

$$(M_E)(\delta\theta_E) - (P)(a\,\delta\theta_E) - (Pa)(\delta\theta_E) = 0$$

or
$$M_E = 2Pa$$

Note that in this case $V_E$ is inactive in work.

From Example 3-9 we see that the value of the principle of virtual work in solving problems of statics is that no free bodies need be taken and that solving simultaneous equations can be avoided.

## 3-4. RELATIONSHIPS BETWEEN LOAD, SHEAR, AND BENDING MOMENT

There exist at any cross section of a loaded beam certain relationships between load, shear, and bending moment that are tremendously helpful in constructing the shear and bending moment curves.

Consider a portion of a beam of any type subjected to transverse loading and moment loading, such as the one shown in Fig. 3-15(a). To investigate the relationships between load, shear, and bending moment in a beam, we may classify the beam segments in the following way [as partly illustrated in Fig. 3-15(a)]:

1. Segment under no load
2. Segment under distributed load
3. Segment under concentrated load
4. Segment under moment load

We shall deal with each of these four cases as follows:

1. *Segment under no load.* As indicated in Fig. 3-15(a), a segment between two concentrated loads is an example of a segment under no load. Let us take

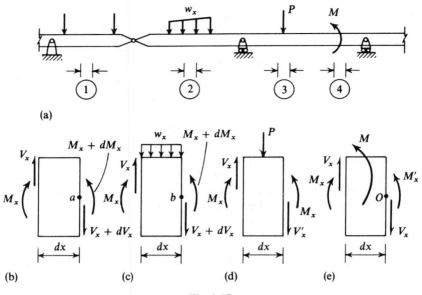

Fig. 3-15

an element cut out by two adjacent cross sections distance $dx$ apart, as shown in Fig. 3-15(b). On the left-hand face of this element, we represent the shear force and bending moment by $V_x$ and $M_x$ and on the right-hand face of the element by $V_x + dV_x$ and $M_x + dM_x$, in which $dV_x$ and $dM_x$ are changes of shear and moment in $dx$. We assume that $x$ increases from left to right. Since the element is in equilibrium, we have from $\sum F_y = 0$

$$V_x - (V_x + dV_x) = 0$$

that is,

$$dV_x = 0$$

or

$$V_x = \text{constant} \qquad (3\text{-}6)$$

Also from $\sum M_a = 0$,

$$M_x + V_x dx - (M_x + dM_x) = 0$$

Reducing and using Eq. 3-6, we arrive at

$$\frac{dM_x}{dx} = \text{constant} \qquad (3\text{-}7)$$

Equation 3-6 states that no change of shear takes place, and Eq. 3-7 states that the rate of change of bending moment at any point with respect to $x$ is constant.

2. *Segment under distributed load.* Let us take an element subjected to a distributed load cut out by two adjacent cross sections distance $dx$ apart, as shown in Fig. 3-15(c). Assume a downward distributed load in a positive direction. From $\sum F_y = 0$,

$$V_x - (V_x + dV_x) - w_x\, dx = 0$$
$$dV_x = -w_x\, dx$$

or
$$\frac{dV_x}{dx} = -w_x \qquad (3-8)$$

From $\sum M_b = 0$,

$$M_x + V_x \, dx - w_x \, dx \frac{dx}{2} - (M_x + dM_x) = 0$$

Neglecting the small term $w_x(dx)^2/2$ and reducing, we find that

$$\frac{dM_x}{dx} = V_x \qquad (3-9)$$

Equation 3-8 states that the rate of change of shear with respect to $x$ at any point is equal to the intensity of the load at that point but with the opposite sign. Equation 3-9 states that the rate of change of bending moment with respect to $x$ at any point is equal to the shear force at that point.

3. *Segment under concentrated load.* Figure 3-15(d) shows an element subjected to a concentrated load $P$. Now $P$ is assumed to be acting at a point. As the distance between two adjacent sections becomes infinitesimal, there will be no moment difference between the sections to the immediate left of $P$ and to the immediate right of $P$. However, an abrupt change in the shear force equal to $P$ between the two sections takes place, since from $\sum F_y = 0$,

$$V_x - P - V_x' = 0$$

or
$$V_x' = V_x - P \qquad (3-10)$$

as indicated in Fig. 3-15(d). Accordingly, there will be an abrupt change in the derivative $dM_x/dx$ at the point of application of concentrated force.

4. *Segment under moment load.* In Fig. 3-15(e) is shown an element subjected to a couple of $M$. Now $M$ is assumed to be acting at a point. As the distance between the two adjacent sections becomes infinitesimal, there will be no shear difference between the sections to the immediate left of $M$ and to the immediate right of $M$. However, there will be an abrupt change of moment equal to $M$ between the two sections, for from $\sum M_o = 0$ we have

$$M_x - M - M_x' = 0$$

or
$$M_x' = M_x - M \qquad (3-11)$$

as indicated in Fig. 3-15(e).

Construction of the shear and moment diagrams is facilitated by the relationships previously stated. For instance, the equation

$$\frac{dV_x}{dx} = -w_x$$

implies that the slope of the shear curve at any point is equal to the negative value of the ordinate of load diagram applied to the beam at that point. There are cases worth noting:

1. For a segment under no load, the slope of the shear curve is zero (i.e.,

parallel to the beam axis). The shear curve is therefore a line parallel to the beam axis.

2. For a segment under a uniform load of intensity $w$, the slope of the shear curve is constant. The shear curve is therefore a sloping line.

3. At a point of concentrated load, the intensity of the load is infinite, and the slope of the shear curve will thus be infinite (i.e., vertical to the beam axis). There will be a discontinuity in the shear curve, and a change process of shear equal to the applied force occurs between the two sides of the loaded point.

4. Under distributed load the change in shear between two cross sections a differential distance $dx$ apart is

$$dV_x = -w_x \, dx$$

Thus, the difference in the ordinates of the shear curve between any two points $a$ and $b$ is given by

$$V_b - V_a = - \int_{x_a}^{x_b} w_x \, dx$$
$$= -(\text{area of load diagram between } a \text{ and } b) \tag{3-12}$$

Suppose that there are additional concentrated forces $\sum P$ acting between $a$ and $b$. The result of the shear difference between the two points must include the effect due to $\sum P$:

$$V_b - V_a = - \int_{x_a}^{x_b} w_x \, dx - \sum P$$
$$= -(\text{area of load diagram between } a \text{ and } b + \sum P) \tag{3-13}$$

in which $\sum P$ has been assumed to act downward.

Similarly, from the equation

$$\frac{dM_x}{dx} = V_x$$

the slope of the bending moment curve at any point equals the ordinate of the shear curve at that point. We note the following:

1. If the shear is constant in a portion of the beam, the bending moment curve will be a straight line in that portion.

2. If the shear varies in any manner in a portion of the beam, the bending moment curve will be a curved line.

3. At a point where a concentrated force acts there will be an abrupt change in the ordinate of shear curve and, therefore, an abrupt change in the slope of the bending moment curve at the point. In fact, the moment curve will have two different slopes at that point.

4. Maximum and minimum bending moments occur at the points where a shear curve goes through the $x$ axis—the maximum where shear changes from positive (at the left) to negative (at the right); the minimum in the reverse manner.

5. For a concentrated force system the maximum bending moment must occur under a certain concentrated force, since change of shear from positive to negative must occur at a certain point where a concentrated force is applied.

6. Referring to the equation $dM_x/dx = V_x$, we find that under transverse loading the change in bending moment between two cross sections a differential distance $dx$ apart is given by

$$dM_x = V_x \, dx$$

Therefore, the difference in the ordinates of the bending moment curve between any two points $a$ and $b$ is given by

$$M_b - M_a = \int_{x_a}^{x_b} V_x \, dx$$

$$= \text{area of shear diagram between } a \text{ and } b \tag{3-14}$$

If there are external moments $\sum M$ acting between $a$ and $b$, then the result of the moment difference between the two points must include the effect due to these moments:

$$M_b - M_a = \int_{x_a}^{x_b} V_x \, dx - \sum M$$

$$= (\text{area of shear diagram between } a \text{ and } b) - \sum M \tag{3-15}$$

in which $\sum M$ has been assumed to act in a counterclockwise direction.

### 3-5. NUMERICAL EXAMPLES

The following examples will serve to illustrate the construction of shear and bending moment diagrams of transversely loaded beams by using the principles stated in Sec. 3-4.

**Example 3-10.** Consider the beam shown in Fig. 3-16(a). From $\sum M_b = 0$ and $\sum M_d = 0$, the support reactions are found to be

$$R_b = 22 \text{ kN} \qquad R_d = 14 \text{ kN}$$

We may now regard the beam as being in equilibrium under the balanced system of applied loads and reactions and present the load diagram as shown in Fig. 3-16(b).

A freehand sketch of the shear diagram can then be drawn, as in Fig. 3-16(c). In connection with this diagram, we note the following facts:

1. The shear at $a$ goes from 0 dropping to $-5$ kN; also, the shear at $e$ is 0. Recall that the shear curve always starts at zero and ends up at zero.

2. There will be constant shear in portion $ab$ since it is not loaded. As a result, the shear curve in this portion is a horizontal line parallel to the beam axis.

3. Except abrupt changes in shear at $b$, $c$, and $d$ corresponding to the concentrated forces acting at these points, the shear curves from $b$ to $e$ are sloping

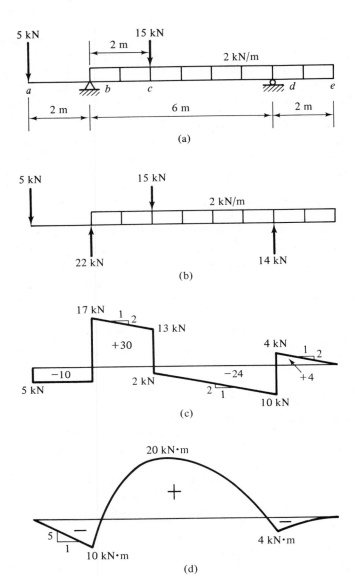

**Fig. 3-16**

lines, the slope being given by

$$\frac{dV}{dx} = -w = -2$$

that is, 2 : 1 downward to the right, as indicated in Fig. 3-16(c).

A freehand sketch of a bending moment diagram can be drawn, as in Fig. 3-16(d). In connection with it, we note the following:

1. Moments at $a$ and $e$ are null. The moment curve from $a$ to $b$ is a sloping line with the slope given by

$$\frac{dM}{dx} = V = -5$$

that is, 5 : 1 downward to the right, as indicated in Fig. 3-16(d).

2. There are extreme values of moment at points $b$, $c$, and $d$ where the shear curve goes through the $x$ axis. Minimum bending moments occur at $b$ and $d$ since abrupt changes in the slope of moment curve from negative to positive take place at these points, corresponding to the abrupt changes in shear from negative to positive. Maximum bending moment occurs at $c$, where an abrupt change in the slope of the moment curve from positive to negative takes place, corresponding to the shear change from positive to negative at $c$.

3. Since the shear curve between $bc$ or $cd$ or $de$ decreases from left to the right, the slope of the moment curve in each portion also decreases from left to right. This means that the moment curve in each portion is concave downward.

4. One way to obtain the ordinates of the moment diagram at $b$, $c$, and $d$ is to compute the areas of the shear diagram [see the values indicated in Fig. 3-16(c)], from which we may find the moment difference between any two points:

$$M_b - M_a = -10 \text{ kN·m} \qquad M_c - M_b = 30 \text{ kN·m}$$
$$M_d - M_c = -24 \text{ kN·m} \qquad M_e - M_d = 4 \text{ kN·m}$$

From the above and using $M_a = M_e = 0$, we find that

$$M_b = -10 \text{ kN·m} \qquad M_c = 20 \text{ kN·m} \qquad M_d = -4 \text{ kN·m}$$

as indicated in Fig. 3-16(d).

It may be noted that the algebraic sum of the total area of the shear diagram for the beam is zero in this example, since $M_a = M_e = 0$ and there is no moment force acting between $a$ and $e$.

**Example 3-11.** Figure 3-17(a) shows a simple beam carrying a distributed load that varies linearly from 2 kips/ft at the left end to 0 at the right end.

Applying $\sum M_b = 0$ and $\sum M_a = 0$, we obtain the vertical reaction at $a$ equal to 16 kips and that at $b$ equal to 8 kips, respectively. The load diagram for the beam is shown in Fig. 3-17(b).

In connection with the sketch of the shear diagram shown in Fig. 3-17(c), we note:

1. The shear ordinates at $a$ and $b$ are from zero to 16 kips and $-8$ kips, respectively.

2. The shear curve has a slope of $-(2 : 1)$ at $a$ and zero at $b$; the slope of the shear curve increases continuously from left to right, which corresponds to a continuously decreasing load intensity from left to right. Hence, the shear curve is concave upward.

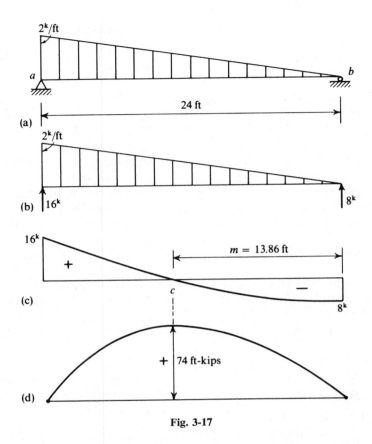

**Fig. 3-17**

In connection with the sketch of the moment diagram shown in Fig. 3-17(d), we note:

1. $M_a = M_b = 0$. Between $a$ and $b$ the moment curve is concave downward corresponding to the continuously decreasing shear (i.e., the continuously decreasing slope of the moment curve from the left end to the right end).

2. There will be a maximum bending moment corresponding to the zero shear, the location of which may be determined by writing the expression for shear at point $c$ in terms of its distance $m$ from end $b$ [see Fig. 3-17(c)], equating this shear to zero, and then solving for $m$. Thus,

$$V_c = \left(\frac{1}{2}\right)\left(\frac{2m}{24}\right)(m) - 8 = 0$$

$$m = 8\sqrt{3} = 13.86 \text{ ft}$$

In this particular case the maximum bending moment can be more conveniently obtained by conventional calculation than from the area of shear diagram.

Using the portion to the right of $c$ as free body, we obtain

$$M_{max} = (8)(13.86) - \frac{(2)(13.86)^3}{(2)(3)(24)} = 74 \text{ ft-kips}$$

## PROBLEMS

**·3-1.** In each part of Fig. 3-18 qualitative loadings are shown. Draw the shear and moment diagrams consistent with these loadings; give the equation for each curve.

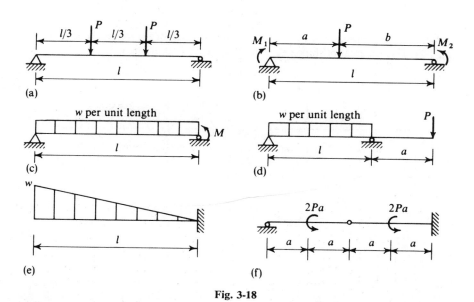

**Fig. 3-18**

**3-2.** Calculate the reactions at supports $A$, $B$, and $D$, the moment at $B$, and the shear at $C$ of the compound beam shown in Fig. 3-19 by the method of virtual work.

**Fig. 3-19**

**3-3.** Use the method of virtual work to find the reactions induced at supports $A$ and $E$ and the shear and moment at $B$ of the compound beam shown in Fig. 3-20.

Fig. 3-20

**3-4.** Sketch the shear and moment diagrams for each of the loaded beams shown in Fig. 3-21. Use the relationships between load, shear, and moment.

Fig. 3-21

56

# 4

## STATICALLY DETERMINATE TRUSSES

### 4-1. GENERAL

A *truss*, such as the one shown in Fig. 4-1, may be defined as a plane structure composed of a number of members joined together at their ends by smooth pins so as to form a rigid framework the external forces and reactions of which are assumed to lie in the same plane and to act only at the pins. Furthermore, we assume that the centroidal axis of each member coincides with the line connecting the joint centers at the ends of member and that the weight of each member is negligible in comparison to the other external forces acting on the truss. From these conditions it follows that each member in a truss is a *two-force* member and is subjected only to direct axial forces (tension or compression). A truss is completely analyzed when the internal axial forces in all members have been determined. It is customary to use a plus sign to designate tension and a minus sign to designate compression. Thus +5 kips means a tension of 5 kips, and −5 kips a compression of 5 kips.

A modern truss made of bolted or welded joints is not really a truss by a strict interpretation of this definition. However, since a satisfactory stress analysis may usually be worked out by assuming that such a structure acts as if it were pin-connected, it may still be called a truss.

Common trusses may be classified according to their formations as *simple*, *compound*, and *complex*.

1. Simple truss. A rigid plane truss can always be formed by beginning with three bars pinned together at their ends in the form of a triangle and then extending from this two new bars for each new joint, as explained in Sec. 2-6. Of course, the new joint and the two joints to which it is connected should never lie along the same straight line, to avoid geometric instability. Trusses the members of which have been so arranged are called *simple trusses*, for they are the simplest type of bar arrangement encountered in practice.

The trusses shown in Fig. 4-2 are all simple trusses. The shaded triangle *abc*

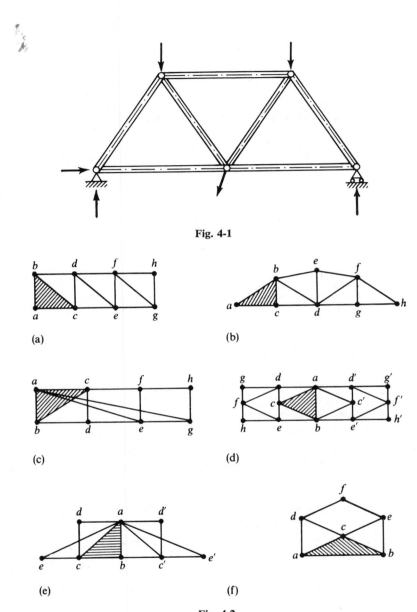

Fig. 4-1

(a)

(b)

(c)

(d)

(e)

(f)

Fig. 4-2

in each truss diagram is the base figure from which we extended the form by using two additional bars to connect each of the new joint in alphabetical order.

It can easily be shown that there exists a very definite relationship between the number of bars $b$ and the number of joints $j$ in a simple truss. Since the base triangle of a simple truss consists of three bars and three joints, the additional bars and joints required to complete the truss are $(b - 3)$ and $(j - 3)$, respectively. These two numbers should be in a 2 : 1 ratio. Thus,

$$b - 3 = 2(j - 3)$$

or

$$b + 3 = 2j$$

Comparing the above equation with the necessary condition for a statically determinate truss (see Sec. 2-6) given by

$$b + r = 2j$$

we find that, if the supports of a simple truss are so arranged that they are composed of three elements of reaction neither parallel nor concurrent, then the structure is stable and statically determinate under general conditions of loading.

2. Compound truss. If two or more simple trusses are connected together to form one rigid framework, the composite truss is called a *compound truss*. One simple truss can be rigidly connected to another simple truss at certain joints by three links neither parallel nor concurrent or by the equivalent of this type of connection. Additional simple trusses can be joined in a similar manner to the framework already constructed to obtain a more elaborate compound truss.

Figure 4-3 shows several examples of compound trusses in which the simple

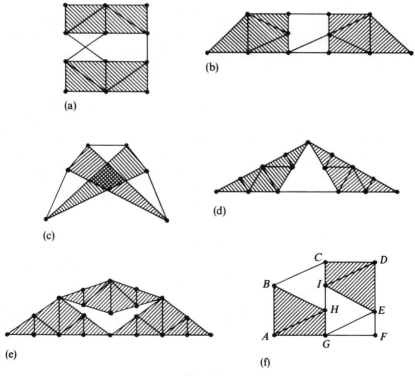

(a)

(b)

(c)

(d)

(e)

(f)

**Fig. 4-3**

trusses that have been connected together are crosshatched. In Fig. 4-3(a)–(c) two simple trusses are connected by three links. In Fig. 4-3(d) two simple trusses are connected by one hinge and one link, which constitute three elements of connection. In Fig. 4-3(e) two simple trusses may be connected by a hinge and a link (bar) similar to that of Fig. 4-3(d); however, the link is replaced by a third simple truss. Sometimes a compound truss is not apparent. The truss shown in Fig. 4-3(f) is such a case. This truss is composed of two simple trusses *ABHG* and *CDEI* connected by two bars *BC* and *HI* and a third truss *EFG*; therefore, it is a compound truss.

Note that the relationship between the number of bars and the number of joints for a simple truss,

$$b + 3 = 2j$$

still holds for a compound truss. Therefore, a compound truss may be supported in the same way as a simple truss to obtain a stable and statically determinate structure.

3. Complex truss. Trusses that cannot be classified as either simple or compound are called *complex trusses*.

A simple truss is shown in Fig. 4-4(a). Rearranging the bars results in a compound truss such as the one shown in Fig. 4-4(b). However, the truss shown in Fig. 4-4(c), made of the same number of bars and joints, does not belong to either of the above categories and may be termed a complex truss. Similarly, we find that the truss shown in Fig. 4-5(a) is a simple truss, that in Fig. 4-5(b) a compound truss, and that in Fig. 4-5(c) a complex truss.

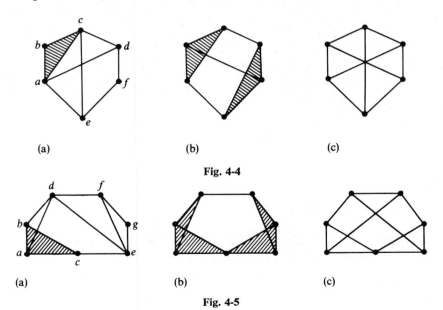

(a)          (b)          (c)

**Fig. 4-4**

(a)          (b)          (c)

**Fig. 4-5**

## 4-2. ANALYSIS OF SIMPLE TRUSSES: METHOD OF JOINT AND METHOD OF SECTION

The method of joint and the method of section are the most fundamental tools in the stress analysis of trusses. These procedures may be explained by considering a specific example, such as the simple truss shown in Fig. 4-6.

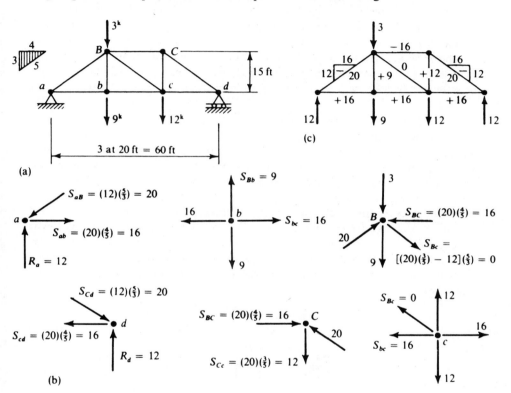

Fig. 4-6

*Method of joint.* The reactions

$$R_a = R_d = 12 \text{ kips}$$

are first obtained by taking the whole truss as a free body.

The two equations of equilibrium $\sum F_x = 0$ and $\sum F_y = 0$ are then applied to the *free body of each joint* in such an order that not more than two unknowns are involved in each free body. This can always be done for a simple truss. In this example we start with joint $a$ at the left end and proceed in succession to joints $b$ and $B$; then we turn to joint $d$ at the right end and proceed to joints $C$ and $c$. We thus provide three checks for the analysis by obtaining the internal forces in members $BC$, $Bc$, and $bc$ from two directions.

The stress analysis for each joint is given briefly in Fig. 4-6(b). Usually, when the slopes of the members are in simple ratios, the solution for unknowns can readily be obtained by inspection rather than by using equations. The arrows in each free body of the joint indicate the directions of the member forces acting on the joint, not the actions of the joint on the member. Note that the internal force in the member is a tensile force if it acts outward such as $S_{ab}$ and that the internal force in the member is a compressive force if it acts toward the joint such as $S_{aB}$.

The *answer diagram* [Fig. 4-6(c)] gives the results obtained from the preceding analysis together with the horizontal and vertical components. A plus sign indicates a tensile force, and a minus sign indicates a compressive force.

*Method of section.* Sometimes when only the stresses in certain members are desired or when the method of joint is less convenient for solving stresses, it is expedient to use the method of section, which involves isolating a portion of the truss by cutting certain members and then solving the stresses on these members with the equilibrium equations. Consider the truss in Fig. 4-6(a). Let us determine the internal forces in the members $BC$, $Bc$, and $bc$.

We start by passing a section $m$–$m$ through these members and treating either side of the truss as a free body (see Fig. 4-7). Note that the sense of the unknown force in each cut member is assumed to be tensile and if this is done, a plus sign in the answer indicates that the assumed sense is correct, and therefore tension; whereas a minus sign indicates that the assumed sense is incorrect, and therefore compression.

**Fig. 4-7**

Since in each free body only three unknowns are involved, the unknowns can be solved by three equilibrium equations. In this example, it is convenient to solve $S_{BC}$ by $\sum M_c = 0$; $S_{bc}$ by $\sum M_B = 0$; and $S_{Bc}$ by $\sum F_y = 0$. Thus, if we consider the left portion of Fig. 4-7 as a free body, we have

$$\sum M_c = 0 \qquad (12)(40) - (9 + 3)(20) + 15S_{BC} = 0$$
$$S_{BC} = -16 \text{ kips (compression)}$$
$$\sum M_B = 0 \qquad (12)(20) - 15S_{bc} = 0$$
$$S_{bc} = +16 \text{ kips (tension)}$$
$$\sum F_y = 0 \qquad 12 - (9 + 3) - V_{Bc} = 0$$
$$V_{Bc} = 0 \quad \text{or} \quad S_{Bc} = 0$$

in which $V_{Bc}$ represents the vertical component of $S_{Bc}$. Since $S_{Bc} = (5/3)V_{Bc}$, the zero value of $V_{Bc}$ evidently implies the nonexistence of $S_{Bc}$.

In applying the method of section, we note that by proper choice of moment centers we can often determine the stresses on certain members, such as the members $BC$ and $bc$ of Fig. 4-7, directly from the moment equations and avoid solving simultaneous equations. This technique is called the *method of moment* and can best be illustrated in the following example.

**Example 4-1.** In Fig. 4-8(a) is shown a simple nonparallel chord truss. Find the stresses in chord members $cd$ and $CD$ and in the diagonal $Cd$.

First, from $\sum M_i = 0$ for the entire structure, the reaction at $a$ is found to be

$$R_a = \frac{(5)(60) + (4 + 3 + 2 + 1)(90)}{8} = 150 \text{ kips}$$

(a)

(b)          (c)          (d)

**Fig. 4-8**

Next, to find the internal force in member $cd$, we pass a section $m$–$m$ through members $CD$, $Cd$, and $cd$, as indicated by the dashed line in Fig. 4-8(a), and take the left portion of the truss as a free body, as shown in Fig. 4-8(b). From $\sum M_C = 0$,

$$S_{cd} = \frac{(150)(50)}{30} = 250 \text{ kips (tension)}$$

To find the internal force in member $CD$, we use the same free body and resolve $S_{CD}$ into a vertical component $V_{CD}$ and a horizontal component $H_{CD}$ at $D$, as shown in Fig. 4-8(c). From $\sum M_d = 0$,

$$H_{CD} = -\frac{(150)(75)}{33} = -341 \text{ kips} \qquad \text{(compression)}$$

Thus, $\qquad S_{CD} = (-341)\left(\frac{25.2}{25}\right) = -344 \text{ kips} \qquad \text{(compression)}$

Similarly, to find the internal force in member $Cd$, we resolve $S_{Cd}$ into a vertical component $V_{Cd}$ and a horizontal component $H_{Cd}$ at $d$, as shown in Fig. 4-8(d). Note that the moment center is chosen at $o$, where the extending lines of members $CD$ and $cd$ intersect. The distance of $oa$ is found to be 200 ft. From $\sum M_o = 0$,

$$V_{Cd} = \frac{(150)(200)}{275} = 109 \text{ kips} \qquad \text{(tension)}$$

Thus, $\qquad S_{Cd} = (109)\left(\frac{7.8}{6}\right) = 142 \text{ kips} \qquad \text{(tension)}$

In general, no truss is analyzed by one method alone. Instead, it is often analyzed by a mixed method based on knowledge from both the joint and section methods combined as illustrated in the following example.

**Example 4-2.** Consider the K truss loaded as shown in Fig. 4-9(a). The reactions are found to be

$$R_a = 150 \text{ kN} \qquad \text{and} \qquad R_g = 125 \text{ kN}$$

To obtain chord member stresses, such as the internal force in member $cd$, we pass a section $m$–$m$ through four bars, as indicated by the dashed line, use either side of the truss as a free body, and apply $\sum M_C = 0$. Thus, if the left portion of the truss is taken as the free body,

$$S_{cd} = \frac{(150)(8)}{6} = 200 \text{ kN} \qquad \text{(tension)}$$

Similarly, we apply $\sum M_c = 0$ to obtain the internal force in chord member $CD$

$$S_{CD} = -\frac{(150)(8)}{6} = -200 \text{ kN} \qquad \text{(compression)}$$

(a)

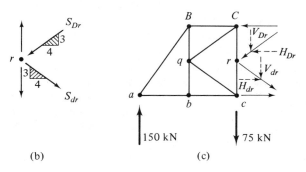

(b)                    (c)

**Fig. 4-9**

These results can be checked by $\sum F_x = 0$; in this case the internal forces in the chord members $cd$ and $CD$ must be equal and opposite.

Suppose that we also wish to find $S_{Dr}$ and $S_{dr}$ in the truss of Fig. 4-9(a). We first fix our attention on the free body of joint $r$, as shown in Fig. 4-9(b). Assume that $S_{Dr}$ is compression; then obviously $S_{dr}$ is tension, as indicated. Furthermore,

$$S_{Dr} = S_{dr} \quad \text{or} \quad V_{Dr} = V_{dr}$$

since the slopes of members $Dr$ and $dr$ are equal and their horizontal components of bar forces must be balanced.

Next, by passing a vertical section through bars $CD$, $Dr$, $dr$, and $cd$, and by taking the left portion of truss as a free body, as shown in Fig. 4-9(c), we obtain from $\sum F_y = 0$,

$$150 - 75 - V_{Dr} - V_{dr} = 0$$

65

Using $V_{Dr} = V_{dr}$ and reducing, we arrive at

$$V_{Dr} = V_{dr} = +37.5 \text{ kN}$$

or                  $$S_{Dr} = S_{dr} = (37.5)\left(\frac{5}{3}\right) = +62.5 \text{ kN}$$

The positive sign indicates that $S_{Dr}$ is compression and $S_{dr}$ is tension, as assumed.

### 4-3. ANALYSIS OF COMPOUND TRUSSES: MIXED METHOD

A compound truss is formed by interconnecting two or more simple trusses usually by three elements of connection neither parallel nor concurrent as explained in Sec. 4-1. Such a truss, if adequately sustained by three elements of support, will result in a stable and statically determinate structure. However, it frequently happens that a compound truss cannot be completely analyzed by the method of joint alone. For instance, Fig. 4-10(a) shows a loaded compound truss that consists of two triangles connected by three bars. As a first step in the analysis of this truss, we apply the equilibrium equations to determine the reactions at $A$ and $B$ by taking the entire structure as a free body. After this, we find that the method of joint cannot be applied at the very beginning because there is no joint to which less than three bars are connected. In order to proceed we must resort to the method of section, and we cut through the three connecting members as indicated by the dashed line in Fig. 4-10(a), thus isolating triangle $ABC$. From the three equilibrium equations we obtain the internal forces for members $Aa$, $Bb$, and $Cc$. As soon as this is done, the rest of the analysis can be carried out by the method of joint.

As another illustration, let us consider the compound truss in Fig. 4-10(b).

After the reactions are found, we start the stress analysis from the left end joint $a$ by the method of joint. This done, each of the joints next met has more than two unknowns which cannot be solved by the method of joint, and the

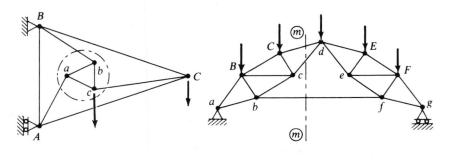

(a)                              (b)

**Fig. 4-10**

difficulty is not overcome by going to the other side of truss. However, we can remove the trouble by taking section $m$–$m$, as shown, and by the method of moment we can easily obtain $S_{bf}$ from $\sum M_d = 0$. After that, we go to joint $b$ and proceed successively to joints $B$, $c$, $C$, $d$, ... without difficulty.

A specific application of the foregoing procedure of the mixed method is illustrated in the following example.

**Example 4-3.** In Fig. 4-11(a) we have a compound truss consisting of two simple trusses (shaded) connected by bar $IJ$ and two triangle trusses $BCD$ and $FGH$. The truss is subjected to a vertical load of 60 kN at joint $E$.

The first step in the analysis is to obtain the reactions at $A$ and $G$ by considering the entire truss as a free body. Thus,

$$H_A = 120 \text{ kN} \qquad H_G = 120 \text{ kN} \qquad V_G = 60 \text{ kN}$$

as indicated. Next, from joint $G$, we find that

$$S_{FG} = -120 \text{ kN} \qquad S_{GH} = -60 \text{ kN}$$

and from joint $C$,

$$S_{BC} = S_{CD} = 0$$

(a)

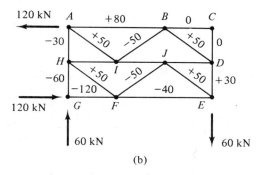

(b)

**Fig. 4-11**

After this the method of joint fails, since each remaining joint involves more than two unknowns. It also appears at first glance that it is not possible to apply the method of section, since we cannot take any section that cuts only three bars that are not concurrent. However, if we pass section $m$–$m$ through five bars, as indicated in Fig. 4-11(a), we can easily obtain from $\sum M_D = 0$ that

$$S_{AH} = -\frac{(120)(3)}{12} = -30 \text{ kN}$$

by taking either portion of the truss as a free body and by assuming that $S_{AH}$ acts in a positive direction. Similarly, from $\sum M_H = 0$ or $\sum F_x = 0$, we obtain that

$$S_{DE} = 30 \text{ kN}$$

Having done this, we can solve the stresses for the remaining bars by the method of joint without difficulty. The analysis may be facilitated by observing that in this particular case

$$S_{AI} = -S_{BI} = S_{BD} = S_{EJ} = -S_{FJ} = S_{FH} = 50 \text{ kN}$$

An answer diagram for the analysis is given in Fig. 4-11(b).

## 4-4. ANALYSIS OF COMPLEX TRUSSES:
### METHOD OF SUBSTITUTE MEMBER

In analyzing a complex truss, we frequently find that the method of joint and the method of section, described in previous sections, are not directly applicable. For example, let us consider the loaded complex truss shown in Fig. 4-12(a). After the reactions at $A$ and $E$ are found, we observe that no further progress can be made by either the method of joint or the method of section. One way to handle this is to substitute for the bar $AD$ a bar $AC$ and thus obtain a stable simple truss, as in Fig. 4-12(b), which can be completely analyzed by the method of joint for the given loading. Next, let the same simple truss be loaded with two equal and opposite forces $X$ at $A$ and $D$ along $AD$, as shown in Fig. 4-12(c). Again a complete analysis can be carried out by the method of joint such that the internal force for each member will be expressed in terms of the unknown $X$.

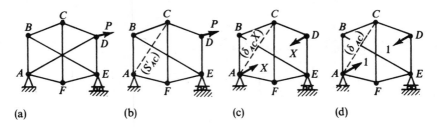

(a)            (b)            (c)            (d)

**Fig. 4-12**

Or for convenience sake, we put a pair of unit forces in place of the $X$'s, as given in Fig. 4-12(d). It is apparent that the stresses obtained from Fig. 4-12(d) times $X$ will give those of Fig. 4-12(c).

Now the analysis of Fig. 4-12(a) can be made equivalent to the superposing effects of Fig. 4-12(b) and (c) if we let the bar force of $AC$ obtained from (b) and (c), or from (b) and (d) times $X$, be equal to zero. Thus, if we let $S'_i$ denote the force in any bar of (b) and $\delta_i$ the corresponding bar force of (d), then the corresponding internal force $S_i$ in any bar of (a) is expressed by

$$S_i = S'_i + \delta_i X$$

in which $X$ denotes the bar force in member $AD$ solved by

$$S_{AC} = S'_{AC} + \delta_{AC} X = 0$$

or

$$X = -\frac{S'_{AC}}{\delta_{AC}}$$

With the value of $X$ determined, the force in any other bar of the given truss can be obtained without difficulty.

The following example will illustrate the application of the *substitute-member method* to the analysis of a complex truss.

**Example 4-4.** In analyzing the loaded complex truss shown in Fig. 4-13(a), we begin with the complete analyses of the corresponding simple trusses of Fig. 4-13(b) and (c). These can be done by the method of joints, and the results are recorded in columns (2) and (3), respectively, of Table 4-1. Next, setting

$$S_{AC} = S'_{AC} + \delta_{AC} X$$
$$= -10.89 + 0.465X = 0$$

we obtain

$$X = +23.42 \text{ kips}$$

which is the value of the bar force in $AD$.

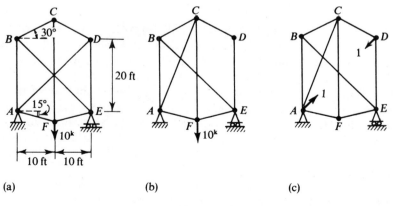

(a)                    (b)                    (c)

**Fig. 4-13**

Having found this, we may now fill in column (4) of Table 4-1 and then complete the stress analysis for the complex truss, as recorded in column (5) of Table 4-1. Note that in this particular case we obtain a special check by the condition of symmetry of the structure.

TABLE 4-1

| (1) Member | (2) $S_i'$ | (3) $\delta_i$ | (4) $\delta_i \cdot X$ | (5) $S_i(= S_i' + \delta_i \cdot X)$ |
|---|---|---|---|---|
| AB | +6.22 | −1.382 | −32.36 | −26.1 (kips) |
| BC | +4.55 | −1.010 | −23.65 | −19.1 |
| CD | 0 | −0.816 | −19.11 | −19.1 |
| DE | 0 | −1.115 | −26.11 | −26.1 |
| EF | +4.08 | −0.910 | −21.31 | −17.2 |
| FA | +4.08 | −0.910 | −21.31 | −17.2 |
| BE | −5.58 | +1.240 | +29.0 | +23.4 |
| CF | +7.89 | +0.471 | +11.03 | +18.9 |
| AC | −10.89 | +0.465 | +10.89 | 0 |

It should be noted that complex trusses may often be arranged so as to be geometrically unstable. However, it is not always possible to see a critical form just by inspection. For example, the complex truss shown in Fig. 4-14 has a

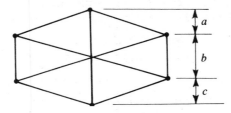

**Fig. 4-14**

critical form if $a = c$; otherwise, it is a stable form. In such cases detection is based on the principle that if the analysis for the truss yields a unique solution then the truss is stable and determinate; on the other hand, if the analysis fails to yield a unique solution, then the truss has a critical form.

The principle described above also suggests a convenient way of detecting critical form known as the *zero-load test*. For an unloaded truss, one possible solution satisfying equilibrium is that all bars of the truss have zero stress. Hence, if with no loads on a truss we can find a set of solutions for the bar stresses other than zero that will also satisfy the condition of equilibrium for the truss, then it has a critical form.

As an illustration, let us consider the complex truss shown in Fig. 4-15.

Without external forces acting on the truss, we assume a tension $S$ in the member $BC$:

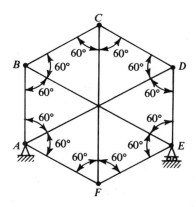

**Fig. 4-15**

$$S_{BC} = S$$

From joint $B$,

$$S_{AB} = S \quad \text{and} \quad S_{BE} = -S$$

and from joint $C$,

$$S_{CD} = S \quad \text{and} \quad S_{CF} = -S$$

Because of the identity in each joint, we can easily see that the set of solution

$$S_{AB} = S_{BC} = S_{CD} = S_{DE} = S_{EF} = S_{FA} = S \qquad S_{AD} = S_{BE} = S_{CF} = -S$$

will satisfy the condition of equilibrium at all joints. Since $S$ may be assigned any value different from zero, this indicates that the truss has a critical form. In such cases no matter how the truss is actually loaded, the truss will have an infinite number of solutions that can satisfy the equilibrium requirements of statics. Therefore, it is statically indeterminate.

## 4-5. A GENERAL METHOD FOR ANALYZING STATICALLY DETERMINATE TRUSSES

Theoretically, we can always solve any statically stable and determinate truss by $2j$ simultaneous equilibrium equations for $j$ joints of the system. The method is perfectly general but must be done with the aid of a modern computer.

As a simple illustration of this process, let us consider the three-hinged truss shown in Fig. 4-16. The unknown elements involved in this truss are the reaction components $H_A$ and $V_A$ at joint $A$, $H_B$ and $V_B$ at joint $B$, and the bar forces $S_a$ and $S_b$. The six unknowns can be solved by six equilibrium equations, two for each of three discrete joints. Thus,

Joint $A$:

$$\sum F_x = 0 \qquad H_A + 0.6 S_a = 0$$
$$\sum F_y = 0 \qquad V_A + 0.8 S_a = 0$$

71

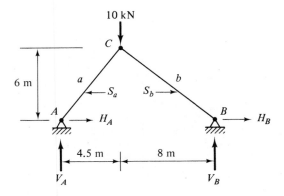

**Fig. 4-16**

Joint $B$:

$$\sum F_x = 0 \qquad H_B - 0.8S_b = 0$$
$$\sum F_y = 0 \qquad V_B + 0.6S_b = 0$$

Joint $C$:

$$\sum F_x = 0 \qquad 0.6S_a - 0.8S_b = 0$$
$$\sum F_y = 0 \qquad 0.8S_a + 0.6S_b = -10$$

Collecting the preceding six equations in matrix form gives

$$\begin{bmatrix} 1 & 0 & 0 & 0 & 0.6 & 0 \\ 0 & 1 & 0 & 0 & 0.8 & 0 \\ 0 & 0 & 1 & 0 & 0 & -0.8 \\ 0 & 0 & 0 & 1 & 0 & 0.6 \\ 0 & 0 & 0 & 0 & 0.6 & -0.8 \\ 0 & 0 & 0 & 0 & 0.8 & 0.6 \end{bmatrix} \begin{Bmatrix} H_A \\ V_A \\ H_B \\ V_B \\ S_a \\ S_b \end{Bmatrix} = \begin{Bmatrix} 0 \\ 0 \\ 0 \\ 0 \\ 0 \\ -10 \end{Bmatrix}$$

$$\qquad\qquad D \qquad\qquad\qquad X \qquad\quad E$$

or
$$DX = E$$

The unknown values are found by

$$X = D^{-1}E$$

The problem is now to find $D^{-1}$. In this case,

$$D^{-1} = \begin{bmatrix} 1 & 0 & 0 & 0 & -0.36 & -0.48 \\ 0 & 1 & 0 & 0 & -0.48 & -0.64 \\ 0 & 0 & 1 & 0 & -0.64 & 0.48 \\ 0 & 0 & 0 & 1 & 0.48 & -0.36 \\ 0 & 0 & 0 & 0 & 0.6 & 0.8 \\ 0 & 0 & 0 & 0 & -0.8 & 0.6 \end{bmatrix}$$

72

After the performance of matrix multiplication, we obtain

$$X = \begin{Bmatrix} H_A \\ V_A \\ H_B \\ V_B \\ S_a \\ S_b \end{Bmatrix} = \begin{Bmatrix} 4.8 \\ 6.4 \\ -4.8 \\ 3.6 \\ -8 \\ -6 \end{Bmatrix} \text{ kN}$$

It should be noted that the unknowns $X$ are to be uniquely determined by $X = D^{-1}E$ under the condition of the nonsingularity of the square matrix $D$; that is, the determinant containing the same elements as $D$ is not zero. This principle provides a general way of detecting the stability of a structural system; that is, if

$$|D| \neq 0$$

then the system has a unique solution, which indicates that the system is stable; on the other hand, if

$$|D| = 0$$

then the system is unstable, since many solutions are possible. Refer to Fig. 4-16. If we replace the hinge support at $B$ with a roller ($H_B = 0$), then

$$D = \begin{bmatrix} 1 & 0 & 0 & 0 & 0.6 & 0 \\ 0 & 1 & 0 & 0 & 0.8 & 0 \\ 0 & 0 & 0 & 0 & 0 & -0.8 \\ 0 & 0 & 0 & 1 & 0 & 0.6 \\ 0 & 0 & 0 & 0 & 0.6 & -0.8 \\ 0 & 0 & 0 & 0 & 0.8 & 0.6 \end{bmatrix}$$

Apparently,

$$|D| = 0$$

implying that the truss is unstable.

Although the illustration we have considered concerns a simple truss, the principle and the method of detecting a critical form described above can be applied to other types of structures, such as beams and rigid frames.

## 4-6. GENERAL DESCRIPTION OF BRIDGE AND ROOF TRUSS FRAMEWORKS

Figure 4-17 shows a typical *through trussed bridge*. The word *through* indicates that the trains (or vehicles) actually travel through the bridge. If the bridge is installed under the floor or deck, then the bridge is called a *deck bridge*. If the trains pass between trusses but the depth is insufficient to allow the use of a top chord bracing system, the bridge is called *half-through*.

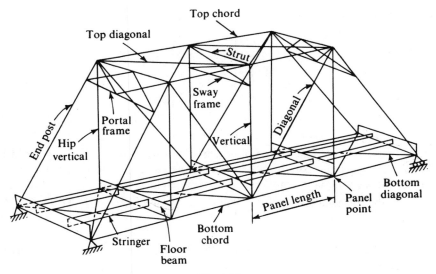

**Fig. 4-17**

Referring to Fig. 4-17, we place the road surface (or the rail and tie system in railways) on the short longitudinal beams called *stringers*, assumed simply supported on the *floor beams* which in turn are supported by the two *main trusses*. The moving loads on bridge are transmitted to the main trusses through the system of the connection of road surface (or rail and tie), stringer, and floor beam.

The top series of truss members parallel to the stringer are called *top chords;* while the corresponding bottom series of members are called *bottom chords*. The members connecting the top and bottom chords form the web system and are referred to as *diagonals* and *verticals*. The end diagonals are called *end posts*, and the side verticals are called *hip verticals*. The point at which web members connect to a chord is called a *panel point*, and the length between two adjacent panel points on the same chord is called the *panel length*.

The cross struts at corresponding top-chord panel points, together with the top diagonals connecting the adjacent struts, make up the top-chord lateral system. The bottom-chord lateral system is composed of the floor beams and the bottom diagonals connecting the adjacent floor beams.

The two main trusses are also cross braced at each top-chord panel point by *sway frames*. The frame in the plane of each pair of end posts is called a *portal frame*.

A typical roof truss framework supported by columns is shown in Fig. 4-18.

A roof truss with its supporting columns is called a *bent*. The space between adjacent bents is called a *bay*. *Purlins* are longitudinal beams that rest on the top chord and preferably at the joints of the truss, in accordance with the definition of truss. The *roof covering* may be laid directly on the purlins for very short

74

**Fig. 4-18**

bay lengths but usually is laid on the wood *sheathings* that, in turn, rest either on the purlins or on the *rafters* (if provided). Rafters are the sloping beams extending from the *ridge* to the *eaves* and are supported by the purlins.

For a symmetrical roof truss the ratio of its *rise*, center height, to its *span*, the horizontal distance between the center lines of the supports, is called *pitch*.

The truss consists of top-chord, bottom-chord, and *web members*. Although the purlins act to strengthen the longitudinal stability, additional bracing is always necessary. The *bracing members* may run from truss to truss longitudinally or diagonally and may be installed in the plane of the bottom chord, the top chord, or both. Surface loads are transmitted from covering, sheathing, rafter, purlin and distributed to adjacent trusses.

### 4-7. CONVENTIONAL TYPES OF BRIDGE AND ROOF TRUSSES

The members of a main truss may be arranged in many different ways. However, the principal types of trusses encountered in bridges and buildings are shown in Figs. 4-19 and 4-20.

Among these types, the Pratt, Howe, and Warren trusses are more commonly used. We may note that in the Pratt truss the diagonals, except the end posts, are stressed in tension and that the verticals, except the hip verticals, are stressed in compression under dead load. On the other hand, in a Howe truss the diagonals are in compression and the verticals are in tension. Note also that, of all the trusses shown in Fig. 4-19 under dead load, the upper chords are in compression and bottom chords in tension.

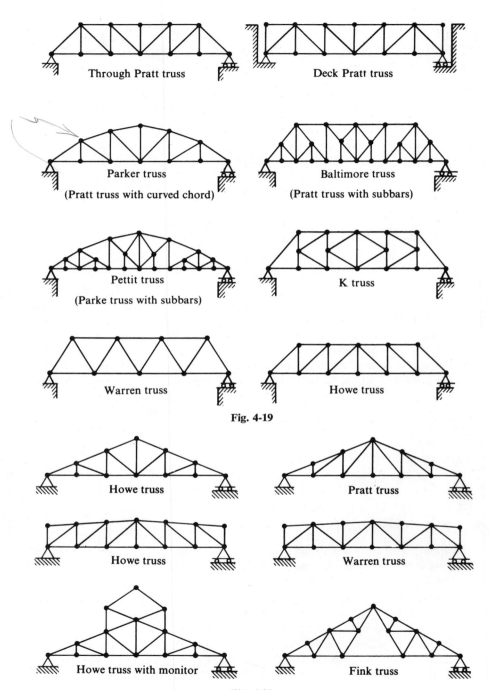

Through Pratt truss

Deck Pratt truss

Parker truss

(Pratt truss with curved chord)

Baltimore truss

(Pratt truss with subbars)

Pettit truss

(Parke truss with subbars)

K truss

Warren truss

Howe truss

**Fig. 4-19**

Howe truss

Pratt truss

Howe truss

Warren truss

Howe truss with monitor

Fink truss

**Fig. 4-20**

76

Saw-tooth truss

Three-hinged arch

**Fig. 4-20** (cont.)

## PROBLEMS

**4-1.** Classify the trusses of Fig. 4-21 as being simple, compound, or complex.

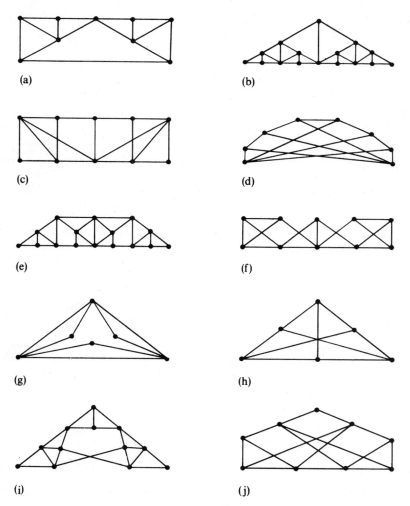

(a)

(b)

(c)

(d)

(e)

(f)

(g)

(h)

(i)

(j)

**Fig. 4-21**

**4-2.** Determine the bar force in each member of the trusses shown in Fig. 4-22 by the method of joint.

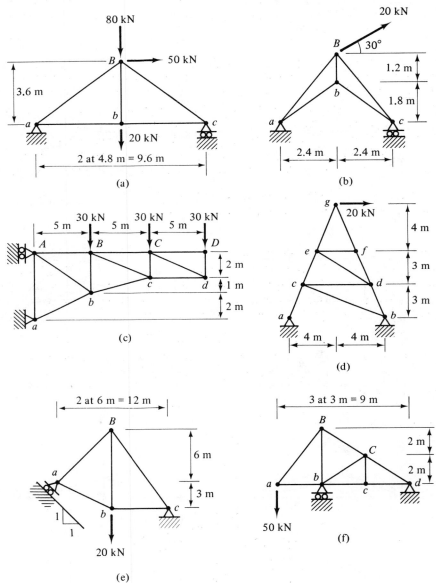

Fig. 4-22

**4-3.** By the method of section compute the bar forces in the lettered bars of the trusses shown in Fig. 4-23.

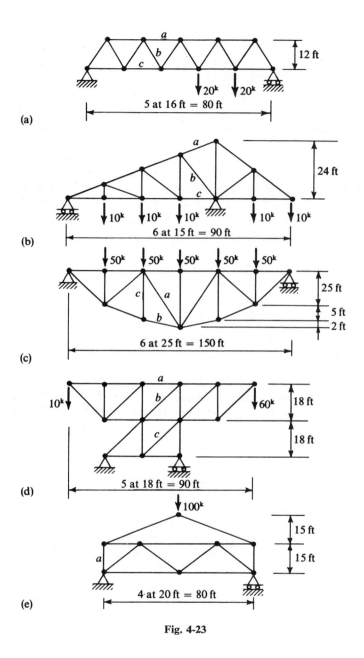

Fig. 4-23

**4-4.** By the mixed method of joint and section determine the bar forces in the lettered bars of the K truss shown in Fig. 4-24.

79

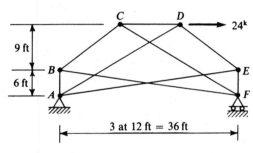

**Fig. 4-24**

**4-5.** Make a complete analysis of the compound truss shown in Fig. 4-25.

**Fig. 4-25**

**4-6.** Use the substitute-member method to make a complete analysis of the complex truss shown in Fig. 4-26. Repeat it by solving 12 joint equations if a digital computer is available.

**Fig. 4-26**

80

# 5

# STATICALLY DETERMINATE RIGID
# FRAMES AND COMPOSITE STRUCTURES

## 5-1. RIGID JOINT

A *rigid frame* may be defined as a structure composed of a number of members connected together by joints some or all of which are rigid, that is, capable of resisting both force and moment as distinguished from a pin-connected joint which offers no moment resistance. In steel structures, rigid joints may be formed by certain types of riveted, bolted, or welded connections. In reinforced concrete structures, the materials in the joined members are mixed together in one unit so as to be substantially rigid. In the analysis of rigid frames, we assume that the centroidal axis of each member coincides with the line connecting the joint centers of the ends of the member. The so-called joint center is therefore the concurrent point of all centroidal axes of members meeting at the joint. With the joint rigid the ends of all connected members must not only translate but also rotate identical amounts at the joint.

## 5-2. ANALYSIS OF STATICALLY DETERMINATE RIGID FRAMES

Rigid frames are usually built to be highly statically indeterminate. The discussion of determinate rigid frames in this chapter is primarily of academic interest rather than of practical use and serves as a prelude to the analysis of indeterminate frames.

To analyze a statically determinate rigid frame, we start by finding the reaction components from statical equations for the entire structure. This done, we are able to determine the shear, moment, and axial force at any cross section of the frame by taking a free body cut through that section and by using the equilibrium equations. Based on the centroidal axis of each member, we can plot the shear, bending moment, and the direct force diagrams for the rigid frame. However, it is the bending moment diagram with which we are mainly concerned in the analysis of a rigid frame.

The following numerical examples will serve to illustrate the procedure.

**Example 5-1.** Analyze the rigid frame in Fig. 5-1(a). Let $H_a$, $V_a$, and $M_a$, respectively, denote the horizontal, vertical, and rotational reaction components at support $a$, and let $V_e$ be the vertical reaction at support $e$.

Fig. 5-1

From the condition equation $M_d = 0$, we find that $V_e = 30$ kN; from $\sum F_x = 0$, $H_a = 6$ kN; from $\sum F_y = 0$, $V_a = 42$ kN; and from $\sum M_a = 0$, $M_a = 78$ kN·m, as indicated in Fig. 5-1(a).

After all the external forces acting on the rigid frames are determined, the internal forces at each end of the members can easily be obtained by taking each member as a free body [Fig. 5-1(b)]. Take member $ac$, for instance. At end $c$ of $ac$, we find the shear force equal to zero by applying $\sum F_x = 0$; axial force equal to 42 kN (down) from $\sum F_y = 0$; and the resisting moment equal to 90 kN·m (clockwise) from $\sum M_c = 0$. By inspection, we see that the shear and moment at end $c$ of the overhanging portion $ac$ are 4.5 kN (up) and 4.5 kN·m (clockwise), respectively. Finally, we use the equilibrium of joint $c$ to obtain the end forces at $c$ of member $ce$ as

shear $= 37.5$ kN    (up)    moment $= 94.5$ kN·m    (counterclockwise)

With all the end forces for each member found, the shear, bending moment, and axial force in any section of the frame can be obtained by simple statics.

The moment diagrams for beam $be$ and column $ac$ are shown separately in Fig. 5-1(c).

**Example 5-2.** Analyze the simply supported gable frame, shown in Fig. 5-2(a), which is composed of two columns and two sloping members.

From $\sum F_x = 0$, $\sum M_e = 0$, and $\sum F_y = 0$ for the entire frame, the reaction elements are found to be

$$H_a = 8 \text{ kips} \qquad V_a = 11 \text{ kips} \qquad V_e = 21 \text{ kips}$$

as shown in Fig. 5-2(a).

Next, we take member $ab$ as a free body. With the end forces known at $a$, we can readily obtain those at the other end $b$ from the equilibrium conditions:

shear $= 8$ kips    moment $= 80$ ft-kips    axial force $= 11$ kips

acting as indicated in Fig. 5-2(b).

Following this, we sketch the free-body diagram for joint $b$, as shown in Fig. 5-2(c). Note that the joint is shown in an exaggerated manner, since theoretically it should be represented by a point and all forces acting on the joint should be concurrent at this point.

Next, let us take member $bc$ as a free body subjected to the external load of 2 kips per horizontal unit length. With the internal forces known at end $b$, we can apply $\sum F_x = 0$, $\sum F_y = 0$, and $\sum M_c = 0$ to obtain the internal forces at end $c$ as

horizontal force $= 0$    vertical force $= 5$ kips    moment $= 104$ ft-kips

These act as indicated in the upper sketch of Fig. 5-2(d). To determine the stresses in each section of the member, we resolve all the indicated forces into components normal and tangential to the member section, as shown in the lower sketch of Fig. 5-2(d). For instance, at end $b$ we have

$$\text{normal force (axial force)} = (11)(\tfrac{3}{5}) = 6.6 \text{ kips}$$
$$\text{tangential force (shear)} = (11)(\tfrac{4}{5}) = 8.8 \text{ kips}$$

Similarly, at end $c$ we have

$$\text{normal force (axial force)} = (5)(\tfrac{3}{5}) = 3 \text{ kips}$$
$$\text{tangential force (shear)} = (5)(\tfrac{4}{5}) = 4 \text{ kips}$$

**Fig. 5-2**

The total uniform load on member $bc$ is 16 kips, of which

$$(16)(\tfrac{4}{5}) = 12.8 \text{ kips acting transversely to the member axis}$$

$$(16)(\tfrac{3}{5}) = 9.6 \text{ kips acting axially to the member axis}$$

thus giving a uniform load of intensity:

$$\frac{12.8}{10} = 1.28 \text{ kips/ft acting transversely to the member axis}$$

$$\frac{9.6}{10} = 0.96 \text{ kip/ft acting axially to the member axis}$$

With these determined, the shear, bending moment, and direct force in any section of member $bc$ can readily be obtained, as shown in Fig. 5-3.

**Fig. 5-3**

In this manner we may proceed from member $bc$ to joint $c$, then to member $cd$ and joint $d$, and finally to member $de$. However, it seems more convenient to analyze $de$ now and then to turn to joint $d$ and member $cd$, and to leave the joint $c$ as a final check, as shown in Fig. 5-2(e), (f), (g), and (h), respectively. The bending moment diagram for the whole frame is plotted in Fig. 5-2(i).

**Example 5-3.** Consider the three-hinged frame loaded as in Fig. 5-4(a). The four reaction elements at supports $a$ and $e$ are first obtained by solving simultaneous equations, three from equilibrium and one from construction.

$$\sum F_x = 0 \qquad H_a - H_e = 0$$
$$\sum F_y = 0 \qquad V_a + V_e - 12 = 0$$
$$\sum M_e = 0 \qquad 12V_a - (12)(10) = 0$$
$$M_c = 0 \qquad 6V_e - 8H_e = 0$$

which give

$$V_a = 10 \text{ kips} \qquad V_e = 2 \text{ kips} \qquad H_a = H_e = 1.5 \text{ kips}$$

The free-body diagrams for members *ab*, *bd*, *de* are then drawn as in Fig. 5-4(b). From these we plot the moment diagram for the frame, as shown in Fig. 5-4(c).

**Fig. 5-4**

It is interesting to note that in this particular case the portion to the right of hinge *c* (i.e., *cde*) carries no external load and is therefore a two-force member if isolated. The line of reaction at support *e*, called $R_e$, must be through points *e* and *c* and must meet the action line of the applied load at some point *o*, as shown in Fig. 5-5(a). Now if we take the whole frame as a free body, we see that the system constitutes a three-force member subjected to the applied load and support reactions. Thus, the line of reaction at support *a*, called $R_a$, must be through points *a* and *o* so that the three forces are concurrent at point *o* as required by equilibrium. The vectors $R_a$ and $R_e$ can then easily be determined by the equilibrium triangle, as shown in Fig. 5-5(b).

In the case where loads are placed both to the left and to the right of the

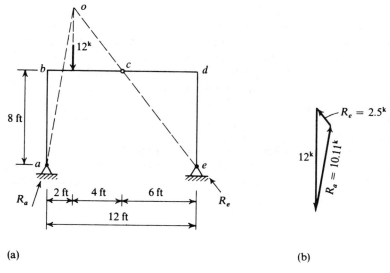

(a)

(b)

Fig. 5-5

connecting hinge of a three-hinged frame, one way to analyze this is to use the method of superposition. This is illustrated in Fig. 5-6, in which the case shown in part (a) can be made equivalent to the sum of effects of (b) and (c), each analyzed by the method discussed previously.

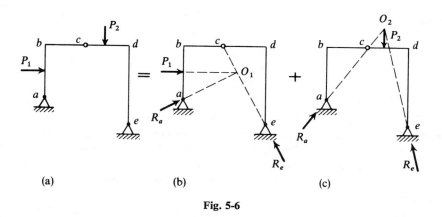

(a)

(b)

(c)

Fig. 5-6

## 5-3. ANALYSIS OF STATICALLY DETERMINATE COMPOSITE STRUCTURES

There are structures in which some members are two-force members subjected to axial forces only and others are not. Such structures may be called *beam trusses* or *truss beams* but are more frequently called *composite structures*.

87

In discussing a composite structure, we begin by investigating the *stability* and *determinacy* of the structure. The most fundamental approach is to separate the composite structure into a number of free bodies of beam and truss through connections or by cutting certain two-force members. After that, we may be in a position to see if the total number of equations of statics equals the total number of unknown elements as a *necessary* condition for statical determinacy. A final decision that the structure is both determinate and stable should be based on the criteria given in Chapter 2.

For example, Fig. 5-7(a) shows a *mill bent* (bent of a mill building) which consists of a roof truss supported at two ends *a* and *g* by two columns hinged to the foundation at *A* and *G*. In this case all truss members are two-force members while the columns are not.

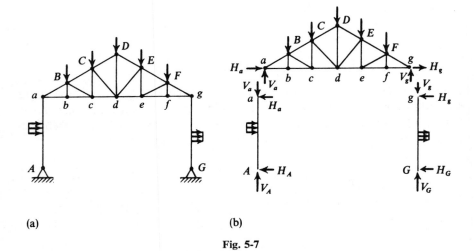

(a)                                     (b)

**Fig. 5-7**

To investigate the stability and determinacy of this composite structure, we separate it into three free bodies through connections at *a* and *g* as well as at *A* and *G*, as shown in Fig. 5-7(b). The total number of unknown elements involved is eight, whereas the total number of equations of statics is nine, since each free body generally provides three equilibrium equations. Therefore, this is an unstable structure.

However, the stability of the structure can be secured by adding two-force members *a'b* and *fg'* called *knee braces*, as shown in Fig. 5-8. The introduction of two links adds two unknowns to the system and makes the structure statically indeterminate to the first degree.

As another example, consider the trussed beam shown in Fig. 5-9(a), in which the simple beam *AD* is strengthened by the addition of trussing. To investigate the determinacy of the structure, we separate the structure into two parts by cutting through the two-force members *Ab*, *Bb*, *Cc*, and *Dc*, as shown

**Fig. 5-8**

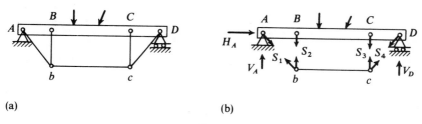

(a)  (b)

**Fig. 5-9**

in Fig. 5-9(b). The total number of unknowns involved in two free bodies is seven, four from bar stresses and three from reactions, as indicated in Fig. 5-9(b); the total number of statical equations is six. Therefore, it is statically indeterminate to the first degree.

To illustrate the stress analysis of a statically determinate composite structure, let us consider the system shown in Fig. 5-10(a), in which members $ab$ and $df$ are two-force members since they are pin-connected at ends and carry no load. If we cut through these two members and take two free bodies, as shown in Fig. 5-10(b), we at once observe that the structure is statically determinate since there is a total of six unknowns involved in the two free bodies and six independent equations of statics to solve them.

Using the free body of beam $ce$ and applying equilibrium equations, we obtain the unknown forces $S_{df}$, $H_c$, and $V_c$ as follows:

$$\sum M_c = 0 \qquad (\tfrac{3}{5}S_{df})(12) - (12)(15) = 0 \qquad S_{df} = 25 \text{ kips}$$
$$\sum F_x = 0 \qquad \tfrac{4}{5}S_{df} - H_c = 0 \qquad H_c = 20 \text{ kips}$$
$$\sum F_y = 0 \qquad \tfrac{3}{5}S_{df} - 12 - V_c = 0 \qquad V_c = 3 \text{ kips}$$

Next, using the free body of member $bg$ and applying equilibrium equations, we obtain all the remaining unknown forces as follows:

$$\sum M_g = 0 \qquad 15S_{ab} - (20)(12) + (\tfrac{4}{5})(25)(3) = 0 \qquad S_{ab} = 12 \text{ kips}$$
$$\sum F_x = 0 \qquad H_g - S_{ab} + 20 - (\tfrac{4}{5})(25) = 0 \qquad H_g = 12 \text{ kips}$$
$$\sum F_y = 0 \qquad V_g + 3 - (\tfrac{3}{5})(25) = 0 \qquad V_g = 12 \text{ kips}$$

**Fig. 5-10**

The results can be checked by taking the whole structure as a free body and applying $\sum M_g = 0$, $\sum F_x = 0$, and $\sum F_y = 0$.

After these unknowns are solved, we can determine the shear, the moment, and the axial force at any section of the structure without difficulty. The bending moment diagrams for members $bg$ and $ce$ are plotted, as shown in Fig. 5-10(c) and (d), respectively.

## 5-4. APPROXIMATE ANALYSIS FOR STATICALLY INDETERMINATE STRUCTURES

As previously mentioned, the rigid frames of present-day construction are highly indeterminate. It will be seen in the later chapters, which deal with statically indeterminate structures, that to obtain the solution for a building frame based on more exact analyses is often tedious and time consuming. In many cases, we cannot obtain the solution without the aid of modern electronic computers. For this reason empirical rules and approximate methods were often used in the past by structural and architectural engineers in designing various kinds of indeterminate structures. In order to do this, as many independent equations of statics as there are independent unknowns must be available. The additional equations of statics are worked out by reasonable assumptions based on experience and knowledge of the more exact analyses.

To illustrate, consider a frame subjected to uniform floor loads such as the one shown in Fig. 5-11(a). The frame is indeterminate to the 24th degree since eight cuts in the girders would render the frame into three stable and determinate

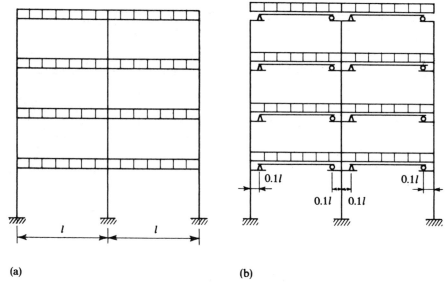

(a)                                        (b)

**Fig. 5-11**

parts and since each cut involves the removal of three elements of restraint (i.e., bending moment, shear, and axial force). A preliminary survey of stresses may be performed by assuming the following so that the indeterminate frame can be solved by a determinate approach, that is, by equations of statics alone:

1. The axial force in each girder is small and can be neglected.
2. A point of inflection (zero moment) occurs in each girder at a point one-tenth of the span length from the left end of the girder.
3. A point of inflection occurs in each girder at a point one-tenth span length from the right end of the girder.

This would render the frame equivalent to the one shown in Fig. 5-11(b), which is statically determinate.

Another case that may also be worth brief mention, without going into details, is the approximate analysis for wind stresses in building frames. Consider a frame subjected to lateral forces (equivalent wind) acting at the joints such as the one shown in Fig. 5-12(a). The frame is statically indeterminate to the 27th degree. There are several methods available for dealing with the problem. The method chosen to illustrate this is called the *cantilever method* and is based on the following assumptions:

1. A point of inflection exists at the center of each girder.
2. A point of inflection exists at the center of each column.
3. The unit axial stresses in the columns of a story vary as the horizontal

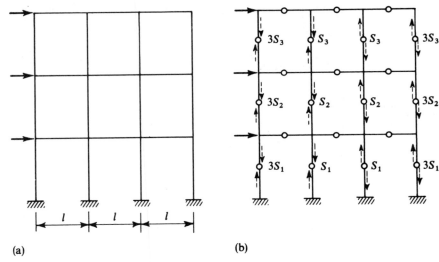

(a)                                        (b)

Fig. 5-12

distances of the columns from the center of gravity of the bent. It is usually further assumed that all columns are identical in a story, so that the axial forces of the columns in a story will vary in proportion to the distances from the center gravity of the bent.

This would lead the frame to appear as the form shown in Fig. 5-12(b). Note that the last assumption virtually puts the column axial forces in one story in terms of a single unknown [see the dashed arrows in Fig. 5-12(b)]. It is therefore equivalent to making $(n - 1)$ additional assumptions for each story, $n$ being the number of columns in one story. In this case, there are three for each story, or nine in total regarding column axial forces. As a result, the total number of additional equations is 30 (9 from column axial forces and 21 from inserting pins), which is three more than are necessary. However, it happens that a statical analysis for the frame can be carried out without inconsistency on the basis of the foregoing assumptions.

Like building frames, statically indeterminate composite structures are often analyzed by a determinate approach. Take, for instance, the mill bent acted on by lateral loads, as shown in Fig. 5-13(a). It is statically indeterminate to the third degree. An approximate analysis of stresses may be carried out on the basis of the following assumptions:

1. The horizontal reactions at the bases of the columns are equal.

2. A point of inflection occurs midway between the base $A$ and the end $B$ of the knee brace of the left column.

3. A point of inflection occurs midway between the base $D$ and the end $C$ of the knee brace of the right column.

92

As indicated in Fig. 5-13(b), the six reaction elements can be determined by the three equilibrium equations together with the foregoing three assumptions. After these are determined, the rest of analysis can be worked out easily.

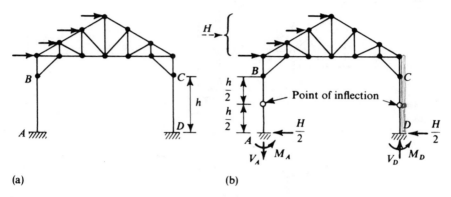

(a)                                                (b)

Fig. 5-13

## PROBLEMS

**5-1.** Analyze each of the frames shown in Fig. 5-14, and draw the bending moment diagram.

(a)

(b)

Fig. 5-14

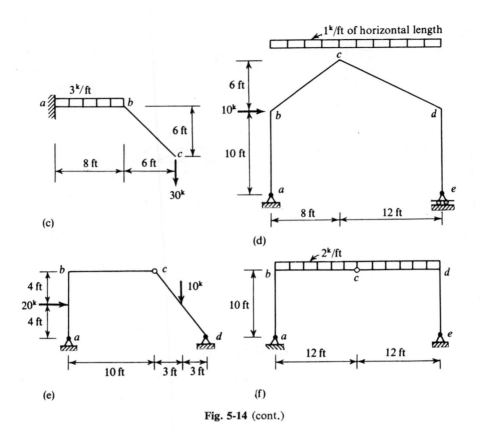

Fig. 5-14 (cont.)

**5-2.** Analyze each of the frames shown in Fig. 5-15 and draw the shear, moment, and axial force diagrams for member *bc*.

Fig. 5-15

**5-3.** Determine the forces acting on members *ad* and *ae* of the composite structure of Fig. 5-16.

94

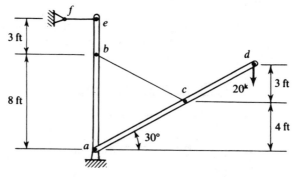

**Fig. 5-16**

**5-4.** Analyze the composite structure shown in Fig. 5-17. Draw the shear, moment, and axial force diagrams for member *ab*.

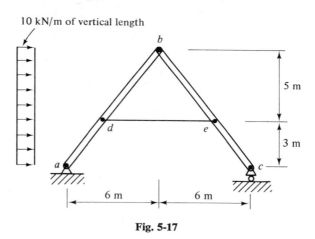

**Fig. 5-17**

**5-5.** Analyze the frame of Fig. 5-18 by the cantilever method. Assume constant *EI* for all members.

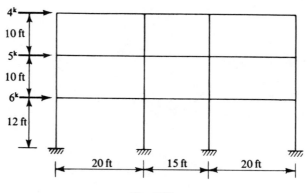

**Fig. 5-18**

**95**

**5-6.** Analyze the portal frame of Fig. 5-19 by assuming that the horizontal reactions are equal and that a point of inflection occurs halfway between the base and the end of the knee brace for each column.

Fig. 5-19

# 6

## INFLUENCE LINES FOR STATICALLY
## DETERMINATE STRUCTURES

### 6-1. THE CONCEPT OF INFLUENCE LINE

In the design of a structure, as discussed in Sec. 1-2, the loading conditions for the structure must be established before the stress analysis can be made. For a static structure, we are mainly concerned with two kinds of load, dead load and live load (the impact load being a fraction of the live load). The former remains stationary with the structure; whereas the latter, either the moving or the movable load, may vary in position on the structure. When one is designing any specific part of a structure, particular attention should be paid to the placement of the live load so that it will cause the maximum live stresses for the part considered. The part of the structure and the type of stress may be the reaction of a support, the shear or moment of a beam section, or the bar force in a truss. In this connection we should note that the position of the load that causes the maximum bending moment at a section will not necessarily cause the maximum shear at the same section or that the condition of loading which causes the maximum axial force at one member may not cause the maximum axial force at some other member. This and other considerations with respect to the relationship between the stress and the corresponding critical position of live load lead to the construction of the influence line.

The concept of the influence line may be illustrated by considering a beam under a single concentrated load, as follow,:

1. Suppose that a single concentrated load is placed transversely on a beam. Each part or each beam section is then affected by the load acting on the beam. For instance, each beam section will have certain shear and bending moment, and the supports will undergo certain reactions. When the load is fixed in position, a different section has different stresses. When the position of the load is varied, the corresponding value of stress at a fixed section varies.

2. As mentioned, we have touched on three distinct things—the value of the stress, the position of the beam section, and the position of the load. It should

be noted that a plane curve is possible only when two variables are related, not three. In this respect, we may express the value of the stress as a function of the position of the beam section, if we let the position of the load be fixed. Throughout the preceding chapters, involving the analysis of beams, the discussion was devoted to finding the value of the stresses on various sections, while the position of the applied load was assumed definite. The familiar shear or moment curve is nothing but the graphic representation of the relationship

$$y = f(x)$$

in which $x$ is taken along the beam axis to indicate the variable position of beam section and $y$ the corresponding value of the function (shear or moment) under fixed load.

3. On the other hand, we may express the value of stress in a fixed section as a function of variable load position. For simplicity and elegance, we use a single unit load to represent the moving live load. As the unit load travels across the span, we fix our attention on its effect upon a certain section of the beam and record the variation of stress (shear or moment) for that section by

$$y = g(x)$$

in which $x$ is taken along the beam axis to indicate the variable position of the unit load and $y$ the corresponding value of function (shear or moment) for the fixed section. The graphic representation of $y = g(x)$ is called the *influence line* for the value of the function under consideration.

4. A generalized definition of the influence line is: *An influence line is a curve the ordinate* ($y$ *value*) *of which gives the value of the function* (shear, moment, reaction, bar force, etc.) *in a fixed element* (member section, support, bar in truss, etc.) *when a unit load is at the ordinate.*

As an illustration, let us begin with the basic approach as to how to draw a bending moment influence line for the midspan section of a simple beam 10 ft long [Fig. 6-1(a)]. We may first divide the span into equal segments, say 10 segments $AB$, $BC$, . . . , $JK$, to indicate the position of load. As the unit load moves continuously from the left to the right, we focus our attention on the midspan section $F$ and compute the bending moment at $F$ for each 1-ft interval. The results are plotted in Fig. 6-1(b). This gives the bending moment influence line for section $F$. The abscissa coincides with the beam axis, indicating the position of the load, and the ordinate gives the corresponding moment at $F$ due to the single unit load placed at the ordinate. For instance, the ordinate at $D$ is 1.5, which is the value of the moment at $F$ caused by a unit load at $D$.

Of course, we need not always plot the influence line in the above-indicated fashion, since it is time-consuming. In most cases we can find an equation $y = g(x)$ expressing the desired stress $y$ at the given section in terms of the load position $x$. The plane curve represented by the equation gives the desired influence line. To illustrate this technique, let us use the same problem previously

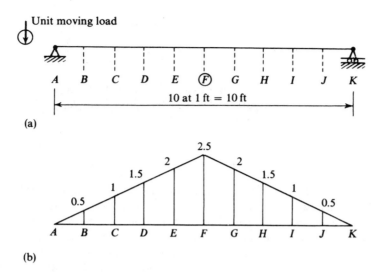

(a)

(b)

**Fig. 6-1**

mentioned but picture it in a different way, as shown in Fig. 6-2(a). The unit load is placed at a distance $x$ from left end $A$. The reactions at ends $A$ and $K$ are expressed as functions of $x$,

$$R_A = \frac{(10 - x)(1)}{10} = 1 - \frac{x}{10} \qquad R_K = \left(\frac{x}{10}\right)(1) = \frac{x}{10}$$

respectively. When the moving load is confined to the left of section $F$ (as shown), the bending moment at $F$ may be found from $R_K$:

$$M_F = 5R_K = \frac{x}{2}$$

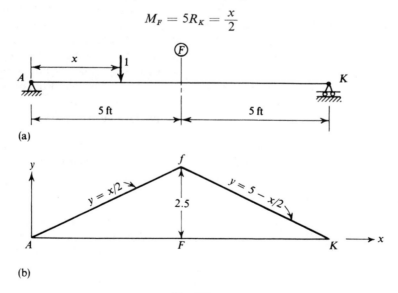

(a)

(b)

**Fig. 6-2**

99

or
$$y = \frac{x}{2}$$

where $y$ denotes $M_F$. As the moving load is confined to the right of section $F$ (not shown), the bending moment at $F$ may be found from $R_A$:

$$M_F = 5R_A = (5)\left(1 - \frac{x}{10}\right) = 5 - \frac{x}{2}$$

or
$$y = 5 - \frac{x}{2}$$

Selecting the coordinate axes as shown in Fig. 6-2(b), we plot $y = x/2$ and $y = 5 - (x/2)$ by two straight lines. The curve $AfK$ given in Fig. 6-2(b) is the desired *bending moment influence line* for section $F$, and the corresponding diagram $AFKf$ is called the *bending moment influence diagram* for section $F$.

Although in this particular case the influence diagram of Fig. 6-2(b) is identical to the moment diagram for the same beam under a unit load at midspan, we must not confuse the influence diagram with a bending moment diagram for the beam. Whereas the latter shows by each ordinate the bending moment at the corresponding section due to a fixed load, the influence diagram shows by each ordinate the bending moment at a fixed section due to a unit load placed at the point where the ordinate stands.

### 6-2. USE OF THE INFLUENCE LINE

By definition an influence line indicates the effect of a unit load moving across a span. It should be pointed out that the unit load for which the influence line is drawn is usually vertical although it may be horizontal or inclined. The construction of influence lines is closely associated with the stress analysis of bridge structures, which are frequently subjected to the action of various systems of moving loads. However, the usefulness of influence lines is not limited to bridge structures. They are also important in the determination of the maximum stresses in other engineering structures subjected to live loads.

An influence line is a useful tool in stress analysis in two ways:

1. It serves as a criteria in determining the maximum stress—a guide to determine just what portion of the structure should be loaded in order to cause the maximum effect on the part under consideration.
2. It simplifies the computation.

To illustrate, consider a simple beam 10 ft long subjected to the passage of a moving uniform load of 1 kip/ft without limit in length and a movable concentrated load of 10 kips that may be placed at any point of the span [see Fig. 6-3(a)]. Determine the maximum bending moment at the midspan section $C$.

We start by drawing the bending moment influence line for section $C$, as in Fig. 6-3(b). It is apparent from the influence line that, to obtain the maximum

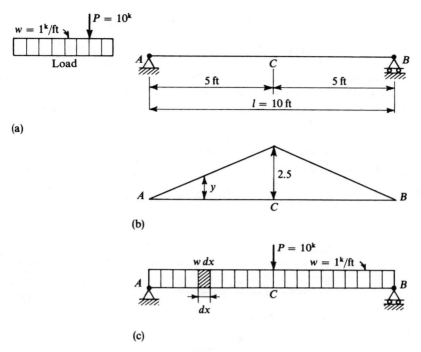

(a)

(b)

(c)

**Fig. 6-3**

$M_C$, the single concentrated load of 10 kips should be placed at the midpoint of the span, where the maximum ordinate of the influence line occurs, and the uniform load should be spread over the entire span.

Next, to compute the bending moment at $C$ due to the live loads so placed, we simply multiply each load by the corresponding influence ordinate and add. Referring to Fig. 6-3(b) and (c), we obtain

$$M_C = P(2.5) + \sum (w\,dx)y$$

Note that the uniform load is treated as a series of infinitesimal concentrated loads each of magnitude $w\,dx$ and the total effect of the uniform load is the summation

$$\sum (w\,dx)y$$

Now,

$$\sum (w\,dx)y = \int_0^l wy\,dx = w \int_0^l y\,dx$$

$$= \text{(load intensity)} \times \text{(area of influence diagram)}$$

Therefore, the total bending moment at $C$ is

$$M_C = (10)(2.5) + (1)\frac{(2.5)(10)}{2} = 25 + 12.5 = 37.5 \text{ ft-kips}$$

This value may be checked by the conventional method of computing $M_C$:

$$M_C = \left(\frac{10}{2}\right)(5) + \frac{(1)(10)^2}{8} = 25 + 12.5 = 37.5 \text{ ft-kips}$$

In this simple case, such a conclusion may be drawn without the aid of the influence diagram; but for more complicated moving load systems, we find that the influence diagram can be of substantial help, as discussed in Chapter 7.

## 6-3. INFLUENCE LINES FOR STATICALLY DETERMINATE BEAMS

**Example 6-1.** In Fig. 6-4(a) is shown a simple beam. Draw the influence lines for

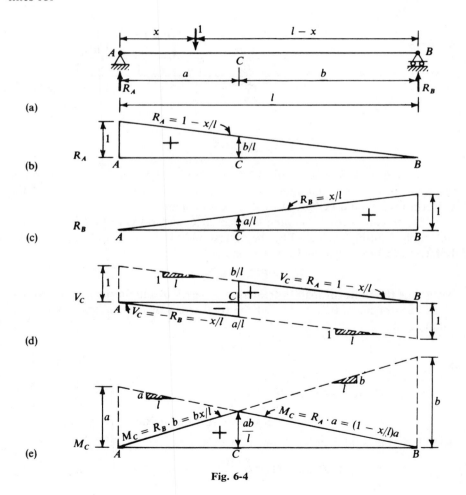

**Fig. 6-4**

1. The reaction at $A$ $(R_A)$ and the reaction at $B$ $(R_B)$
2. The shear at any section $C$ $(V_C)$
3. The moment at any section $C$ $(M_C)$

To draw the influence line for $R_A$, we place a unit load distance $x$ from $A$ and express $R_A$ as the function of $x$:

$$R_A = \left(\frac{l-x}{l}\right)(1) = 1 - \frac{x}{l}$$

which represents a straight line with unit ordinate at $A$ and zero at $B$ as shown in Fig. 6-4(b). Similarly, we draw the influence line for $R_B$ as shown in Fig. 6-4(c) based on

$$R_B = \frac{x}{l}$$

To draw the influence line for $V_C$, we note that as long as the unit load is applied at any position to the left of section $C$, $0 \leqslant x < a$, $V_C$ is found to be equal to $R_B$ but with the opposite sign. When the unit load is applied at any section to the right of $C$, $a < x \leqslant l$, $V_C$ is found to be equal to $R_A$. Therefore, the influence line $V_C$ in the portion of $AC$ is the same as that of $R_B$ but with negative sign, and the influence line $V_C$ in the portion $CB$ is the same as that of $R_A$. These two lines are parallel with equal slopes of $-1/l$, and there is an abrupt change of unity when the unit load passes from the left of $C$ to the right of $C$ [see Fig. 6-4(d) for the influence line of $V_C$].

To draw the influence line for $M_C$, we note that as long as the load is to the left of $C$, we can use segment $CB$ as a free body and find that $M_C = R_B b$. When the load passes $C$ to the right, we can use segment $AC$ as a free body and find that $M_C = R_A a$. Both are straight lines but with different slopes, as shown in Fig. 6-4(e).

In connection with the influence lines of Fig. 6-4, we note the following:

1. The maximum reaction due to a single concentrated load occurs when the load is at the support.

2. The maximum reaction due to a uniform load occurs when the beam is fully loaded.

3. The maximum shear at any section $C$ due to a single concentrated load occurs when the load is just to the right or left of the section and is on the longer of the two segments into which $C$ divides the beam.

4. The maximum shear at any section $C$ due to a uniform load occurs when the load extends from $C$ to the more distant support. Referring to Fig. 6-4(d), if $b > a$, we find the load is distributed over $BC$ to cause a maximum $V_C$.

5. The maximum moment at any section $C$ due to a single concentrated load occurs when the load is at $C$.

6. The maximum moment at any section $C$ due to a uniform load occurs when the beam is fully loaded.

By comparing the influence lines for the same function at different points, we note the following:

1. The maximum shear for a simple beam due to a single concentrated load occurs at the end and equals the load.

2. The maximum shear for a simple beam due to a uniform load occurs at the end when the beam is fully loaded and equals one-half the total load.

3. The maximum moment for a simple beam due to a single concentrated load $Q$ occurs at the midspan section and equals $Ql/4$.

4. The maximum moment for a simple beam due to a uniform load of intensity $q$ occurs at the midspan section when the beam is fully loaded and equals $ql^2/8$.

**Example 6-2.** Consider the simple beam with an overhang shown in Fig. 6-5(a). To construct the influence line for $R_B$, we place a unit load distance $x$ from end $A$ and apply $\sum M_C = 0$ to obtain

$$R_B = \frac{20 - x}{16}$$

The linear expression represents a straight line with the maximum ordinate $\frac{5}{4}$ at $A$ and the minimum ordinate 0 at $C$, as shown in Fig. 6-5(b). Note that when the unit load is placed at $B$, the influence ordinate for $R_B$ should be equal to unity.

The influence line for $R_C$ may be found by applying $\sum F_y = 0$:

$$R_C = 1 - R_B = 1 - \frac{20 - x}{16} = \frac{x - 4}{16}$$

which is also a linear function of $x$ representing a straight line, as shown in Fig. 6-5(c). As a check, the ordinate at $C$ should be equal to unity; and that at $B$ to zero.

The influence line for the shear at the section just to the left of $B$, called $(V_B)_L$, is given in Fig. 6-5(d). As long as the unit load is on the overhanging portion of beam, $(V_B)_L = -1$; as load passes $B$ to the right, $(V_B)_L = 0$.

The influence line for the shear at the section just to the right of $B$, called $(V_B)_R$, is shown in Fig. 6-5(e). As long as the unit load is on the overhanging portion of beam, $(V_B)_R$ equals $R_C$ but with opposite sign. As the load is on the simple-beam portion, $(V_B)_R$ equals $R_B$.

By a similar approach, we construct the influence line for shear at section $D$, as shown in Fig. 6-5(f). As a check, when the unit load passes $D$ from the left to the right, the shear at $D$ increases suddenly from $-\frac{1}{4}$ to $+\frac{3}{4}$ (i.e., there is an abrupt change of shear equal to unity at $D$).

The influence line for the moment at $D$ is shown in Fig. 6-5(g). We note that as long as the unit load is confined to the portion $AD$, the moment at $D$ may be found from $R_C$,

$$M_D = 12R_C = (12)\left(\frac{x - 4}{16}\right) = \frac{3x - 12}{4} \qquad (0 \leqslant x \leqslant 8)$$

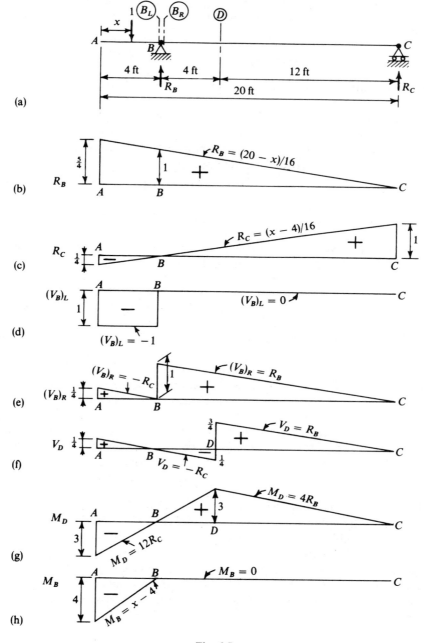

Fig. 6-5

which represents a straight line from $A$ to $D$ with the ordinates equal to $-3$ at $A$ and $+3$ at $D$. When the load passes $D$ to the right, the moment at $D$ may be found from $R_B$:

$$M_D = 4R_B = (4)\left(\frac{20 - x}{16}\right) = \frac{20 - x}{4} \qquad (8 \leqslant x \leqslant 20)$$

which represents a straight line from $D$ to $C$ with ordinates $+3$ at $D$ and $0$ at $C$.

Finally, we construct the influence line for the moment at $B$, as shown in Fig. 6-5(h). When the load is placed at $A$, $M_B$ has its greatest negative value of 4. As the load travels from $A$ to $B$, the moment varies linearly from $-4$ to $0$. As the load enters the portion $BC$, there will be no moment at $B$.

### 6-4. BEAM INFLUENCE LINES BY VIRTUAL WORK

By the principle of virtual work we can derive a very simple and elegant method for constructing the influence lines for beams. It can be stated as follows:

1. To obtain an influence line for the reaction of any statically determinate beam, remove the support and make a positive unit displacement of its point of application. The deflected beam is the influence line for the reaction.

2. To obtain an influence line for the shear at a section of any statically determinate beam, cut the section and induce a unit relative transverse sliding displacement between the portion to the left of the section and the portion to the right of the section keeping all other constraints (both external and internal) intact. The deflected beam is the influence line for the shear at the section.

3. To obtain the influence line for the moment at a section of any statically determinate beam, cut the section and induce a unit rotation between the portion to the left of the section and the portion to the right of the section keeping all other constraints (both external and internal) intact. The deflected beam is the influence line for the moment at the section.

To prove the theory, we take the case of a simple beam. The procedure of proof is generally applicable to more complicated beams. Figure 6-6(a) shows a simple beam subjected to a single unit moving load. To find the reaction at $A$ by the method of virtual work, we remove the constraint at $A$, substitute $R_A$ for it, and let $A$ travel a small virtual displacement $\delta s_A$ along $R_A$. We then have a deflected beam $A'B$, as shown in Fig. 6-6(b), where $y$ indicates the transverse displacement at the point of unit load. Applying the virtual-work equation, we obtain

$$(R_A)(\delta s_A) - (1)(y) = 0$$

from which
$$R_A = \frac{y}{\delta s_A}$$

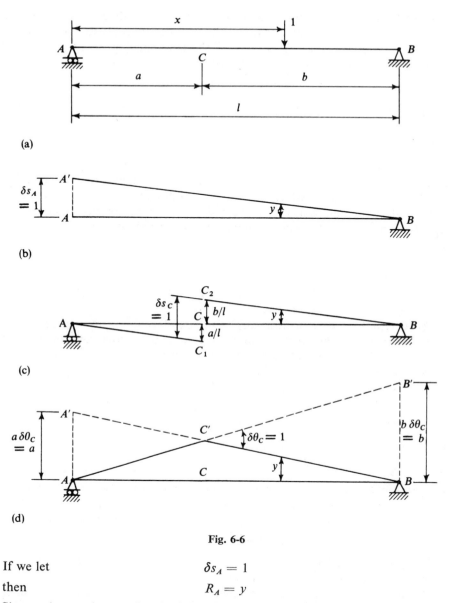

**Fig. 6-6**

If we let
$$\delta s_A = 1$$
then
$$R_A = y$$

Since $y$ is, on the one hand, the ordinate of the deflected beam at the point where the unit load stands and is, on the other hand, the value of function $R_A$ due to the unit moving load (i.e., the influence ordinate at the point), we conclude that the deflected beam $A'B$ of Fig. 6-6(b) is the influence line for $R_A$ if $\delta s_A$ is set to be unity.

To determine the shearing force at any beam cross section $C$, we cut the beam at $C$ and let the two portions $AC$ and $CB$ have a relative virtual transverse displacement $\delta s_C$ at $C$ without causing relative rotation between the two portions. This is equivalent to rotating $AC$ and $BC$ the same small angle about $A$ and $B$ respectively. Applying the virtual-work equation, we obtain

$$(V_C)(\delta s_C) - (1)(y) = 0$$

from which
$$V_C = \frac{y}{\delta s_C}$$

If we let
$$\delta s_C = 1$$

then
$$V_C = y$$

This proves that the deflected beam $AC_1C_2B$ of Fig. 6-6(c) is the influence line for $V_C$. It should be pointed out that the virtual displacement is supposed to be vanishingly small and that, when we say $\delta s_C = 1$, we do not mean that $\delta s_C = 1$ ft or 1 in., but one unit of very small distance for which the expressions

$$CC_1 = \frac{a}{l}$$

$$CC_2 = \frac{b}{l}$$

shown in Fig. 6-6(c) are justified.

To determine the moment at any beam cross section $C$ by the method of virtual work, we cut the beam at $C$ and induce a relative virtual rotation between the two portions $AC$ and $CB$ at $C$ without producing relative transverse sliding between the two. Thus, by virtual work,

$$(M_C)(\delta\theta_C) - (1)(y) = 0$$

from which
$$M_C = \frac{y}{\delta\theta_C}$$

If we let
$$\delta\theta_C = 1$$

then
$$M_C = y$$

This proves that the deflected beam $AC'B$ of Fig. 6-6(d) is the influence line for $M_C$. Note that, when we say $\delta\theta_C = 1$, we do not mean that $\delta\theta_C = 1$ radian. One unit of $\delta\theta_C$ may be as small as $\frac{1}{100}$ radian, for which it is justified to write

$$AA' = a \cdot \delta\theta_C = a \text{ units} \qquad BB' = b \cdot \delta\theta_C = b \text{ units}$$

as indicated in Fig. 6-6(d). Compare the influence lines of Fig. 6-6 with those obtained in Fig. 6-4; they are identical.

To illustrate the rules given above, we take a simple beam with an overhang and a compound beam so that we may clearly see that the method based on virtual work can be used to advantage in constructing influence lines for beams.

**Example 6-3.** Figure 6-7(a) shows a simple beam with an overhang the same as that in Fig. 6-5(a). We wish to obtain the influence lines for $R_B$, $R_C$, $(V_B)_L$, $(V_B)_R$, $V_D$, $M_D$, and $M_B$ by the virtual-work method.

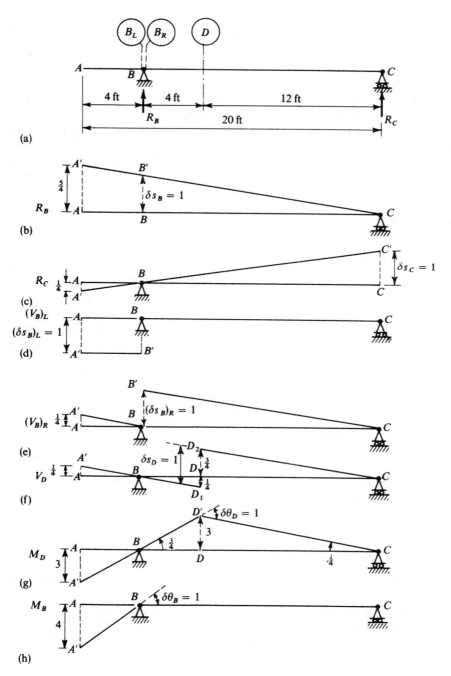

Fig. 6-7

To construct the influence line for $R_B$, we remove the support at $B$ and let it move a unit distance upward. The deflected beam $A'B'C$, shown in Fig. 6-7(b), is the desired influence line. The influence line for $R_C$ is obtained in a similar manner, as shown in Fig. 6-7(c) by $A'BC'$.

To construct the influence line for $(V_B)_L$, we cut the section immediately to the left of $B$ and let the left portion have a unit relative transverse displacement with respect to the right portion at the cut point without causing relative rotation between the two. Since the portion to the right of the cut section is a simple beam, it remains stable and rigid. Only the portion to the left of the cut section can go down a unit distance. Thus, the influence line for $(V_B)_L$ is represented by $A'B'BC$ in Fig. 6-7(d).

To construct the influence line for $(V_B)_R$, we cut the section immediately to the right of $B$ and let the two portions of the beam have a unit relative transverse displacement without producing relative rotation. To do this, end $B$ of $BA$ cannot move because of the presence of the support at $B$; however, it is allowed to rotate about the hinge at $B$. We, therefore, first raise the end $B$ of $BC$ a unit distance to $B'$ and then turn $BA$ to $BA'$ parallel to $B'C$. The deflected beam so arrived at, as shown in Fig. 6-7(e) by $A'BB'C$, is the influence line for $(V_B)_R$.

To construct the influence line for $V_D$, we cut the beam at $D$ and let the two portions $DA$ and $DC$ have a relative transverse displacement equal to unity. The two are kept parallel to each other so that no relative rotation is introduced. Then the deflected beam $A'D_1D_2C$ of Fig. 6-7(f) is the desired influence line. The slope of the deflected beam is found to be $-1/16$, with which the influence ordinates can easily be determined, as indicated in Fig. 6-7(f).

The influence line for $M_D$ is shown in Fig. 6-7(g) by the deflected beam $A'D'C$. It results from cutting the beam at $D$ and letting $DA$ and $DC$ have a unit rotation at $D$ without allowing relative translation between the two. Referring to Fig. 6-7(g), we find that the rotation of $DC$ about support $C$ is $\frac{1}{4}$ and the rotation of $DA$ about support $B$ is $\frac{3}{4}$. Thus, the influence ordinate at $D$ is found to be $(12)(\frac{1}{4})$ or 3.

Similarly, we construct the influence line for $M_B$, as shown in Fig. 6-7(h) by $A'BC$, by cutting the section just to the left of $B$ and letting the left portion of the beam have a unit relative rotation with respect to the right portion of the beam at $B$.

**Example 6-4.** Figure 6-8(a) shows a compound beam. Draw influence lines for $R_A$, $V_D$, $M_D$, $V_E$, $M_E$.

To construct the influence line for $R_A$, we remove support $A$ and move end $A$ up a unit distance. The deflected beam $A'CB$ shown in Fig. 6-8(b) is the influence line for $R_A$. Note that portion $CB$ is a cantilever and will remain unmoved.

To construct the influence line for $V_D$, we cut the beam through $D$ and let the left portion of beam have a relative transverse displacement equal to unity with respect to the right portion of beam at $D$ without causing relative rotation

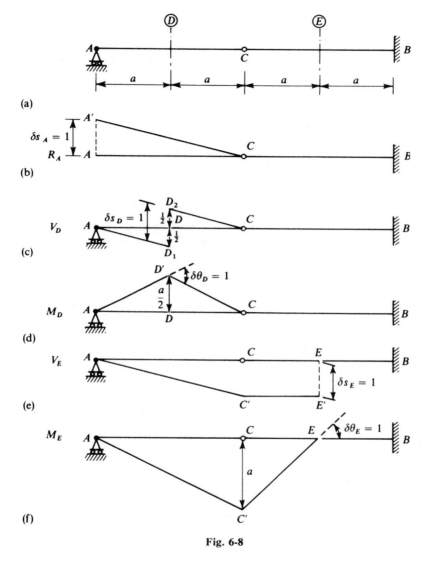

**Fig. 6-8**

between the two. The deflected beam $AD_1D_2CB$, shown in Fig. 6-8(c), is the influence line for $V_D$.

To construct the influence line for $M_D$, we cut the beam through $D$ and let the left portion of beam rotate a unit angle with respect to the right portion at $D$. The deflected beam $AD'CB$ of Fig. 6-8(d) is the influence line for $M_D$.

The influence line for $V_E$ is shown in Fig. 6-8(e) by $AC'E'EB$, which results from cutting the beam through $E$ and moving the left portion of beam down a unit distance with respect to the right portion of beam at $E$, while keeping the deflected portion $C'E'$ parallel to $BE$.

The influence line for $M_E$ is shown in Fig. 6-8(f) by $AC'EB$, which results from cutting the beam through $E$ and rotating the left portion of beam a unit angle with respect to the right portion of beam at $E$. Point $E$ is kept fixed in the original position.


### 6-5. STRINGER AND FLOOR-BEAM SYSTEM

In the case where a long-span girder is used in bridge construction, the live loads are usually not applied directly to the girder but are transmitted by a *stringer and floor-beam system* to the main girder. A similar arrangement is used in a trussed bridge where the moving wheels are carried by the bridge floor to the stringers, which are supported by the floor beams. The floor beams then carry the stringer reactions to the panel points of the main trusses. Regardless of actual details of construction, we may picture this system as in Fig. 6-9, where the main girder $AB$ carries a number of floor beams that, in turn, support stringers running parallel to the girder. Those portions of the main girder between the floor beams, such as 0–1, 1–2, and so on, are called *panels*, and the end points of the panels, such as 0, 1, 2, etc., are called *panel points*. It is assumed that the stringers are each simply supported on the floor beams. Consequently, a load $P$ applied to any stringer will be transmitted to the main girder only at the two corresponding panel points. The girder (likewise the truss) is therefore *panel point loaded* for any condition of loading on the stringer. It is obvious that the division of the load between two panel points has no effect on the reactions at $A$ and $B$. However, the internal shearing force in the main girder is made constant throughout any one panel by such a division. Thus, it is customary to say *panel shear* rather than shear at a particular section.

**Fig. 6-9**

Note that it is not necessary for the stringers in every panel to be simply supported by the adjacent floor beams, as shown in Fig. 6-9. For instance, stringers may be arranged as in Fig. 6-10(a) or (b). However, the type in Fig. 6-9 is most commonly used and will be discussed.

To illustrate the analysis of the girder, let us consider the system shown in Fig. 6-11(a). When the loads are transmitted through stringers and floor beams,

(a)                                              (b)

Fig. 6-10

Fig. 6-11

113

we find that the girder is subjected to forces as shown in Fig. 6-11(b). The reaction at $A$ is found to be 17.25 kips from $\sum M_B = 0$ by using either the loading shown in Fig. 6-11(a) or that shown in Fig. 6-11(b). Similarly, the reaction at $B$ is found to be 21.75 kips.

Based on the load diagram of Fig. 6-11(b), the shear diagram is plotted as in Fig. 6-11(c) and the moment diagram as in Fig. 6-11(d). Since the girder is panel point loaded, the shear has the same value at any section in a given panel of the girder.

## 6-6. INFLUENCE LINES FOR GIRDERS
### WITH FLOOR SYSTEMS

Consider the girder shown in Fig. 6-12(a), which consists of four equal panels each 10 ft long.

The influence lines for $R_A$ and $R_B$ are readily drawn as shown in Fig. 6-12(b) and (c), respectively. They are constructed in the same way as the influence lines for reactions of a simple beam 40 ft long, since end reactions of the girder are not affected by the presence of the floor system.

The influence line for the shear in panel 1–2, denoted by $V_{1-2}$, is shown in Fig. 6-12(d). When the load is confined to 0–1, we observe that $V_{1-2}$ is equal to $R_B$ but with a minus sign. When the load is confined to 2–4, $V_{1-2}$ is equal to $R_A$. As the load travels from panel point 1 to panel point 2, the value of $V_{1-2}$ has a linear variation. For, in general, in a structure where the loads are transmitted by members of a floor system to panel points, the influence line between consecutive panel points will be a straight line provided that the stringers act as simple beams spanning the adjacent floor beams. This can be proved as follows:

Consider a girder with a floor system, shown partially in Fig. 6-13(a). Let a unit load travel along a panel $m$–$n$. When the unit load is at $m$, we let $y_m$ be the corresponding influence ordinate for it; when the unit load is at $n$, we let $y_n$ be the corresponding influence ordinate for it [see Fig. 6-13(b)]. Now when the unit load is in any intermediate position, say distance $x$ from $m$, it will be transmitted to the girder through the floor beams at $m$ and $n$ with amounts $(l - x)/l$ and $x/l$, respectively. The effect of the load is found by multiplying each of these values by the corresponding value of the influence ordinate, and adding. This gives

$$(1)(y) = \left(\frac{l-x}{l}\right)(y_m) + \left(\frac{x}{l}\right)(y_n)$$

in which $y$ denotes the influence ordinate where the unit load is located as indicated in Fig. 6-13(b). This, being a linear expression of $x$, specifies the influence ordinate for the general intermediate position in the panel.

Similarly, we draw the influence line for shear in panel 0–1 as shown in Fig. 6-12(e).

Fig. 6-12

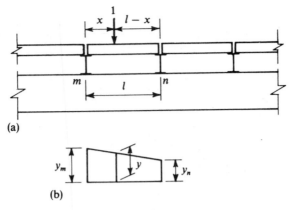

(a)

(b)

Fig. 6-13

To construct the influence line for the moment at panel point 1, denoted by $M_1$, we note that when the unit load is at 0, $M_1 = 0$; when the unit load is located between panel points 1 and 4, $M_1 = 10R_A$, which represents a straight line with maximum ordinate 7.5 at panel point 1 and 0 at panel point 4. Connecting the influence ordinates at panel points 0 and 1 by a straight line, we complete the influence line for $M_1$, as shown in Fig. 6-12(f).

Similarly, we draw the influence line for moment at panel point 2 as shown in Fig. 6-12(g).

It should be pointed out that in this girder, in which each stringer has a length of one panel, the influence line for the moment at any panel point is the same as the influence line for the moment at the corresponding point in a simple beam.

To draw the moment influence line at any intermediate section, say section $m$ in the second panel, we note that as the load is confined to 0–1, $M_m = 28R_B$, which represents a straight line with 0 ordinate at 0 and a maximum ordinate 7 at panel point 1. As the load is confined to 2–4, $M_m = 12R_A$, which also represents a straight line with 0 ordinate at panel point 4 and a maximum ordinate 6 at panel point 2. It remains for us to connect the influence ordinates at panel points 1 and 2 by a straight line to complete the influence line for $M_m$, as shown in Fig. 6-12(h).

## 6-7. INFLUENCE LINES FOR STATICALLY DETERMINATE BRIDGE TRUSSES

Influence lines for bridge trusses are drawn as a unit load travels across the loaded chord (i.e., the chord to which the floor systems are connected). Throughout our discussion we shall assume that the stringers act as simple beams between the adjacent floor beams so that the influence line will be a straight line between any two adjacent panel points as described in Sec. 6-6.

116

**Example 6-5.** Figure 6-14(a) shows a Pratt truss with parallel top and bottom chords. For convenience in further analysis, we may begin with the influence lines for the end reactions at $a$ and $g$, which can readily be drawn as shown in Fig. 6-14(b). The dashed line gives the influence line for reaction at $g$.

From this, we construct the influence line for the bar force in the end post $aB$ as shown in Fig. 6-14(c). Note that with the unit load at $a$, $S_{aB} = 0$; with the unit load confined in the range $bg$, $S_{aB} = (-\sqrt{2})(R_a)$, the minus sign indicates a compression (see joint $a$). Thus, the influence diagram for $S_{aB}$ from $b$ to $g$ is in direct proportion to the influence diagram for $R_a$. As the load travels from $a$ to $b$, the influence line will be a straight line varying from 0 at $a$ to $(-\sqrt{2})(\frac{5}{6})$ at $b$.

To construct the influence line for the bar force in the hip verticle $Bb$, we note that, unless the unit load is on stringer $ab$ or stringer $bc$, there will be no force transmitted to panel point $b$ and therefore no force in member $Bb$. Now since $S_{Bb} = 0$ when the unit load is at $a$, $S_{Bb} = 1$ when the unit load is at $b$, and $S_{Bb} = 0$ when the unit load is at $c$, it is obvious that the influence line for $S_{Bb}$ is the one shown in Fig. 6-14(d), where the plus sign indicates tension in the member.

The construction of an influence line for the bar force in a diagonal member is illustrated by Fig. 6-14(e), where the vertical component of the bar force in member $Cd$, denoted by $V_{Cd}$, is considered.

As the unit load is at, or to the left of, panel point $c$, we find by the method of section (using the right portion of truss as a free body) that $V_{Cd} = -R_g$ (compression). Hence, the influence line is a straight line from 0 at $a$ to $-\frac{1}{3}$ at $c$. Similarly, when the unit load is at or to the right of panel point $d$, we find that $V_{Cd} = R_a$ (tension). Hence, the influence line is a straight line from $\frac{1}{2}$ at $d$ to 0 at $g$. As the unit load travels from $c$ to $d$, the influence line is straight, varying from $-\frac{1}{3}$ to $\frac{1}{2}$.

If we wish to construct the influence line for the bar force in $Cd$, we simply multiply each influence ordinate of $V_{Cd}$ by $\sqrt{2}$.

The influence line for the bar force in member $Cc$ is shown in Fig. 6-14(f) and can be constructed by the method of section as described for the truss diagonal. In this particular case, we observe that $S_{Cc} = -V_{Cd}$ by using joint $C$ as a free body.

To construct the influence line for a chord member, such as bar $cd$, we apply the method of moment about joint $C$. With the unit load at, or to the left of, panel point $c$, we find that $S_{cd} = 100R_g/25 = 4R_g$. Hence, the influence line is a straight line varying from 0 at $a$ to $\frac{4}{3}$ at $c$. With the unit load at, or to the right of, panel point $c$, we find that $S_{cd} = 50R_a/25 = 2R_a$. Hence, the influence line is a straight line varying from $\frac{4}{3}$ at $c$ to 0 at $g$. Thus, we complete the influence diagram for $S_{cd}$. As shown in Fig. 6-14(g), it is a triangle with its apex under the moment center $C$.

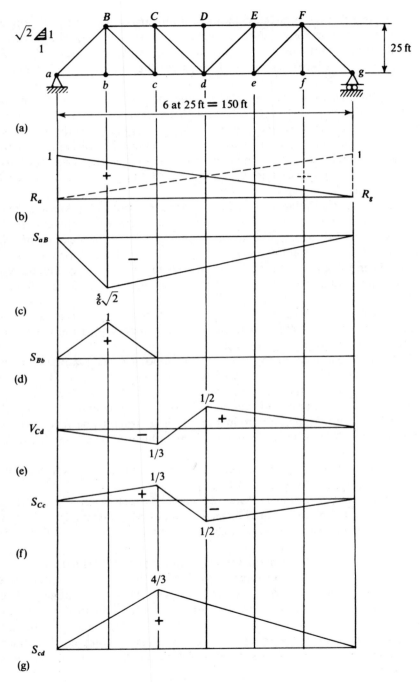

$\sqrt{2}\,\angle\!1\atop1$

B   C   D   E   F

25 ft

a
b   c   d   e   f   g

6 at 25 ft = 150 ft

(a)

1                           1

$R_a$   +                   ---   $R_g$

(b)

$S_{aB}$

−

$\frac{5}{6}\sqrt{2}$

(c)

1

$S_{Bb}$   +

(d)

1/2

$V_{Cd}$   −   +

1/3

(e)

1/3

$S_{Cc}$   +

−

1/2

(f)

4/3

$S_{cd}$   +

(g)

Fig. 6-14

118

**Example 6-6.** In Fig. 6-15(a) is shown a K truss. Let us draw the influence lines for the bar forces (or components) in members $dp$, $dq$, and $Dp$. To do this, we may either directly compute influence ordinates for each successive panel point or draw influence lines for members other than those under consideration and from these deduce the desired influence lines. The latter procedure will be used in this problem.

We begin with the influence diagram for the bar force of the lower chord $cd$, as shown in Fig. 6-15(b), the apex of which is at panel point $c$ with the ordinate equal to $\frac{6}{7}$, found by taking section $m$–$m$ and using $C$ as the moment center. Similarly, we can draw the influence line for the bar force of the lower chord $de$, as shown in the same figure, by taking section $n$–$n$ and using $D$ as the moment

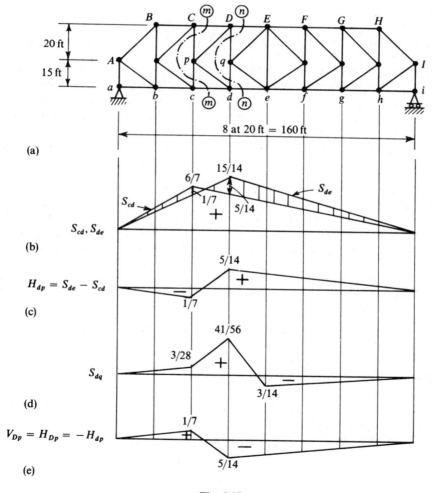

(a)

(b)

(c)

(d)

(e)

**Fig. 6-15**

center. The influence diagram is also a triangle with its apex at panel point $d$ and the ordinate equal to $\frac{15}{14}$. Both influence lines can be drawn without difficulty.

Now by use of the equation $\sum F_x = 0$ at joint $d$, we obtain the horizontal component of the bar force in $dp$ equal to the difference between $S_{de}$ and $S_{cd}$. That is,

$$H_{dp} = S_{de} - S_{cd}$$

The influence diagram for $H_{dp}$ can thus be obtained from the shaded portion of Fig. 6-15(b) rearranged as in Fig. 6-15(c).

To construct the influence line for the bar force in $dq$, we note, from $\sum F_y = 0$ at joint $d$, that

$$S_{dq} = -V_{dp} = -\tfrac{3}{4}H_{dp}$$

when the unit load is placed at any panel points other than $d$, and that

$$S_{dq} = 1 - V_{dp} = 1 - \tfrac{3}{4}H_{dp} = 1 - (\tfrac{3}{4})(\tfrac{5}{14}) = \tfrac{41}{56}$$

when the unit load is placed at $d$. See Fig. 6-15(d) for the influence line for $S_{dq}$.

The influence line for the vertical component of the bar force in $Dp$ is shown in Fig. 6-15(e), which is the same figure for $H_{dp}$ but with opposite sign, since

$$V_{Dp} = H_{Dp} = -H_{dp}$$

by using $\sum F_x = 0$ at joint $p$.

**Example 6-7.** Let us consider the statically determinate long-span bridge truss of Fig. 6-16(a). The difficulty with the influence line for such a truss disappears, since we know that it is always possible to obtain the ordinate to the influence line of any element for a unit load at each panel point of the truss.

Suppose that we wish to construct the influence lines for the reaction at support $c$, denoted by $R_c$, and the bar force in member $BC$, denoted by $S_{BC}$, of the indicated truss. It is advantageous to construct first the influence line for the force in the hanger $Ee$, as shown in Fig. 6-16(b). This can easily be verified by cutting the hanger and taking the moments about $d$ ($M_d = 0$) of the forces acting on the suspended span $de$.

Next, we construct the influence line for $R_c$, as shown in Fig. 6-16(c). Note that by taking a free body through the hanger $Ee$ and the panel $bc$, we observe that

$$R_c + S_{Ee} = 1$$

as long as the unit load is confined to portion $ce$.

The influence line for the bar force in member $BC$ is then drawn, with the force computed by taking the moments about $d$ ($M_d = 0$) of the forces acting on the part of structure between panel $bc$ and $d$ [see Fig. 6-16(a)]. For instance, for a unit load at point $d$, $R_c = 1$; by taking $M_d = 0$, we obtain

$$S_{BC} = \frac{(1)(60)}{40} = \frac{3}{2}$$

The influence line for $S_{BC}$ is shown in Fig. 6-16(d).

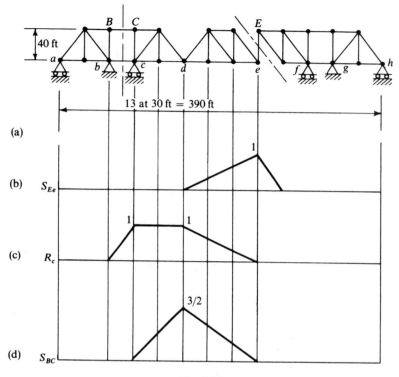

(b) $S_{Ee}$

(c) $R_c$

(d) $S_{BC}$

**Fig. 6-16**

## PROBLEMS

**6-1.** Given a simple beam 24 ft long, construct the influence lines for the shear and bending moment at a section 8 ft from the left end, and obtain the maximum shear and bending moment for the section resulting from a moving uniform load of 3 kips/ft and a movable concentrated load of 50 kips.

**6-2.** A cantilever beam 20 ft long is fixed at the right end. Construct the shear and moment influence lines for sections 5 ft, 10 ft, and 20 ft from the free end. Using the same loadings given in Prob. 6-1, compute maximum shears and moments at these sections.

**6-3.** In Fig. 6-17 is shown a simple beam with an overhang. Draw the influence lines for $R_B$, $R_C$, $V_D$, $M_B$, and $M_D$.

**Fig. 6-17**

**121**

**6-4.** Given a compound beam such as that shown in Fig. 6-18, construct the influence lines for $R_A$, $R_C$, $R_E$, $V_B$, $M_B$, and $M_C$. Compute the maximum value for each of them due to a moving uniform load of 20 kN/m.

**Fig. 6-18**

**6-5.** Solve Prob. 6-3 by the method of virtual work.

**6-6.** Construct the influence lines for Prob. 6-4 by the method of virtual work.

**6-7.** For the girder of Fig. 6-19, construct influence lines for (a) the reaction at end $A$, (b) the reaction at end $B$, (c) the shear in panel 0–1, (d) the shear in panel 1–2, (e) the moment at panel point 2, and (f) the moment at a point 15 ft from the left end.

**Fig. 6-19**

**6-8.** For the trusses shown in Fig. 6-20, construct the influence lines for the bar force (or component) in each of the lettered bars.

(a)

**Fig. 6-20**

(b)

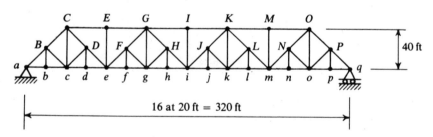

(c)

**Fig. 6-20** (cont.)

**6-9.** For the subdivided Warren truss, shown in Fig. 6-21, draw the influence lines for the bar forces (or components) in members $fF$, $ef$, $eF$, $FG$, and $gG$.

**Fig. 6-21**

**6-10.** For the long-span truss shown in Fig. 6-22, draw the influence lines for the bar forces in the lettered bars; also draw the influence lines for the reactions at $a$ and $c$.

**Fig. 6-22**

123

**6-11.** For the satically determinate rigid frame shown in Fig. 6-23, draw the influence lines for $M_e$ and the axial force of member $ce$ as the unit load travels across beam $ad$.

**Fig. 6-23**

**6-12.** Consider the statically determinate composite system in Fig. 6-24, which consists of a beam supported by a three-hinged truss. Draw the influence lines for reaction at $c$ and bar force in $be$ due to the passage of a unit load over beam $ac$.

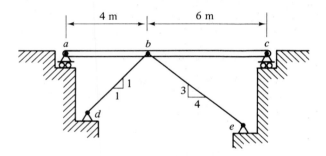

**Fig. 6-24**

# 7

## MOVING CONCENTRATED LOADS:

## CRITERIA FOR MAXIMA

### 7-1. GENERAL

As mentioned in Sec. 6-1, when designing any specific part of a structure, we should know where to place the live load in order to produce the maximum live stresses for the part considered. Under a single concentrated load or a uniform load the critical position causing certain maximum live stress can be spotted at once by inspection at the influence line. For more complicated conditions of loading, which are of various magnitude and spacing, such as a series of moving wheels on a locomotive, we cannot tell the critical position by just looking at the influence line. The method that should be followed for such cases is essentially one of trial aided by the use of criteria based on the influence line, in order to minimize computations.

For a different type of influence line, there will be a corresponding different criterion for maxima. In this chapter we discuss the criteria for maxima derived from the more common types of influence diagrams.

### 7-2. CRITERION FOR THE MAXIMUM VALUE OF A FUNCTION HAVING THE INFLUENCE DIAGRAM

This category includes the influence line for the end reaction in a simple beam and that for the end reaction of a simply supported girder or truss.

Consider a simple beam subjected to the passage of a series of wheel loads of fixed spacing and magnitude, the resultant of which is denoted by $P$ located at a distance $p$ from the right end $B$ [see Fig. 7-1(a)].

To find the reaction at end $A$, we first construct the influence line for $R_A$, as shown in Fig. 7-1(b), from which

$$R_A = Py = P(p)\left(\frac{1}{l}\right)$$

where $1/l$ is the slope of the influence line, a constant. The reaction at $A$ is,

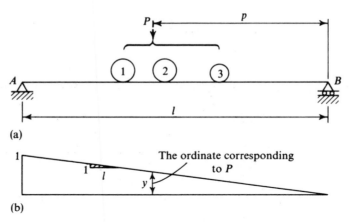

**Fig. 7-1**

therefore, in direct proportion to $P$ and $p$. If there is no change in the value of $P$ (i.e., there is no load entering or leaving the span during the movement), then the maximum reaction will occur when wheel 1 is placed over the support in such a way that $p$ will be the longest possible distance. If several systems of concentrated loads pass over a beam, the one which has the greatest value of $Pp$ will cause the maximum reaction; and this reaction will always occur when one load is placed over the support.

**Example 7-1.** Find the maximum reaction that will occur in a girder of 60-ft span under the car loads shown in Fig. 7-2(a).

Using the criterion established in the foregoing discussion, we make three trials for the left-end reaction, as shown in Fig. 7-2(b), (c), and (d), respectively. Brief computations are also given in each figure. Note that in a simple case, such as this, it is not always necessary to try all three loadings. They serve, here, to illustrate the method of trial only.

By comparing the results, we conclude that the maximum reaction is 35 kips when wheel 3 is directly over the left-end support.

## 7-3. CRITERION FOR THE MAXIMUM VALUE OF A FUNCTION HAVING THE INFLUENCE DIAGRAM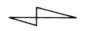

The influence line for the shear in a simple beam section is of this type. Consider the simple beam in Fig. 7-3(a) subjected to the passage of a group of wheel loads the resultant of which is $P$. We wish to find the maximum shear at section $C$, the influence line of which is shown in Fig. 7-3(b).

A glance at the influence line suggests that the first possible position for producing the maximum $V_C$ is to place wheel 1 directly over $C$, as shown in Fig. 7-3(c). In this case, the shear at $C$, called $(V_C)_1$, equals the reaction at $A$:

126

(a)

(b)

(c)

(d)

Fig. 7-2

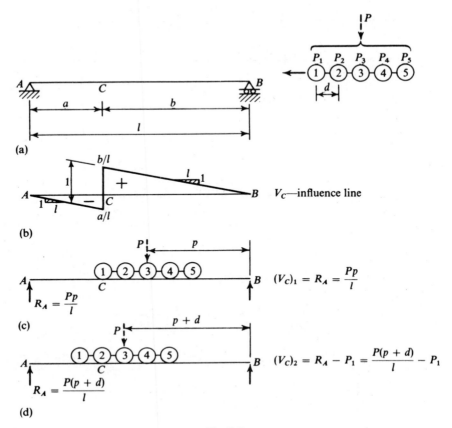

Fig. 7-3

$$(V_C)_1 = \frac{Pp}{l}$$

If the group moves toward the left, we observe that as soon as wheel 1 passes $C$, there will be a sudden drop of $V_C$ equal to $P_1$. Since the influence ordinate increases toward the left, another possible maximum shear will not be reached until wheel 2 reaches $C$ [see Fig. 7-3(d)]. In this case, the shear at $C$, called $(V_C)_2$, equals the reaction at $A$ minus $P_1$:

$$(V_C)_2 = \frac{P(p+d)}{l} - P_1$$

where $d$ is the distance between wheel 1 and wheel 2—the distance the group moves in the second trial.

Now let $\Delta V_C$ be the difference between the second result and the first result:

$$\Delta V_C = (V_C)_2 - (V_C)_1 = \frac{P(p+d)}{l} - P_1 - \frac{Pp}{l} = \frac{Pd}{l} - P_1$$

Then if

$$\frac{Pd}{l} > P_1 \left( \text{or} \ \frac{P}{l} > \frac{P_1}{d} \right) \qquad (V_C)_2 \text{ will be greater}$$

and if

$$\frac{Pd}{l} < P_1 \left( \text{or} \ \frac{P}{l} < \frac{P_1}{d} \right) \qquad (V_C)_1 \text{ will be greater}$$

These criteria will hold provided that there is no load entering or leaving the span. Note that the term $Pd/l$ is the increase in $V_C$, whereas $P_1$ is the decrease in $V_C$ during the movement. In general, if the increment in the value of the function is greater than the decrement, we let the system move; otherwise, no progress should be made.

**Example 7-2.** Consider the group of wheels shown in Fig. 7-4(a), which will pass over a simple beam of 50-ft span in the assumed direction. Let it be desired to find the maximum shear $(+)$ and the minimum shear $(-)$ at the midspan section.

To do this, we construct the shear influence line for the midspan section as shown in Fig. 7-4(a).

**Fig. 7-4**

1. Maximum shear $(+)$. The first trial position to find maximum positive shear at the midspan section is shown in Fig. 7-4(b). If the criterion

$$\frac{Pd}{l} \gtrless P_1$$

is applied to the first wheel, we find that

$$\frac{(45)(8)}{50} = 7.2 > 5$$

Therefore, we let the group move forward 8 ft, as shown in Fig. 7-4(c). The second position produces the maximum shear since applying the foregoing criterion for the second wheel gives

$$\frac{(45)(5)}{50} = 4.5 < 10$$

which indicates that the loss in shear will be greater than the gain if the wheels move 5 ft farther to the left.

When the load is in the position in Fig. 7-4(c), we compute the maximum shear for the midspan section as

$$\frac{(45)(9.22 + 10)}{50} - 5 = 12.3 \text{ kips}$$

2. Maximum shear $(-)$. The position of loading for the minimum negative shear at the midspan section is obvious, as shown in Fig. 7-4(d), under which the shear is given by

$$- \frac{(45)(13.78 + 2)}{50} = - 14.2 \text{ kips}$$

## 7-4. CRITERION FOR THE MAXIMUM VALUE OF A
## FUNCTION HAVING THE INFLUENCE DIAGRAM

The influence line for the bending moment in a simple beam section, that for the bending moment at the panel point of a simple girder bridge, and that for the bar force in a chord member of certain bridge trusses are of this type.

Consider the simple beam with a system of concentrated loads entering from the right-hand side shown in Fig. 7-5(a). Ascertain the critical loading position for the maximum bending moment in section $C$.

We start by constructing the influence line for $M_C$, as shown in Fig. 7-5(b). The system of loads takes the position shown, with the loads on the span divided into two groups, $G_1$ moving over the left segment length $a$; and $G_2$ over $b$. The bending moment at $C$ under this loading is found to be

$$(M_C)_1 = G_1 y_1 + G_2 y_2$$

[see Fig. 7-5(b)]. If the system moves a small distance $\Delta x$ to the left, the bending moment at $C$ will be

$$(M_C)_2 = G_1 \left[ y_1 - \left(\frac{b}{l}\right)\Delta x \right] + G_2 \left[ y_2 + \left(\frac{a}{l}\right)\Delta x \right]$$

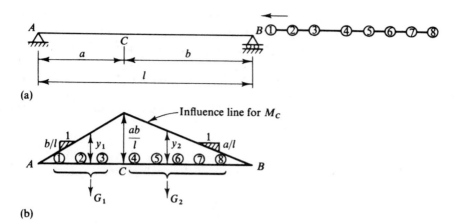

(a)

(b)

**Fig. 7-5**

As a result, the change in the bending moment is

$$\Delta M_C = (M_C)_2 - (M_C)_1 = (G_2)\left(\frac{a}{l}\right)\Delta x - (G_1)\left(\frac{b}{l}\right)\Delta x$$

Thus, the rate of change of $M_C$ with respect to the movement $\Delta x$ is

$$\frac{\Delta M_C}{\Delta x} = G_2\left(\frac{a}{l}\right) - G_1\left(\frac{b}{l}\right)$$

for which we note the following:

1. As long as

$$G_2\left(\frac{a}{l}\right) > G_1\left(\frac{b}{l}\right) \qquad \text{or} \qquad \frac{G_1}{a} < \frac{G_2}{b}$$

the bending moment at $C$ will increase as a result of the progressive movement of the loads to the left.

In order to obtain the maximum bending effect at $C$, the movement to the left must be continued until some one in group $G_2$ arrives at the peak ordinate.

Further movement to the left will change the value of $G_1$ and $G_2$. Unless the inequality still holds for the new arrangement, no further progress should be made.

2. Similarly, if

$$G_2\left(\frac{a}{l}\right) < G_1\left(\frac{b}{l}\right) \qquad \text{or} \qquad \frac{G_1}{a} > \frac{G_2}{b}$$

the bending moment at $C$ will increase as a result of the movement to the right.

In order to obtain the maximum $M_C$, we should let the system move to the right until one of $G_1$ reaches the peak ordinate. Further movement to the right will depend on whether the foregoing inequality is still valid for the new arrangement.

3. In both cases, the magnitude of the effect of the load system on $M_C$ increases as the movement to the left, or right, is continued until some load is at the peak ordinate. Loads entering or leaving the span will increase the growth of $M_C$ with this movement.

4. Thus, the criterion for maximum bending moment at $C$ is *to place some load at the peak ordinate* so that:

   a. When this load is just to the right of the peak ordinate (i.e., included in $G_2$), we have

$$\frac{G_1}{a} < \frac{G_2}{b}$$

   b. When this load is just to the left of the peak ordinate (i.e., included in $G_1$), we have

$$\frac{G_1}{a} > \frac{G_2}{b}$$

**Example 7-3.** For the beam and loading in Fig. 7-6(a), find the maximum bending moment at the midspan section $C$.

In this case, the influence diagram for $M_C$ is an isosceles triangle with the peak ordinate at the midspan, as shown in Fig. 7-6(b). We then place wheel 3 at $C$ and find that

$$\frac{G_1}{a} = \frac{15}{25} < \frac{G_2}{b} = \frac{30}{25}$$

if wheel 3 moves to the right of the midspan and that

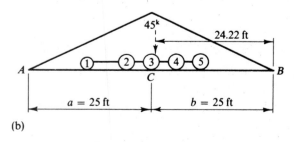

(b)

**Fig. 7-6**

$$\frac{G_1}{a} = \frac{25}{25} > \frac{G_2}{b} = \frac{20}{25}$$

if wheel 3 moves to the left of the midspan.

Therefore, the load position in Fig. 7-6(b) is the position that causes the maximum bending moment at $C$, which is found to be

$$M_C = \frac{(45)(24.22)}{50}(25) - (10)(5) - (5)(8 + 5) = 429.95 \text{ ft-kips}$$

## 7-5. CRITERION FOR THE MAXIMUM VALUE OF A FUNCTION HAVING THE INFLUENCE DIAGRAM ⟋▽⟍

This category includes the influence lines for the shear in an interior panel of a girder bridge and the bar force in the web member of certain bridge trusses.

Consider the girder bridge subjected to the passage of the group of wheel loads shown in Fig. 7-7(a). Find the maximum shear in the second panel from the left, whose influence line is shown in Fig. 7-7(b).

Fig. 7-7

The best approach is to try several loading positions and to compare the changes in the value of the function because of the movement. Let us try the first loading position, shown in Fig. 7-7(c), with wheel 1 at the peak ordinate. Next, let the system move to the left until wheel 2 reaches the peak ordinate, as shown in Fig. 7-7(d). The computations to the right of Fig. 7-7(c) and (d) show that the movement results in an increase in the value of the function. Next, let this system move further to the left until wheel 3 reaches the peak ordinate, as shown in Fig. 7-7(e). The computations to the right of Fig. 7-7(d) and (e) show that this gives a decrease in the value of the function. Thus, the second position of loading, shown in Fig. 7-7(d), produces the maximum shear in the second panel of the girder.

By using the influence diagram, we find the maximum value of $V_{b-c}$ to be

$$V_{b-c} = (10)\left(\frac{20 + 15 + 10 + 5}{40}\right) - (5)\left(\frac{1}{4}\right)\left(\frac{3.33 - 2}{3.33}\right)$$

$$= 12.5 - 0.5 = 12 \text{ kips}$$

Note that this method of locating a load system for the maximum effect is perfectly general and may always be employed in cases with more complicated influence diagrams.


### 7-6. ABSOLUTE MAXIMUM BENDING MOMENT

Under a single concentrated load or a uniform load the maximum bending moment in a simple beam occurs at the midspan section. However, when a simple beam is subjected to a group of concentrated loads, the maximum bending moment does not usually occur at midspan.

Consider a simple beam subjected to the passage of a series of wheel loads, as shown in Fig. 7-8. The maximum moment must occur under a certain load $P_x$, where the shear of the beam changes sign since the necessary condition for the maximum bending moment is $dM/dx = V = 0$.

Now let the resultant of the group be $P$; the resultant of the wheels to the left of $P_x$ be $P_L$; the distance between $P_x$ and $P_L$ be $c$; and that between $P_x$ and

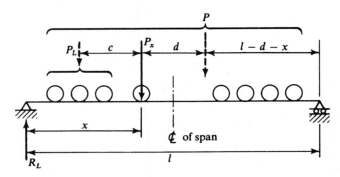

**Fig. 7-8**

$P$ be $d$. Also let the distance from the left end to $P_x$ be $x$, as shown. Taking $\sum M = 0$ about the right end gives the left end reaction as

$$R_L = \frac{P(l - d - x)}{l}$$

The moment under $P_x$ will be

$$M_x = (R_L)(x) - (P_L)(c)$$
$$= \frac{(P)(l - d - x)(x)}{l} - (P_L)(c)$$

Now setting

$$\frac{dM_x}{dx} = 0$$

we obtain the expression

$$x = l - d - x$$

which means that in order to obtain the absolute maximum bending moment for a simple beam under a set of moving concentrated loads, the distance from one end of the beam to wheel load $P_x$ should be equal to the distance from the other end of the beam to resultant $P$. In other words, the center line of the span bisects the distance between $P_x$ and $P$.

Usually the absolute maximum bending moment occurs under the load which is closest to the resultant of all the wheel loads.

**Example 7-4.** Find the absolute bending moment for the beam and loading in Fig. 7-9(a).

To do this, we place the loads as shown in Fig. 7-9(b), in which the center line of the span bisects wheel 3 and the resultant $P$. The absolute maximum bending moment will occur under wheel 3 and is found to be

$$M = \frac{(45)(24.61)^2}{50} - (10)(5) - (5)(8 + 5) = 430.09 \text{ ft-kips}$$

**Fig. 7-9**

It is interesting to compare these results with the maximum bending moment at midspan due to the same loading. The latter has been found to be 429.95 ft-kips (see Example 7-3), which is very close to, but still smaller than, the absolute maximum bending moment.

## PROBLEMS

**7-1.** A simple beam 45 ft long carries moving loads of 5 kips, 10 kips, and 10 kips spaced 5 ft apart. Calculate (a) the maximum left reaction, and (b) the maximum shear and bending moment at a section 15 ft from the left end.

**7-2.** Find the maximum shear for panel 1–2 and the maximum bending moment at panel point 2 for the girder in Fig. 7-10 resulting from the passage of a set of moving wheels from the right.

Fig. 7-10

**7-3.** For a simple beam of 60-ft span subjected to the loading given in Fig. 7-10 (let the set of wheels be from either of two directions), compute (a) the maximum end shear, (b) the maximum shear at the quarter point, (c) the maximum bending moment at the center, and (d) the absolute maximum bending moment in the beam.

**7-4.** For the truss and the loading shown in Fig. 7-11, compute the maximum forces in bars *a* and *b*. Consider both tension and compression in bar *a*.

Fig. 7-11

**7-5.** For the compound beam and loads shown in Fig. 7-12, find (a) the maximum reaction at *C*, and (b) the maximum moment at *D*.

Fig. 7-12

**7-6.** For the beam and loads shown in Fig. 7-13, find the absolute maximum moment.

Fig. 7-13

**7-7.** Refer to the frame and loads in Fig. 7-14. Find the maximum moment and vertical reaction at support $E$ due to the passage of loads over the beam.

Fig. 7-14

# 8

# ELASTIC DEFORMATIONS

# OF STRUCTURES

## 8-1. GENERAL

The calculation of elastic deformations for structures, both the linear deformations of points and the rotational deformations of lines (slopes) from their original positions, is of great importance in the analysis, design, and construction of structures. For instance, in the erection of a bridge structure, especially when the cantilever method is used, the theoretical elevations of some or all joints must be computed for each stage of the work. In building design the sizes of beams and girders are sometimes governed by the allowable deflections. Most important, the stress analysis for statically indeterminate structures is based largely upon an evaluation of their elastic deformations under load. By a statically indeterminate structure we mean a structure in which the number of unknown elements involved is greater than the number of equations of statics available for solution of the equations. If such is the case, there will be infinite number of solutions that can satisfy the statical equations. In order to reach a *unique* correct solution, the conditions of the *continuity* of structure, which are associated with the geometric and elastic properties of structure, are a necessary supplement.

Numerous methods of computing elastic deformations have been developed. Among them the following are considered the most significant in conventional structural analysis and will, therefore, be discussed in this chapter:

1. The method of virtual work (unit-load method)
2. Castigliano's theorem
3. The conjugate-beam method

## 8-2. CURVATURE OF ELASTIC LINE

The mathematical definition for curvature is *the rate at which a curve is changing direction.* To derive the expression for curvature, we shall consider a curve such as the one shown in Fig. 8-1. The average rate of change of direction between

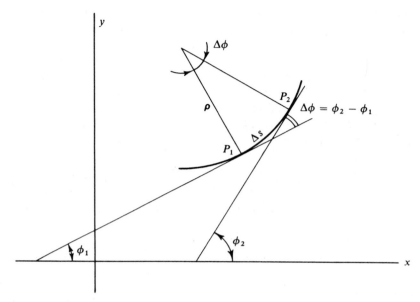

**Fig. 8-1**

points $P_1$ and $P_2$ is $\Delta\phi/\Delta s$. The limiting value of this ratio as $\Delta s$ approaches zero is called *curvature*, and the *radius of curvature* is the reciprocal of the curvature. Thus, if we let $\kappa$ denote the curvature and $\rho$ the radius of curvature, we have

$$\kappa = \frac{1}{\rho} = \lim_{\Delta s \to 0} \frac{\Delta\phi}{\Delta s} = \frac{d\phi}{ds}$$

Now since $\tan \phi = dy/dx$,

$$\frac{d}{dx} \tan \phi = \frac{d^2y}{dx^2}$$

or

$$(1 + \tan^2 \phi) \frac{d\phi}{dx} = \frac{d^2y}{dx^2}$$

This gives

$$\frac{d^2y}{dx^2} = \left[1 + \left(\frac{dy}{dx}\right)^2\right] \frac{d\phi}{dx}$$

whereby

$$\frac{d\phi}{dx} = \frac{d^2y/dx^2}{1 + (dy/dx)^2}$$

Also,

$$\frac{dx}{ds} = \frac{1}{ds/dx} = \frac{1}{[(dx^2 + dy^2)/dx^2]^{1/2}} = \frac{1}{[1 + (dy/dx)^2]^{1/2}}$$

Hence,

$$\kappa = \frac{d\phi}{ds} = \left(\frac{d\phi}{dx}\right)\left(\frac{dx}{ds}\right) = \frac{d^2y/dx^2}{[1 + (dy/dx)^2]^{3/2}} \tag{8-1}$$

For a loaded beam with its longitudinal axis taken as the $x$ axis, we may

**139**

set $dy/dx$ in formula 8-1 equal to zero if the deflection of beam is small. Thus, we obtain

$$\kappa = \frac{d\phi}{ds} \approx \frac{d^2 y}{dx^2} \tag{8-2}$$

In general, except for very deep beams with a short span, the deflection due to the shearing force is negligible and only that due to the bending moment is considered. In order to develop a formula for the curvature due to elastic bending, let us consider a small element of a beam shown in Fig. 8-2. Owing to the action of bending moment $M$, the two originally parallel sections $AB$ and $A'B'$ will change directions. This angle change is denoted by $d\phi$. If the length of the element is $ds$ and the maximum bending stress, which occurs at the extreme fibers, is called $f$, the total elongation at the top or bottom fiber is $c\,d\phi$ (see Fig. 8-2), which equals $f\,ds/E$, $E$ being the modulus of elasticity. Thus,

$$c\,d\phi = \frac{f\,ds}{E}$$

or

$$\frac{d\phi}{ds} = \frac{f}{Ec}$$

Replacing $f$ with $Mc/I$, $I$ being the moment of inertia of the cross-sectional area of the beam about the axis of bending, gives

$$\frac{d\phi}{ds} = \frac{M}{EI} \tag{8-3}$$

which expresses the relationship between the curvature and the bending moment. Now equating Eqs. 8-2 and 8-3, we obtain the approximate curvature for a loaded beam as

$$\frac{d^2 y}{dx^2} = \frac{M}{EI} \tag{8-4}$$

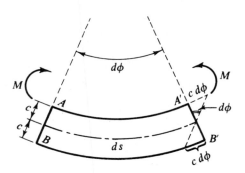

Fig. 8-2

Note that Eq. 8-4 involves four major assumptions:

1. Small deflection of beam

2. Elastic material
3. Only bending moment considered significant
4. Plane section remaining plane after bending

The curvature, established in the coordinate axes of Fig. 8-1, clearly has the same sign as $M$, but the sign may be reversed if the direction of the $y$ axis is reversed. In that case, we have

$$\frac{d^2y}{dx^2} = -\frac{M}{EI} \tag{8-4a}$$

## 8-3. EXTERNAL WORK AND INTERNAL WORK

If a variable force $F$ moves along its direction a distance $ds$, the work done is $F\,ds$. The total work done by $F$ during a period of movement may be expressed by

$$W = \int_{s_1}^{s_2} F\,ds \tag{8-5}$$

where $s_1$ and $s_2$ are the initial and final values of the position.

Consider a load gradually applied to a structure. Its point of application deflects and reaches a value $\Delta$ as the load increases from 0 to $P$. As long as the principle of superposition holds, a linear relationship exists between the load and the deflection as represented by the line $oa$ in Fig. 8-3. The total work performed by the applied load during this period is given by

$$W = \int_0^{\Delta} F\,ds = \int_0^{\Delta} \left(\frac{Ps}{\Delta}\right) ds = \frac{1}{2}\,P\Delta \tag{8-6}$$

which equals the area of the shaded triangle $oab$ in Fig. 8-3.

If further deflection $\delta\Delta$, caused by an agent other than $P$, occurs to the structure in the action line of $P$, then the additional amount of work done by

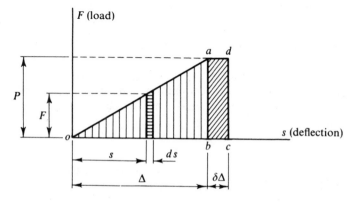

**Fig. 8-3**

the already existing load $P$ will be $P \, \delta\Delta$, which equals the shaded rectangular area *abcd* shown in Fig. 8-3.

Similarly, the work done by a couple $M$ to turn an angular displacement $d\phi$ is $M \, d\phi$. The total work done by $M$ is

$$W = \int_{\phi_1}^{\phi_2} M \, d\phi \tag{8-7}$$

Also, the work performed by a gradually applied couple $C$ accompanied by a rotation increasing from 0 to $\Theta$ is given by

$$W = \tfrac{1}{2} C\Theta \tag{8-8}$$

Now consider a beam subjected to gradually applied forces. As long as the linear relationship between the load and the deflection maintains, all the external work will be converted into internal work or elastic strain energy. Let $dW$ be the strain energy restored in an infinitesimal element of the beam (see Fig. 8-2). We have

$$dW = \tfrac{1}{2} M \, d\phi$$

if only the bending moment $M$ produced by the forces on the element is considered significant. Using Eq. 8-3,

$$\frac{d\phi}{ds} = \frac{M}{EI}$$

or

$$d\phi = \frac{M \, ds}{EI}$$

we have

$$dW = \frac{M^2 \, ds}{2EI}$$

For a loaded beam with its longitudinal axis taken as the $x$ axis, we let $ds \approx dx$ and obtain

$$dW = \frac{M^2 dx}{2EI}$$

The total strain energy restored in the beam of length $l$ is, therefore, given by

$$W = \int_0^l \frac{M^2 \, dx}{2EI} \tag{8-9}$$

For a truss subjected to gradually applied loads, the internal work performed by a member with constant cross-sectional area $A$, length $L$, and internal axial force $S$ is $S^2 L / 2AE$. The total internal work or elastic strain energy for the entire truss is

$$W = \Sigma \frac{S^2 L}{2AE} \tag{8-10}$$

In some special cases deformations of structures can be found by equating external work $W_E$ and internal work (strain energy) $W_I$,

$$W_E = W_I \tag{8-11}$$

For instance, to find the deflection at the free end of the loaded cantilever beam shown in Fig. 8-4, we have

$$W_E = \tfrac{1}{2} P \Delta_b$$

$$W_I = \int_0^l \frac{M^2 \, dx}{2EI}$$

$$= \int_0^l \frac{(-Px)^2 \, dx}{2EI} = \frac{P^2 l^3}{6EI}$$

Setting $W_E = W_I$ gives

$$\Delta_b = \frac{Pl^3}{3EI}$$

Note that the method illustrated is quite limited in application since it is only applicable to deflection at a point of concentrated force. Furthermore, if more than one force is applied simultaneously to a structure, then more than one unknown deformation will appear in one equation, and a solution becomes impossible. Thus, we do not consider this as a general method.

**Fig. 8-4**

## 8-4. METHOD OF VIRTUAL WORK
### (UNIT-LOAD METHOD)

Consider the two cases in Fig. 8-5. Figure 8-5(a) illustrates a deformed elastic structure (be it a beam, a rigid frame, or a truss) subjected to the gradually applied loads $P_1, P_2, \ldots$ which move their points of application the distances $\Delta_1, \Delta_2, \ldots$ respectively. In order to find an expression for the deformation at any point of the structure, say the vertical deflection component at point $C$, we present the case of Fig. 8-5(b), which shows the same structure with all the actual loads removed but a virtual load of unity being gradually applied at point $C$ along the desired deflection. Let $\delta$ denote the distance the unit load moves its point of application. Note that the virtual load is supposed to be vanishingly small and so are the corresponding virtual deformations.

Also shown in Fig. 8-5(a) is one of the typical deformed elements (be it a fiber in a beam or a rigid frame, or a bar in a truss) of length $L$ subjected to internal forces, called $S$, with a corresponding change in length $dL$. In Fig.

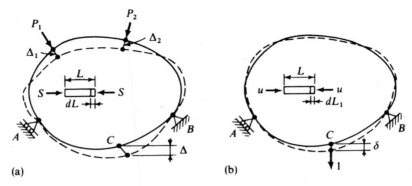

**Fig. 8-5**

8-5(b) the same element is subjected to internal forces, called $u$, with a corresponding change in length $dL_1$.

Since the external work done by the applied loads must equal the internal strain energy of all elements in the structure, we obtain for Fig. 8-5(a),

$$\tfrac{1}{2} P_1\Delta_1 + \tfrac{1}{2} P_2\Delta_2 = \tfrac{1}{2} \sum S \cdot dL \qquad (8\text{-}12)$$

and for Fig. 8-5(b),

$$\tfrac{1}{2}(1)(\delta) = \tfrac{1}{2} \sum u \cdot dL_1 \qquad (8\text{-}13)$$

Now imagine that the case in Fig. 8-5(b) exists first; the actual loads $P_1$ and $P_2$ are then gradually applied to it. Equating the total work done and the total strain energy restored during this period, we have

$$\tfrac{1}{2}(1)(\delta) + \tfrac{1}{2} P_1\Delta_1 + \tfrac{1}{2} P_2\Delta_2 + 1 \cdot \Delta$$
$$= \tfrac{1}{2} \sum u \cdot dL_1 + \tfrac{1}{2} \sum S \cdot dL + \sum u \cdot dL \qquad (8\text{-}14)$$

Since the strain energy and work done must be same whether the loads are applied together or separately, we obtain from subtracting the sum of Eqs. 8-12 and 8-13 from Eq. 8-14,

$$\overset{\text{actual}}{1 \cdot \underset{\text{virtual}}{\Delta = \sum u \cdot dL}} \qquad (8\text{-}15)$$

Note that Eq. 8-15 is the basic equation of the unit-load method. When the rotation of tangent at any point in the structure is desired, we need only replace the unit virtual force with a *unit virtual couple* in the procedure described above, and we obtain

$$\overset{\text{actual}}{1 \cdot \underset{\text{virtual}}{\theta = \sum u \cdot dL}} \qquad (8\text{-}16)$$

where $u$ is the internal force for a typical element caused by the unit couple and $\theta$ is the desired rotation angle.

144

To find a working formula for solving beam deformations, let us consider a statically determinate beam subjected to loads $P_1$ and $P_2$ shown in Fig. 8-6(a); the longitudinal axis of the beam is taken as the $x$ axis. To find the vertical deflection $\Delta$ at an arbitrary point $C$, we place a unit vertical force at $C$, as shown in Fig. 8-6(b), and apply Eq. 8-15:

$$1 \cdot \Delta = \sum u \cdot dL$$

To interpret the terms $dL$ and $u$ involved in the equation above, let us first refer to Fig. 8-6(a) and observe that in the present case $dL$ is the change of length of any fiber having length $dx$ and cross-sectional area $dA$ caused by the actual loads $P_1$ and $P_2$. $dL$ equals unit elongation times $dx$ and can, therefore, be expressed by $My\, dx/EI$ in which $M$ is the bending moment at the section considered resulting from the actual loads, $I$ the moment of inertia of the cross-sectional area of the beam about the axis of bending, $y$ the distance from the fiber to the axis of bending, and $E$ the modulus of elasticity. Next, refer to Fig. 8-6(b) and observe that $u$ in this case is the internal force of the same fiber resulting from a fictitious unit load applied at $C$; $u$ equals the bending stress of the fiber times $dA$, that is, $u = my\, dA/I$, where $m$ is the bending moment at the same section due to the unit load.

Substituting $dL = My\, dx/EI$ and $u = my\, dA/I$ in the basic equation gives

$$1 \cdot \Delta = \sum \left(\frac{my}{I} dA\right)\left(\frac{My}{EI} dx\right)$$

$$= \int_0^l \frac{Mm\, dx}{EI^2} \int_A y^2\, dA$$

(a)

(b)

**Fig. 8-6**

Using $\int_A y^2\, dA = I$, we obtain

$$1 \cdot \Delta = \int_0^l \frac{Mm\, dx}{EI} \tag{8-17}$$

Equation 8-17 is the working formula for the determination of the deflection at any point of a beam. If the rotation of tangent at $C$ is desired, we place a unit couple at $C$ and apply the basic formula

$$1 \cdot \theta = \sum u \cdot dL$$

In a similar manner, we obtain

$$1 \cdot \theta = \int_0^l \frac{Mm\, dx}{EI} \tag{8-18}$$

where $m$ is the bending moment at any section due to a unit couple at $C$.

**Example 8-1.** Find the deflection and slope at the free end of a cantilever beam subjected to a uniform load [Fig. 8-7(a)].

To find $\Delta_b$, we place a unit vertical downward load at $b$ [Fig. 8-7(b)].

$$\Delta_b = \int_0^l \frac{Mm\, dx}{EI} = \int_0^l \frac{(-wx^2/2)(-x)\, dx}{EI} = \frac{wl^4}{8EI}$$

To find $\theta_b$, we place a unit clockwise couple at $b$ [Fig. 8-7(c)].

$$\theta_b = \int_0^l \frac{Mm\, dx}{EI} = \int_0^l \frac{(-wx^2/2)(-1)\, dx}{EI} = \frac{wl^3}{6EI}$$

The positive results indicate that $\Delta_b$ and $\theta_b$ are in the directions assumed.

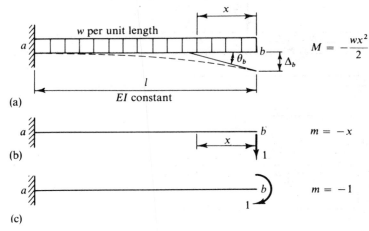

Fig. 8-7

**Example 8-2.** Find $\theta_A$, $\theta_C$, and $\Delta_C$ of the loaded beam in Fig. 8-8(a). Assume constant $EI$.

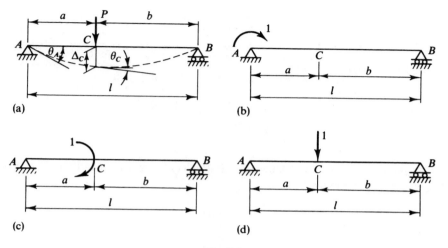

**Fig. 8-8**

To do this, we find it is advantageous to use double origins to perform the integration. That is,

$$\int_0^l \frac{Mm\,dx}{EI} = \int_0^a \frac{Mm\,dx}{EI} + \int_0^b \frac{Mm\,dx}{EI}$$

The terms of $M$ and $m$ in the expression above, solving for $\theta_A$, $\theta_C$, and $\Delta_C$, are evaluated as shown in Table 8-1.

$$\theta_A = \int_0^a \frac{(Pbx/l)[1 - (x/l)]\,dx}{EI} + \int_0^b \frac{(Pax/l)(x/l)\,dx}{EI}$$

$$= \frac{1}{EI}\left(\frac{Pa^2b}{2l} - \frac{Pa^3b}{3l^2} + \frac{Pab^3}{3l^2}\right) = \frac{Pab}{6EIl}\left(3a - \frac{2a^2}{l} + \frac{2b^2}{l}\right) = \frac{Pab(l+b)}{6EIl}$$

$$\theta_C = \int_0^a \frac{(Pbx/l)(-x/l)\,dx}{EI} + \int_0^b \frac{(Pax/l)(x/l)\,dx}{EI}$$

$$= \frac{1}{EI}\left(-\frac{Pa^3b}{3l^2} + \frac{Pab^3}{3l^2}\right) = \frac{Pab(b-a)}{3EIl}$$

$$\Delta_C = \int_0^a \frac{(Pbx/l)(bx/l)\,dx}{EI} + \int_0^b \frac{(Pax/l)(ax/l)\,dx}{EI}$$

$$= \frac{1}{EI}\left(\frac{Pa^3b^2}{3l^2} + \frac{Pa^2b^3}{3l^2}\right) = \frac{Pa^2b^2}{3EIl}$$

If $a = b = l/2$, then

$$\theta_C = 0 \qquad \theta_A = \frac{Pl^2}{16EI} \qquad \Delta_C = \frac{Pl^3}{48EI}$$

147

TABLE 8-1

| Section | Origin | Limit | $M$ Fig. 8-8(a) | $m$ for $\theta_A$ Fig. 8-8(b) | $m$ for $\theta_C$ Fig. 8-8(c) | $m$ for $\Delta_C$ Fig. 8-8(d) |
|---------|--------|-------|------|------|------|------|
| AC | A | 0 to a | $\dfrac{Pb}{l}x$ | $1 - \dfrac{x}{l}$ | $-\dfrac{x}{l}$ | $\dfrac{b}{l}x$ |
| BC | B | 0 to b | $\dfrac{Pa}{l}x$ | $\dfrac{x}{l}$ | $\dfrac{x}{l}$ | $\dfrac{a}{l}x$ |

**Example 8-3.** Find the deflection at the center of the beam in Fig. 8-9. Use $E = 30,000$ kips/in.$^2$.

Fig. 8-9

Refer to Table 8-2 and obtain

$$E\Delta_C = \int_0^l \frac{Mm\,dx}{I}$$

$$= 2\left[\frac{1}{1,000}\int_0^{10}(5x)\left(\frac{x}{2}\right)dx + \frac{1}{1,500}\int_0^{10}\frac{5(10+x)^2}{2}\,dx\right]$$

$$= 2\left\{\frac{1}{1,000}\left[\frac{5}{6}x^3\right]_0^{10} + \frac{1}{1,500}\left(\frac{5}{2}\right)\left[100x + 10x^2 + \frac{x^3}{3}\right]_0^{10}\right\}$$

$$= 2\left(\frac{5}{6} + \frac{70}{18}\right) = 9.44$$

TABLE 8-2

| Section | Origin | Limit (ft) | $M$ (ft-kips) | $m$ (ft-kips)* | $I$ (in.$^4$) |
|---------|--------|-----------|---------------|----------------|---------------|
| ab | a | 0 to 10 | $5x$ | $\dfrac{x}{2}$ | 1,000 |
| bc | b | 0 to 10 | $5(10+x)$ | $\frac{1}{2}(10+x)$ | 1,500 |

*We use a unit load of 1 kip.

Now let us check the dimensions of both sides of the expression above. Note that a unit load of 1 kip must be included in the left side of the expression.

$$30,000 \left(\frac{\text{kips}}{\text{in.}^2}\right)(1 \text{ kip})(\Delta_c) = 9.44 \frac{\text{ft-kips ft-kips ft}}{\text{in.}^4}$$

Thus,
$$\Delta_c = \frac{9.44}{30,000} \frac{\text{ft}^3}{\text{in.}^2}$$

or
$$\Delta_c = \frac{(9.44)(1,728) \text{ in.}}{30,000} = 0.544 \text{ in.} \qquad \text{(down)}$$

In a rigid frame the strain energy due to axial forces and shearing forces is usually much smaller than that due to bending moment and can, therefore, be neglected. The formula

$$\int \frac{Mm \, dx}{EI}$$

previously derived for beam deformations is also good for finding the elastic deformations for a rigid frame, as illustrated in the following examples.

**Example 8-4.** Determine the horizontal, vertical, and rotational deflection components at end $a$ of the rigid frame shown in Fig. 8-10(a). Assume that all members have the same value of $EI$.

Fig. 8-10

To perform the integration for the entire frame denoted by

$$\int_F$$

we must consider each member as a unit, the centroidal axis of the member being taken as the $x$ axis. Thus,

$$\int_F \frac{Mm \, dx}{EI} = \int_{ab} \frac{Mm \, dx}{EI} + \int_{bc} \frac{Mm \, dx}{EI} + \int_{cd} \frac{Mm \, dx}{EI}$$

The terms of $M$ and $m$ in this expression, when we solve for each of the deflection components at $a$, are listed in Table 8-3, in which we use $m_1$ to denote the bending moment at any section due to a unit horizontal force applied at

TABLE 8-3

| Member | Origin | Limit (ft) | $M$ (ft-kips) Fig. 8-10(a) | $m_1$ (ft-kips) Fig. 8-10(b) | $m_2$ (ft-kips) Fig. 8-10(c) | $m_3$ (ft-kips) Fig. 8-10(d) |
|--------|--------|-----------|------|------|------|------|
| $ab$ | $a$ | 0 to 10 | 0 | $-x$ | 0 | 1 |
| $bc$ | $b$ | 0 to 10 | $-\dfrac{1.2x^2}{2}$ | $-10$ | $x$ | 1 |
| $cd$ | $c$ | 0 to 10 | $-60$ | $x-10$ | 10 | 1 |

$a$; $m_2$ that due to a unit vertical force at $a$; and $m_3$ that due to a unit couple at $a$. Note that the bending moment resulting in compression on the outside fibers of the frame is assumed to be positive.

To find the horizontal deflection at $a$, called $\Delta_1$, we apply

$$\Delta_1 = \int_F \frac{M m_1\, dx}{EI}$$

$$= \frac{1}{EI}\left[0 + \int_0^{10}\left(-\frac{1.2x^2}{2}\right)(-10)\, dx + \int_0^{10}(-60)(x-10)\, dx\right]$$

$$= 5{,}000\ \frac{\text{kips-ft}^3}{EI} \qquad \text{(right)}$$

Similarly, we have the vertical deflection at $a$, called $\Delta_2$,

$$\Delta_2 = \int_F \frac{M m_2\, dx}{EI}$$

$$= \frac{1}{EI}\left[0 + \int_0^{10}\left(-\frac{1.2x^2}{2}\right)(x)\, dx + \int_0^{10}(-60)(10)\, dx\right]$$

$$= -7{,}500\ \frac{\text{kips-ft}^3}{EI} \qquad \text{(down)}$$

and the rotational displacement at $a$, called $\Delta_3$,

$$\Delta_3 = \int_F \frac{M m_3\, dx}{EI}$$

$$= \frac{1}{EI}\left[0 + \int_0^{10}\left(-\frac{1.2x^2}{2}\right)(1)\, dx + \int_0^{10}(-60)(1)\, dx\right]$$

$$= -800\ \frac{\text{kips-ft}^2}{EI} \qquad \text{(counterclockwise)}$$

**Example 8-5.** Find the deflection components at $a$ of the same frame for each of three loading cases shown in Fig. 8-10(b)–(d). Let

$\delta_{11}$ = horizontal deflection at $a$ due to a unit horizontal force at $a$

$\delta_{21}$ = vertical deflection at $a$ due to a unit horizontal force at $a$

$\delta_{31}$ = rotational displacement at $a$ due to a unit horizontal force at $a$

Then

$$\delta_{11} = \int_F \frac{(m_1)^2 \, dx}{EI} \qquad \delta_{21} = \int_F \frac{m_1 m_2 \, dx}{EI} \qquad \delta_{31} = \int_F \frac{m_1 m_3 \, dx}{EI}$$

since in this case [see Fig. 8-10(b)] $M = m_1$.

Likewise, if the frame is subjected only to a unit vertical force at $a$ [see Fig. 8-10(c)], the three deflection components at $a$ are found to be

$$\delta_{12} = \int_F \frac{m_2 m_1 \, dx}{EI} \qquad \delta_{22} = \int_F \frac{(m_2)^2 \, dx}{EI} \qquad \delta_{32} = \int_F \frac{m_2 m_3 \, dx}{EI}$$

And if the frame is subjected only to a unit couple at $a$ [see Fig. 8-10(d)], the three deflection components at $a$ are found to be

$$\delta_{13} = \int_F \frac{m_3 m_1 \, dx}{EI} \qquad \delta_{23} = \int_F \frac{m_3 m_2 \, dx}{EI} \qquad \delta_{33} = \int_F \frac{(m_3)^2 \, dx}{EI}$$

Taking numerical values from Table 8-3 and substituting in each of the expressions above, we find

$$\delta_{11} = \frac{1}{EI}\left[ \int_0^{10} x^2 \, dx + \int_0^{10} (-10)(-10) \, dx + \int_0^{10} (x - 10)^2 \, dx \right]$$

$$= 1{,}667 \, \frac{\text{kips-ft}^3}{EI} \qquad \text{(right)}$$

$$\delta_{21} = \frac{1}{EI}\left[ 0 + \int_0^{10} (-10)(x) \, dx + \int_0^{10} (x - 10)(10) \, dx \right]$$

$$= -1{,}000 \, \frac{\text{kips-ft}^3}{EI} \qquad \text{(down)}$$

$$\delta_{31} = \frac{1}{EI}\left[ \int_0^{10} (-x)(1) \, dx + \int_0^{10} (-10)(1) \, dx + \int_0^{10} (x - 10)(1) \, dx \right]$$

$$= -200 \, \frac{\text{kips-ft}^2}{EI} \qquad \text{(counterclockwise)}$$

$$\delta_{12} = \delta_{21} = -1{,}000 \, \frac{\text{kips-ft}^3}{EI} \qquad \text{(left)}$$

$$\delta_{22} = \frac{1}{EI}\left[ 0 + \int_0^{10} x^2 \, dx + \int_0^{10} (10)^2 \, dx \right] = 1{,}333 \, \frac{\text{kips-ft}^3}{EI} \qquad \text{(up)}$$

$$\delta_{32} = \frac{1}{EI}\left[ 0 + \int_0^{10} (x)(1) \, dx + \int_0^{10} (10)(1) \, dx \right] = 150 \, \frac{\text{kips-ft}^2}{EI} \qquad \text{(clockwise)}$$

$$\delta_{13} = -200 \, \frac{\text{kips-ft}^3}{EI} \qquad \text{(left)}$$

$$\delta_{23} = 150 \, \frac{\text{kips-ft}^3}{EI} \qquad \text{(up)}$$

$$\delta_{33} = \frac{1}{EI}\left[ \int_0^{10} dx + \int_0^{10} dx + \int_0^{10} dx \right] = 30 \, \frac{\text{kips-ft}^2}{EI} \qquad \text{(clockwise)}$$

Note that $\delta_{13}$ has the same value as $\delta_{31}$, but they differ by one dimension of length. The same is true for $\delta_{23}$ and $\delta_{32}$.

In matrix form the result shown above is

$$\begin{bmatrix} \delta_{11} & \delta_{12} & \delta_{13} \\ \delta_{21} & \delta_{22} & \delta_{23} \\ \delta_{31} & \delta_{32} & \delta_{33} \end{bmatrix} = \begin{bmatrix} \int_F \dfrac{(m_1)^2\,dx}{EI} & \int_F \dfrac{m_1 m_2\,dx}{EI} & \int_F \dfrac{m_1 m_3\,dx}{EI} \\ \int_F \dfrac{m_1 m_2\,dx}{EI} & \int_F \dfrac{(m_2)^2\,dx}{EI} & \int_F \dfrac{m_2 m_3\,dx}{EI} \\ \int_F \dfrac{m_1 m_3\,dx}{EI} & \int_F \dfrac{m_2 m_3\,dx}{EI} & \int_F \dfrac{(m_3)^2\,dx}{EI} \end{bmatrix}$$

$$= \begin{bmatrix} 1{,}667 \text{ ft} & -1{,}000 \text{ ft} & -200 \text{ ft} \\ -1{,}000 \text{ ft} & 1{,}333 \text{ ft} & 150 \text{ ft} \\ -200 & 150 & 30 \end{bmatrix} \dfrac{\text{kips-ft}^2}{EI}$$

The working formula for deflection of any joint of a loaded truss can be evaluated from the basic equation, Eq. 8-15,

$$1 \cdot \Delta = \sum u \cdot dL$$

by considering each member of the truss as an element. Thus, the term $dL$ is the shortening or lengthening of a bar due to applied loads and can be expressed by $SL/AE$. The equation above becomes

$$1 \cdot \Delta = \sum_1^m \frac{SuL}{AE} \tag{8-19}$$

where $S$ = internal force in any member due to actual loads

  $u$ = internal force in the same member due to a fictitious unit load at the point where the deflection is sought, acting along the desired direction

  $L$ = length of the member

  $A$ = cross-sectional area of the member

  $E$ = modulus of elasticity of the member

  $m$ = total number of members

Equation 8-19 in matrix form is

$$\Delta = \begin{bmatrix} u_1 & u_2 & \cdots & u_m \end{bmatrix} \begin{bmatrix} \dfrac{L_1}{A_1 E} & & & \\ & \dfrac{L_2}{A_2 E} & & \\ & & \ddots & \\ & & & \dfrac{L_m}{A_m E} \end{bmatrix} \begin{Bmatrix} S_1 \\ S_2 \\ \vdots \\ S_m \end{Bmatrix} \tag{8-19a}$$

Sometimes the change of bar length $dL$ is not caused by any external force but is due to the effect of temperature. If this is the case, we let $dL = \alpha t L$ and the working formula for finding deflection due to temperature change is given by

$$1 \cdot \Delta = \sum u \cdot \alpha t L \tag{8-20}$$

where $\alpha$ = coefficient of linear thermal expansion
  $t$ = temperature rise in degrees

**Example 8-6.** Find the vertical deflection of joint $b$ of the loaded truss shown in Fig. 8-11(a). Assume that $L$ (ft)$/A$ (in.²) = 1 and that $E$ = 30,000 kips/in.² for all members.

(a)

(b)

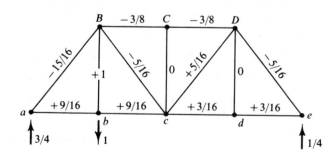

(c)

**Fig. 8-11**

We begin with the evaluation of $S$ and $u$, and the answer diagrams for them are shown in Fig. 8-11(b) and (c), respectively. Next we apply Eq. 8-19. The complete solution is given in Table 8-4.

$$\Delta_v = \Sigma \frac{SuL}{AE} = \frac{+202}{30,000} = +0.00673 \text{ ft} \qquad \text{(down)}$$

TABLE 8-4

| Member | $\dfrac{L}{A}\left(\dfrac{\text{ft}}{\text{in.}^2}\right)$ | $S$ (kips) | $u*$ | $\dfrac{SuL}{A}\left(\dfrac{\text{ft-kips}}{\text{in.}^2}\right)$ |
|--------|:---:|:---:|:---:|:---:|
| ab | 1 | +36 | + 9/16 | + 20.25 |
| bc | 1 | +36 | + 9/16 | + 20.25 |
| cd | 1 | +12 | + 3/16 | + 2.25 |
| de | 1 | +12 | + 3/16 | + 2.25 |
| BC | 1 | −24 | − 3/8 | + 9.0 |
| CD | 1 | −24 | − 3/8 | + 9.0 |
| aB | 1 | −60 | −15/16 | + 56.25 |
| Bb | 1 | +64 | +1 | + 64.0 |
| Bc | 1 | −20 | − 5/16 | + 6.25 |
| Cc | 1 | 0 | 0 | 0 |
| cD | 1 | +20 | + 5/16 | + 6.25 |
| Dd | 1 | 0 | 0 | 0 |
| De | 1 | −20 | − 5/16 | + 6.25 |
|  |  |  | $\Sigma$ | +202.0 |

*We use a fictitious load of 1 (not 1 kip) for determining $u$ values.

**Example 8-7.** For the loaded structure in Example 8-6, find the absolute deflection of joint $b$.

To do this, we have to obtain the horizontal deflection of joint $b$ in addition to the vertical deflection of that joint already found. The vector sum of these two displacement components is the solution.

When a unit horizontal load is applied at joint $b$ to the right, only the member $ab$ is under the stress of tension (i.e., $u = 1$); all other members are unstressed. The horizontal movement, called $\Delta_h$, at joint $b$ is thus given by

$$\Delta_h = \left(\frac{SuL}{AE}\right)_{ab} = \frac{(36)(1)}{30,000} = +0.0012 \text{ ft} \qquad \text{(right)}$$

and the absolute deflection of joint $b$ is given by

$$\Delta = \sqrt{(\Delta_v)^2 + (\Delta_h)^2} = \sqrt{(0.00673)^2 + (0.0012)^2} = 0.00684 \text{ ft}$$

moving down to the right and making an angle $\phi$ with the horizontal direction,

$$\phi = \tan^{-1}\frac{0.00673}{0.00120} = \tan^{-1} 5.72 \approx 80°$$

**Example 8-8.** For the same loaded truss [Fig. 8-11(a)], find the rotation of member $bc$.

Finding the rotation of member $bc$ is equivalent to finding the relative

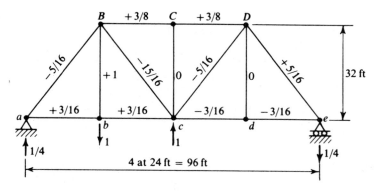

**Fig. 8-12**

displacement between ends $b$ and $c$ (in the direction perpendicular to $bc$) divided by the length of $bc$. Assume counterclockwise rotation. We then apply a pair of unit fictitious loads to joints $b$ and $c$ and evaluate the $u$ value for each member, as shown in Fig. 8-12. The computation leading to the solution of the relative displacement between joints $b$ and $c$ perpendicular to the original line of $bc$ is contained in Table 8-5. The rotation of the member, denoted by $\theta$, is then determined:

$$\theta = \frac{80}{24E} = \frac{80}{(24)(30,000)} = \frac{1}{9,000} \text{ radian}$$

The positive value of the angle indicates a counterclockwise rotation.

TABLE 8-5

| Member | $\dfrac{L}{A}\left(\dfrac{\text{ft}}{\text{in.}^2}\right)$ | $S$ (kips) | $u$ | $\dfrac{SuL}{A}\left(\dfrac{\text{ft-kips}}{\text{in.}^2}\right)$ |
|--------|------|------|------|------|
| ab | 1 | +36 | + 3/16 | +27/4 |
| bc | 1 | +36 | + 3/16 | +27/4 |
| cd | 1 | +12 | − 3/16 | − 9/4 |
| de | 1 | +12 | − 3/16 | − 9/4 |
| BC | 1 | −24 | + 3/ 8 | −9 |
| CD | 1 | −24 | + 3/ 8 | −9 |
| aB | 1 | −60 | − 5/16 | +75/4 |
| Bb | 1 | +64 | +1 | +64 |
| Bc | 1 | −20 | − 15/16 | +75/4 |
| Cc | 1 | 0 | 0 | 0 |
| cD | 1 | +20 | − 5/16 | −25/4 |
| Dd | 1 | 0 | 0 | 0 |
| De | 1 | −20 | + 5/16 | −25/4 |
| | | | $\Sigma$ | +80 |

**Example 8-9.** Find the vertical deflection at joint $b$ resulting from a rise in temperature of 50°F in the top chords $BC$ and $CD$ (Fig. 8-13). $\alpha = 0.0000065$ in./in./1°F.

155

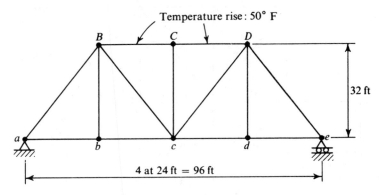

**Fig. 8-13**

On a statically determinate truss, no reactions or internal forces can be developed because of a temperature rise or drop in truss members. However, certain changes of bar length will take place if the temperature rises or drops in a bar. This in turn will cause the distortion of the whole truss.

To find the vertical deflection of joint $b$, we apply Eq. 8-20,

$$\Delta_b = \sum u \cdot \alpha t L$$

Note that, in this problem, only bars $BC$ and $CD$ are involved in computation since the rest of the members undergo no change of length. Now $u = -\frac{3}{8}$ [see Fig. 8-11(c)] and $\alpha t L = (0.0000065)(50)(24) = 0.0078$ for $BC$ and $CD$. Thus,

$$\Delta_b = 2(-\tfrac{3}{8})(0.0078) = -0.00585 \text{ ft}$$

The negative sign indicates an upward movement of joint $b$.

### 8-5. CASTIGLIANO'S THEOREM

In 1876, Alberto Castigliano published a notable paper in which he presented a general method for determining the deformations of linear structures. It can be stated as follows: *The first partial derivative of the total strain energy of the structure with respect to one of the applied actions gives the displacement along that action.*

If we use $P$ to denote the action (force or couple), $\Delta_p$ the corresponding displacement (deflection or rotation) along $P$, and $W$ the total strain energy, the statement can be expressed by

$$\Delta_p = \frac{\partial W}{\partial P} \tag{8-21}$$

To demonstrate the theorem, consider the loaded beam in Fig. 8-14. The deflected position is represented by the dashed line.

If we consider only the internal work resulting from the bending moment, we have the total strain energy of the beam (see Eq. 8-9):

**Fig. 8-14**

$$W = \int_0^l \frac{M^2 \, dx}{2EI}$$

Now let $M_1$ be the bending moment at any section due to the gradually applied load $P_1$, and let $M_2$ be the bending moment at the same section due to the gradually applied load $P_2$. We have the total bending moment at any section given by

$$M = M_1 + M_2 = m_1 P_1 + m_2 P_2$$

where $m_1$ = bending moment at any section due to a unit load in place of $P_1$
 $m_2$ = bending moment of the same section due to a unit load in place of $P_2$

Thus, $\quad \dfrac{\partial W}{\partial P_1} = \dfrac{\partial}{\partial P_1} \int_0^l \dfrac{M^2 \, dx}{2EI} = \int_0^l \dfrac{M(\partial M / \partial P_1) \, dx}{EI} = \int_0^l \dfrac{M m_1 \, dx}{EI} = \Delta_1$

and $\quad \dfrac{\partial W}{\partial P_2} = \dfrac{\partial}{\partial P_2} \int_0^l \dfrac{M^2 \, dx}{2EI} = \int_0^l \dfrac{M(\partial M / \partial P_2) \, dx}{EI} = \int_0^l \dfrac{M m_2 \, dx}{EI} = \Delta_2$

The last step of equality in each of the two expressions above is based on Eq. 8-17 from virtual work.

Let us now turn to the loaded truss in Fig. 8-15. The total strain energy (see Eq. 8-10) is

$$W = \Sigma \frac{S^2 L}{2AE}$$

If we let

$S_1$ = internal force in any bar due to the gradually applied load $P_1$

$S_2$ = internal force in the same bar due to the gradually applied load $P_2$

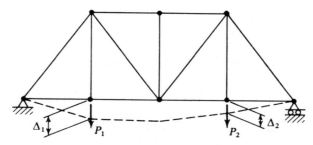

**Fig. 8-15**

157

we have the total internal force in any bar given by

$$S = S_1 + S_2 = P_1 u_1 + P_2 u_2$$

where $u_1$ = internal force in any bar due to a unit load in place of $P_1$
$u_2$ = internal force in the same bar due to a unit load in place of $P_2$

Thus,     $$\frac{\partial W}{\partial P_1} = \frac{\partial}{\partial P_1} \sum \frac{S^2 L}{2AE} = \sum \frac{S(\partial S/\partial P_1)L}{AE} = \sum \frac{Su_1 L}{AE} = \Delta_1$$

and     $$\frac{\partial W}{\partial P_2} = \frac{\partial}{\partial P_2} \sum \frac{S^2 L}{2AE} = \sum \frac{S(\partial S/\partial P_2)L}{AE} = \sum \frac{Su_2 L}{AE} = \Delta_2$$

The last step of equality in each of the expressions above is based on Eq. 8-19 from virtual work.

It is interesting to point out that the Castigliano's theorem basically does not differ from the method of virtual work for the analysis of linear structures subjected to external forces. The difference is only a matter of the arrangement of calculation. Using the method of virtual work, we have from Eqs. 8-17 and 8-19,

$$\Delta = \int \frac{Mm\, dx}{EI} \qquad \text{for a beam or rigid frame}$$

and     $$\Delta = \sum \frac{SuL}{AE} \qquad \text{for a truss}$$

while in the application of Castigliano's theorem, we have

$$\Delta = \int \frac{M(\partial M/\partial P)\, dx}{EI} \qquad \text{for a beam or rigid frame} \qquad (8\text{-}22)$$

and     $$\Delta = \sum \frac{S(\partial S/\partial P)\, L}{AE} \qquad \text{for a truss} \qquad (8\text{-}23)$$

**Example 8-10.** Find the vertical deflection at the free end $b$ of a cantilever beam $ab$ subjected to a concentrated load $P$ at $b$ (see Fig. 8-4).

$$\Delta_b = \frac{\partial W}{\partial P} = \int_0^l \frac{M(\partial M/\partial P)\, dx}{EI} = \frac{1}{EI} \int_0^l M \frac{\partial M}{\partial P}\, dx$$

Now     $$M = -Px \qquad \frac{\partial M}{\partial P} = -x$$

Therefore,     $$\Delta_b = \frac{1}{EI} \int_0^l (-Px)(-x)\, dx = \frac{Pl^3}{3EI} \qquad \text{(down)}$$

**Example 8-11.** For the beam and load shown in Fig. 8-16(a), find the vertical and rotational displacements at the free end $a$. Given $w = 15$ kN/m, $l = 5$ m, $E = 20{,}000$ kN/cm², and $I = 12{,}000$ cm⁴, determine the magnitude of the displacements.

To find the displacements at the free end, we note that in this case no vertical force or moment force actually acts at the free end, and thus Castigliano's theorem cannot be directly applied. In order to carry out the partial derivative, we must first assume the imaginary forces $Q_1$, $Q_2$ corresponding respectively,

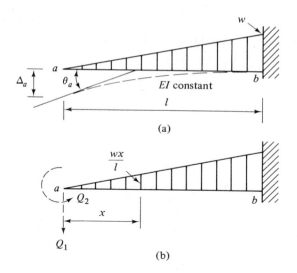

Fig. 8-16

to the vertical deflection $\Delta_a$ and the rotation $\theta_a$ at the free end, and then to set $Q_1 = Q_2 = 0$ in the final operation.

Refer to Fig. 8-16(b). The moment at any section is given by

$$M = -Q_1 x - Q_2 - \frac{wx^3}{6l}$$

It follows that

$$\frac{\partial M}{\partial Q_1} = -x \qquad \text{and} \qquad \frac{\partial M}{\partial Q_2} = -1$$

$$\Delta_a = \int_0^l \frac{M(\partial M/\partial Q_1)\,dx}{EI}$$

$$= \int_0^l \frac{(-wx^3/6l)(-x)\,dx}{EI} = \frac{wl^4}{30EI} \qquad \text{(down)}$$

$$\theta_a = \int_0^l \frac{M(\partial M/\partial Q_2)\,dx}{EI}$$

$$= \int_0^l \frac{(-wx^3/6l)(-1)\,dx}{EI} = \frac{wl^3}{24EI} \qquad \text{(counterclockwise)}$$

Substituting $w = 15$ kN/m, $l = 5$ m, $E = 20,000$ kN/cm², and $I = 12,000$ cm⁴ into the foregoing expressions, we have

$$\Delta_a = \frac{(15)(5)^4(10)^6}{(30)(20,000)(12,000)} = 1.3 \text{ cm}$$

$$\theta_a = \frac{(15)(5)^3(10)^4}{(24)(20,000)(12,000)} = 0.003 \text{ radian}$$

159

**Example 8-12.** Find, by Castigliano's theorem, the horizontal displacement $\Delta_1$ and the rotational displacement $\Delta_2$ at support $c$ for the rigid frame shown in Fig. 8-17(a). Consider the bending effect only.

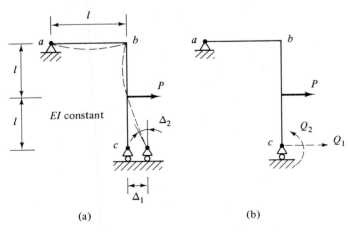

Fig. 8-17

Since no horizontal force or moment actually acts at support $c$, we must assume forces $Q_1$ and $Q_2$ of zero value, corresponding respectively to the desired displacements $\Delta_1$ and $\Delta_2$ at $c$ [Fig. 8-17(b)], in order to carry out the partial derivatives such that

$$\Delta_1 = \int_F \frac{M(\partial M/\partial Q_1)\, dx}{EI}$$

$$\Delta_2 = \int_F \frac{M(\partial M/\partial Q_2)\, dx}{EI}$$

where $\int_F$ indicates the sign of integration carried through the entire frame. The complete solution is given in Table 8-6.

**Example 8-13.** Given the loaded truss in Fig. 8-11(a), find the horizontal deflection at $D$.

Assume that joint $D$ will move to the right. To apply the theorem we place an imaginary horizontal force $Q$ acting at $D$, as shown in Fig. 8-18(a). The bar forces thus obtained are shown in Fig. 8-18(b). The complete solution is shown in Table 8-7.

$$\Delta = \Sigma \frac{S(\partial S/\partial Q)L}{AE} = +\frac{36}{30,000} = +0.0012\,\text{ft} \qquad \text{(right)}$$

TABLE 8-6

| Member | Origin | Limit | $M$ | $\dfrac{\partial M}{\partial Q_1}$ | $\dfrac{\partial M}{\partial Q_2}$ | $\displaystyle\int \dfrac{M(\partial M/\partial Q_1)\,dx}{EI}$ | $\displaystyle\int \dfrac{M(\partial M/\partial Q_2)\,dx}{EI}$ |
|---|---|---|---|---|---|---|---|
| $ab$ | $a$ | $0$ to $l$ | $\left(P + 2Q_1 + \dfrac{Q_2}{l}\right)x$ | $2x$ | $\dfrac{x}{l}$ | $\displaystyle\int_0^l \dfrac{(Px)(2x)\,dx}{EI}$ | $\displaystyle\int_0^l \dfrac{(Px)(x/l)\,dx}{EI}$ |
| $bc$ | $c$ | $l$ to $2l$ | $P(x - l) + Q_1 x + Q_2$ | $x$ | $1$ | $\displaystyle\int_l^{2l} \dfrac{P(x - l)(x)\,dx}{EI}$ | $\displaystyle\int_l^{2l} \dfrac{P(x - l)(1)\,dx}{EI}$ |
| | | | | | | $\Delta_1 = \dfrac{3Pl^3}{2EI}$ | $\Delta_2 = \dfrac{5Pl^2}{6EI}$ |

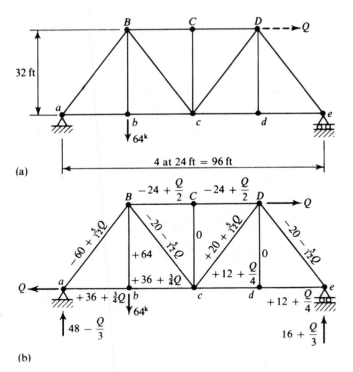

Fig. 8-18

TABLE 8-7

| Member | $\dfrac{L}{A}$ (ft/in.²) | $S$ (kips) | $\dfrac{\partial S}{\partial Q}$ | $\dfrac{S(\partial S/\partial Q)L}{A}$ (ft-kips/in.²) |
|--------|--------------------------|------------|----------------------------------|-------------------------------------------------------|
| $ab$ | 1 | $+36 + 3/4\,Q$ | $+3/4$ | $+27$ |
| $bc$ | 1 | $+36 + 3/4\,Q$ | $+3/4$ | $+27$ |
| $cd$ | 1 | $+12 + 1/4\,Q$ | $+1/4$ | $+\ 3$ |
| $de$ | 1 | $+12 + 1/4\,Q$ | $+1/4$ | $+\ 3$ |
| $BC$ | 1 | $-24 + 1/2\,Q$ | $+1/2$ | $-12$ |
| $CD$ | 1 | $-24 + 1/2\,Q$ | $+1/2$ | $-12$ |
| $aB$ | 1 | $-60 + 5/12\,Q$ | $+5/12$ | $-25$ |
| $Bb$ | 1 | $+64$ | $0$ | $0$ |
| $Bc$ | 1 | $-20 - 5/12\,Q$ | $-5/12$ | $+\ 8.33$ |
| $Cc$ | 1 | $0$ | $0$ | $0$ |
| $cD$ | 1 | $+20 + 5/12\,Q$ | $+5/12$ | $+\ 8.33$ |
| $Dd$ | 1 | $0$ | $0$ | $0$ |
| $De$ | 1 | $-20 - 5/12\,Q$ | $-5/12$ | $+\ 8.33$ |
| | | | $\Sigma$ | $+36.0$ |

$$\Delta = \Sigma \frac{S(\partial S/\partial Q)L}{AE} = +\frac{36}{30,000} = +0.0012 \text{ ft} \qquad \text{(right)}$$

162

## 8-6. CONJUGATE-BEAM METHOD

The purpose of this method is to transform the problem of solving the slopes and deflections of a structure resulting from the applied loads (actual loads) to a problem of solving the shears and moments of a conjugate beam due to the elastic load derived from the angle changes of structural elements. The advantage of this method over the method of virtual work or Castigliano's theorem are:

1. Unlike the previous methods, which are used to find one item of deformation at one point of the structure in an operation, this method enables us to find deformations at many points of the structure in a single setup.

2. It is generally acknowledged that structural engineers prefer to deal with shear and bending moment rather than to do tedious integral calculus.

Consider a typical loaded beam, such as the one shown in Fig. 8-19(a), for which we may plot the moment diagram and, therefore, the $M/EI$ diagram as in Fig. 8-19(b).

We recall that the curvature at any point of the beam of Fig. 8-19(a) is given by Eq. 8-4a:

$$\frac{d^2y}{dx^2} = -\frac{M}{EI}$$

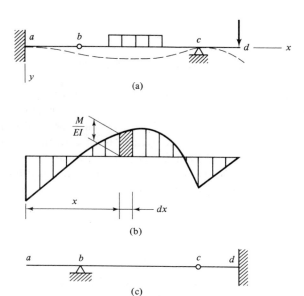

Fig. 8-19

163

Now since the slope at any point of the beam is expressed by

$$\frac{dy}{dx} = \tan \theta \approx \theta$$

for small deformation, we have

$$\frac{d\theta}{dx} = -\frac{M}{EI}$$

or

$$d\theta = -\frac{M\,dx}{EI}$$

Integrate.

$$\theta = -\int \frac{M}{EI}\,dx \tag{8-24}$$

Substituting for $\theta$ by $dy/dx$ and integrating again gives

$$y = \int \theta\,dx = -\iint \frac{M}{EI}\,dx\,dx \tag{8-25}$$

Next, for a beam under a distributed load of intensity $w(x)$, the relationships between the load, the shear, and the bending moment at any point are given by (see Sec. 3-4)

$$\frac{dV}{dx} = -w$$

and

$$\frac{dM}{dx} = V$$

Thus, over a portion of the beam

$$V = -\int w\,dx \tag{8-26}$$

and

$$M = \int V\,dx = -\iint w\,dx\,dx \tag{8-27}$$

Now suppose that we have a beam, called a *conjugate beam*, whose length equals that of the actual beam in Fig. 8-19(a). Let this beam be subjected to the so-called *elastic load* of intensity $M/EI$ given in Fig. 8-19(b). [Elastic load is sometimes referred to as the *angle load*, a term obviously associated with $d\theta = M\,(dx/EI)$.] The integral expressions for the shear and moment over a portion of the conjugate beam, denoted by $\bar{V}$ and $\bar{M}$, respectively, can be obtained by replacing $w$ in Eqs. 8-26 and 8-27 with $M/EI$:

$$\bar{V} = -\int \frac{M}{EI}\,dx \tag{8-28}$$

and

$$\bar{M} = -\iint \frac{M}{EI}\,dx\,dx \tag{8-29}$$

When we compare Eqs. 8-24 and 8-25 with Eqs. 8-28 and 8-29, it follows logically that, with properly prescribed boundary conditions for the conjugate beam, we may reach the following results:

1. The slope at a given section of a loaded beam (actual beam) equals the shear in the corresponding section of the conjugate beam subjected to the elastic load.

2. The deflection at a given section of a loaded beam equals the bending moment in the corresponding section of the conjugate beam subjected to the elastic load.

Thus far we have only stated that the conjugate beam is identical to the actual beam with regard to the length of the beam. In order that the above-stated identities be possible, the setup of the support and connection of the conjugate beam must be such as to induce shear and moment in the conjugate beam in conformity to the slope and deflection induced by the counterparts in the actual beam. This is given in Table 8-8 and can be briefly summarized as follows:

$$\text{fixed end} \longleftrightarrow \text{free end}$$

$$\text{simple end} \longleftrightarrow \text{simple end}$$

$$\text{interior connection} \longleftrightarrow \text{interior support}$$

The symbols between the two groups indicate conjugation to each other.

TABLE 8-8

| Actual Beam Subjected to Applied Load | | | Conjugate Beam Subjected to Elastic Load | |
|---|---|---|---|---|
| Fixed end | $\begin{cases} \theta = 0 \\ y = 0 \end{cases}$ | | $\left.\begin{array}{l} \bar{V} = 0 \\ \bar{M} = 0 \end{array}\right\}$ | Free end |
| Free end | $\begin{cases} \theta \neq 0 \\ y \neq 0 \end{cases}$ | | $\left.\begin{array}{l} \bar{V} \neq 0 \\ \bar{M} \neq 0 \end{array}\right\}$ | Fixed end |
| Simple end (hinge or roller) | $\begin{cases} \theta \neq 0 \\ y = 0 \end{cases}$ | | $\left.\begin{array}{l} \bar{V} \neq 0 \\ \bar{M} = 0 \end{array}\right\}$ | Simple end (hinge or roller) |
| Interior support (hinge or roller) | $\begin{cases} \theta \neq 0 \\ y = 0 \end{cases}$ | | $\left.\begin{array}{l} \bar{V} \neq 0 \\ \bar{M} = 0 \end{array}\right\}$ | Interior connection (hinge or roller) |
| Interior connection (hinge or roller) | $\begin{cases} \theta \neq 0 \\ y \neq 0 \end{cases}$ | | $\left.\begin{array}{l} \bar{V} \neq 0 \\ \bar{M} \neq 0 \end{array}\right\}$ | Interior support (hinge or roller) |

Thus, the conjugate beam for the beam in Fig. 8-19(a) is the one shown in Fig. 8-19(c). If we use the $M/EI$ diagram of Fig. 8-19(b) as the load to put on the beam in Fig. 8-19(c), the resulting shear and bending moment for any section of this beam will give the slope and deflection for the corresponding section of the original beam. Other examples of conjugate beams are shown in Fig. 8-20.

Unlike the actual beams, the conjugate beams may be unstable in themselves. For instance, the conjugate beams shown in Fig. 8-20(c) and (f) are unstable

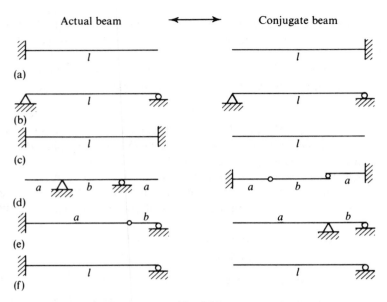

Fig. 8-20

beams. However, they will maintain an unstable equilibrium under the action of an elastic load.

The same figure indicates that the conjugate-beam method is not limited to the analysis of statically determinate beams; in fact, the conjugate-beam method is also applicable to statically indeterminate beams.

The sign convention we use may be stated as follows. The origin of the loaded beam is taken at the left end of the beam with $y$ positive downward and $x$ positive to the right. As a result, a positive deflection means a downward deflection and a positive slope means a clockwise rotation of the beam section. Recall that the derivation from the relationships among load, shear, and bending moment is based on taking the downward load as positive. Therefore, a positive $M/EI$ should be taken as a downward elastic load.

Having defined these, we readily see that a positive shear at a section of the conjugate beam corresponds to a clockwise rotation at the section of the actual beam. A positive moment at a section of the conjugate beam corresponds to a downward deflection at the section of the actual beam.

**Example 8-14.** Find, by the conjugate-beam method, the vertical deflection at the free end $c$ of the cantilever beam shown in Fig. 8-21(a). Assume constant $EI$.

To do this, we place an elastic load on the conjugate beam, as shown in Fig. 8-21(b). The vertical deflection at $c$ of the actual beam is the moment at $c$ of the conjugate beam. Thus,

166

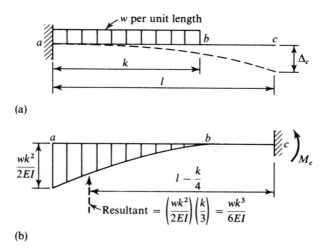

(a)

(b)

**Fig. 8-21**

$$\Delta_c = \left(\frac{wk^3}{6EI}\right)\left(l - \frac{1}{4}k\right) \qquad \text{(down)}$$

**Example 8-15.** Find $\theta_A$, $\theta_C$, and $\Delta_C$ for the loaded beam shown in Fig. 8-22(a) by the conjugate-beam method. Assume constant $EI$. Note that the deformations were solved by the method of virtual work in Example 8-2.

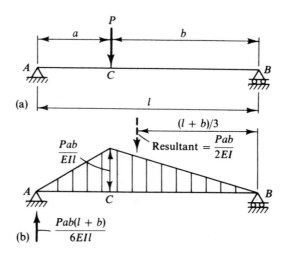

**Fig. 8-22**

The conjugate beam together with the elastic load is shown in Fig. 8-22(b), in which the resultant of the loading is found to be $Pab/2EI$ acting at a distance $(l + b)/3$ from the right end as indicated. Thus,

$$\theta_A = \frac{Pab(l+b)}{6EIl} \qquad \text{(clockwise)}$$

$$\theta_C = \frac{Pab(l+b)}{6EIl} - \left(\frac{Pab}{EIl}\right)\left(\frac{a}{2}\right) = \frac{Pab(b-a)}{3EIl} \qquad \text{(clockwise, if } b > a\text{)}$$

$$\Delta_C = \frac{Pa^2b(l+b)}{6EIl} - \left(\frac{Pab}{EIl}\right)\left(\frac{a}{2}\right)\left(\frac{a}{3}\right) = \frac{Pa^2b^2}{3EIl} \qquad \text{(down)}$$

**Example 8-16.** Use the conjugate-beam method to determine the deflection and rotation at point $b$ in Fig. 8-23(a). $E = 20,000$ kN/cm$^2$.

To do this, we first plot the moment diagram as shown in Fig. 8-23(b). The diagram for the elastic load and the conjugate beam is then given in Fig. 8-23(c).

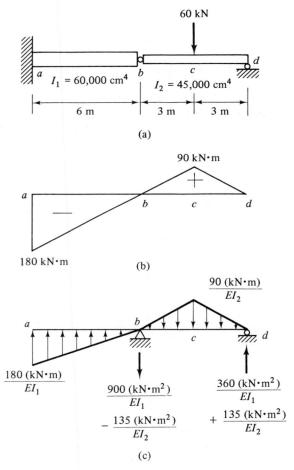

(a)

(b)

(c)

**Fig. 8-23**

The deflection at $b$ of the original beam is the bending moment about $b$ of the conjugate beam subjected to the elastic load. Using the overhanging portion of the conjugate beam, we obtain

$$\Delta_b = \frac{(180)(6)(4)}{2EI_1} = \frac{2,160}{EI_1}$$

$$= \frac{(2,160)(10)^6}{(20,000)(60,000)} = 1.8 \text{ cm} \qquad \text{(down)}$$

Because of the hinge connection provided at $b$, there is a change of slope taking place at that point. In fact, we have different slopes in the immediate left and right of $b$, corresponding to the shearing forces in the conjugate beam. Thus,

$$(Q_b)_{\text{left}} = \frac{(180)(6)}{2EI_1} = \frac{540}{EI_1}$$

$$= \frac{(540)(10)^4}{(20,000)(60,000)} = 0.0045 \text{ radian} \qquad \text{(clockwise)}$$

$$(Q_b)_{\text{right}} = \frac{270}{EI_2} - \left(\frac{360}{EI_1} + \frac{135}{EI_2}\right) = -\frac{360}{EI_1} + \frac{135}{EI_2}$$

$$= -\frac{(360)(10)^4}{(20,000)(60,000)} + \frac{(135)(10)^4}{(20,000)(45,000)}$$

$$= -0.0015 \text{ radian} \qquad \text{(counterclockwise)}$$

The relative rotation between the left and right sides of $b$ is the reaction at support $b$ of the conjugate beam:

$$\frac{900}{EI_1} - \frac{135}{EI_2} = \frac{(900)(10)^4}{(20,000)(60,000)} - \frac{(135)(10)^4}{(20,000)(45,000)} = 0.006 \text{ radian}$$

## PROBLEMS

**8-1.** Use the method of virtual work to determine the vertical deflection at center and the slope at left end for a simply supported beam subjected to a uniform load over the entire span. Assume constant $EI$.

**8-2.** Use the method of virtual work to determine the vertical deflections at the load point and at the center of the beam in Fig. 8-24. $E$ and $I$ are both constant.

**Fig. 8-24**

**8-3.** Use the method of virtual work to determine the slope and deflection at the load point of the beam in Fig. 8-25. Use $E = 30,000$ kips/in.$^2$.

**Fig. 8-25**

**8-4.** By the method of virtual work, find the horizontal, vertical, and rotational displacement components at point $a$ of the frame shown in Fig. 8-26. Use $E = 30,000$ kips/in.$^2$ and $I = 500$ in.$^4$.

**Fig. 8-26**

**8-5.** For the load and beam in Fig. 8-27, use the method of virtual work to determine the slope and deflection at points $b$ and $c$.

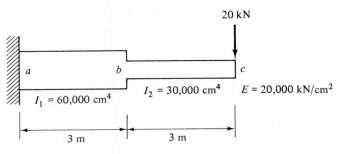

**Fig. 8-27**

**8-6.** For the loads and frame in Fig. 8-28, use the method of virtual work to find the horizontal, vertical, and rotational displacement components at point $a$. $EI = 2 \times 10^8$ kN · cm².

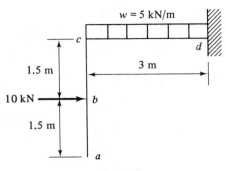

**Fig. 8-28**

**8-7.** For the loads and truss in Fig. 8-29, use the method of virtual work to find the displacement components corresponding to the applied loads at joint $B$. $E = 20{,}000$ kN/cm² and $A = 20$ cm².

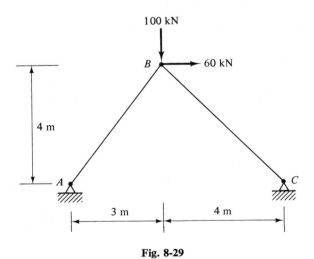

**Fig. 8-29**

**8-8.** For the truss in Fig. 8-30, the area of each bar in square inches equals one-half its length in feet. $E = 30{,}000$ kips/in.². Use the method of virtual work to compute (a) the vertical deflection at point $B$, (b) the horizontal deflection at point $C$, (c) the relative deflection between points $b$ and $C$ along the line joining them, and (d) the rotation of member $bc$.

**Fig. 8-30**

**8-9.** Solve Prob. 8-1 by Castigliano's theorem.

**8-10.** Solve Prob. 8-2 by Castigliano's theorem.

**8-11.** Solve Prob. 8-3 by Castigliano's theorem.

**8-12.** Solve Prob. 8-4 by Castigliano's theorem.

**8-13.** Solve Prob. 8-5 by Castigliano's theorem.

**8-14.** Solve Prob. 8-6 by Castigliano's theorem.

**8-15.** Solve Prob. 8-7 by Castigliano's theorem.

**8-16.** Solve Prob. 8-8 by Castigliano's theorem.

**8-17.** Solve Prob. 8-1 by the conjugate-beam method.

**8-18.** Solve Prob. 8-2 by the conjugate-beam method.

**8-19.** Solve Prob. 8-3 by the conjugate-beam method.

**8-20.** For the load and beam shown in Fig. 8-31, use the conjugate-beam method to find the deflection at $b$ and the rotation at $c$. $E = 20,000$ kN/cm$^2$ and $I = 5,000$ cm$^4$.

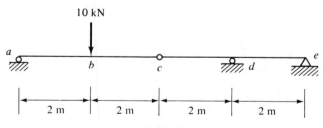

**Fig. 8-31**

# 9

## ANALYSIS OF STATICALLY

## INDETERMINATE STRUCTURES BY THE

## METHOD OF CONSISTENT DEFORMATIONS

### 9-1. GENERAL

Statically indeterminate structures can be analyzed by direct use of the theory of elastic deformations developed in Chapter 8. Any statically indeterminate structure can be made statically determinate and stable by removing the extra restraints called *redundant forces* or simply *redundants*, that is, the force elements which are more than the minimum necessary for the static equiliblium of the structure. The statically determinate and stable structure that remains after removal of the extra restraints is called the *primary structure*. The original structure is then equivalent to the primary structure subjected to the combined action of the original loads plus the unknown redundants. The conditional equations for geometric consistency of the original structure at redundant points called the *compatibility equations*, are then obtained from the primary structure by superposition of the deformations caused by the original loads and redundants. We can have as many compatibility equations as the number of unknown redundants so that the redundants can be determined by solving these simultaneous equations. This method known as *consistent deformations* is generally applicable to the analysis of any structure, whether it is being analyzed for the effect of loads, support settlement, temperature change, or any other case. However, there is only one restriction on the use of this method: the principle of superposition must hold.

As an illustration, consider the loaded continuous beam with nonyielding supports shown in Fig. 9-1(a). It is statically indeterminate to the second degree, that is to say, with two redundants. The first step in the application of the method is to remove, say, the two interior supports and to introduce in these releases the redundant actions called $X_1$ and $X_2$, respectively, and by so doing to reduce or cut back the structure to a condition of determinateness and stability. The original structure is now considered as a simple beam (the primary structure) subjected to the combined action of a number of external forces and two redundants $X_1$ and $X_2$, as shown in Fig. 9-1(b).

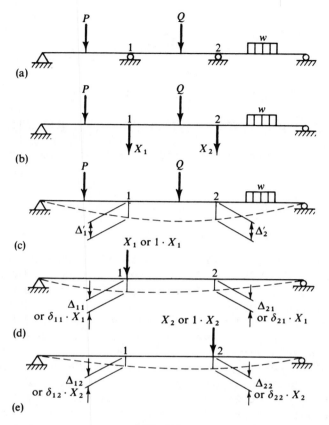

**Fig. 9-1**

The resulting structure in Fig. 9-1(b) can be regarded as the superposition of those shown in Fig. 9-1(c)–(e). Consequently, any deformation of the structure can be obtained by the superposition of these effects.

Referring to Fig. 9-1(b), for unyielding supports, we find that compatibility requires

$$\Delta_1 = 0 \qquad (9\text{-}1)$$

$$\Delta_2 = 0 \qquad (9\text{-}2)$$

where $\Delta_1$ = deflection at redundant point 1 (in the line of redundant force $X_1$)
$\Delta_2$ = deflection at redundant point 2 (in the line of redundant force $X_2$)

By the principle of superposition we may expand Eqs. 9-1 and 9-2 as

$$\Delta_1' + \Delta_{11} + \Delta_{12} = 0 \qquad (9\text{-}3)$$

$$\Delta_2' + \Delta_{21} + \Delta_{22} = 0 \qquad (9\text{-}4)$$

where $\Delta_1'$ = deflection at redundant point 1 due to external loads [see Fig. 9-1(c)]

$\Delta_{11}$ = deflection at redundant point 1 due to redundant force $X_1$ [see Fig. 9-1(d)]

$\Delta_{12}$ = deflection at redundant point 1 due to redundant force $X_2$ [see Fig. 9-1(e)]

The rest are similar.

Equations 9-3 and 9-4 may be expressed in terms of the *flexibility coefficients*. A typical flexibility coefficient $\delta_{ij}$ is defined by

$\delta_{ij}$ = displacement at point $i$ due to a unit action at $j$, all other points being assumed unloaded

Thus, Eqs. 9-3 and 9-4 may be written as

$$\Delta_1' + \delta_{11}X_1 + \delta_{12}X_2 = 0 \tag{9-5}$$

$$\Delta_2' + \delta_{21}X_1 + \delta_{22}X_2 = 0 \tag{9-6}$$

Apparently,

$\delta_{11}$ = deflection at point 1 due to a unit force at point 1 [see Fig. 9-1(d)]

$\delta_{12}$ = deflection at point 1 due to a unit force at point 2 [see Fig. 9-1(e)]

and so on.

Both the deflections resulting from the original external loads and the flexibility coefficients for the primary structure can be obtained by any method described in Chapter 8. The remaining redundant unknowns are then solved by simultaneous equations. This process can be generalized. Thus, for a structure with $n$ redundants, we have

$$\Delta_1' + \delta_{11}X_1 + \delta_{12}X_2 + \cdots + \delta_{1n}X_n = 0$$

$$\Delta_2' + \delta_{21}X_1 + \delta_{22}X_2 + \cdots + \delta_{2n}X_n = 0$$

$$\vdots \tag{9-7}$$

$$\Delta_n' + \delta_{n1}X_1 + \delta_{n2}X_2 + \cdots + \delta_{nn}X_n = 0$$

Equation 9-7 in matrix form is

$$\begin{Bmatrix} \Delta_1' \\ \Delta_2' \\ \cdot \\ \cdot \\ \cdot \\ \Delta_n' \end{Bmatrix} + \begin{bmatrix} \delta_{11} & \delta_{12} & \cdots & \delta_{1n} \\ \delta_{21} & \delta_{22} & \cdots & \delta_{2n} \\ \cdot & & & \\ \cdot & & & \\ \cdot & & & \\ \delta_{n1} & \delta_{n2} & \cdots & \delta_{nn} \end{bmatrix} \begin{Bmatrix} X_1 \\ X_2 \\ \cdot \\ \cdot \\ \cdot \\ X_n \end{Bmatrix} = \begin{Bmatrix} 0 \\ 0 \\ \cdot \\ \cdot \\ \cdot \\ 0 \end{Bmatrix} \tag{9-8}$$

or simply

$$\Delta' + FX = 0 \tag{9-9}$$

In a more general form, we may include the prescribed displacements (other than zeros) occurring at the releases of the original structures. Then these values $\Delta_1, \Delta_2, \ldots$ must be substituted for the zeros on the right-hand side of Eq. 9-8. Thus,

$$\begin{Bmatrix} \Delta_1' \\ \Delta_2' \\ \cdot \\ \cdot \\ \cdot \\ \Delta_n' \end{Bmatrix} + \begin{bmatrix} \delta_{11} & \delta_{12} & \cdots & \delta_{1n} \\ \delta_{21} & \delta_{22} & \cdots & \delta_{2n} \\ \cdot & & & \\ \cdot & & & \\ \cdot & & & \\ \delta_{n1} & \delta_{n2} & \cdots & \delta_{nn} \end{bmatrix} \begin{Bmatrix} X_1 \\ X_2 \\ \cdot \\ \cdot \\ \cdot \\ X_n \end{Bmatrix} = \begin{Bmatrix} \Delta_1 \\ \Delta_2 \\ \cdot \\ \cdot \\ \cdot \\ \Delta_n \end{Bmatrix} \tag{9-10}$$

or simply

$$\Delta' + FX = \Delta \tag{9-11}$$

in which the column matrix $\Delta'$ on the left-hand side represents the displacements at redundant points of the released structure due to the original loads; the square matrix $F$ represents the structure flexibility, each column of which gives various displacements at redundant points due to a certain unit redundant force; and the column matrix $\Delta$ on the right-hand side contains the actual displacements at redundant points of the original structure. Equation 9-11 expresses the compatibility condition at redundant points in terms of unknown redundant forces.

## 9-2. ANALYSIS OF STATICALLY INDETERMINATE BEAMS BY THE METHOD OF CONSISTENT DEFORMATIONS

The method of consistent deformations is quite easy to understand and can be most effectively demonstrated by a series of illustrations. In all the following examples we assume that only the bending distortion is significant.

**Example 9-1.** Analyze the propped beam shown in Fig. 9-2(a), which is statically indeterminate to the first degree. Assume constant $EI$.

*Solution 1* One of the reactions may be considered as being extra. In this case let us first choose the vertical reaction at $b$ as the redundant assumed to be acting downward, as shown in Fig. 9-2(b). By the principle of superposition we may consider the beam as being subjected to the sum of the effects of the original uniform loading and the unknown redundant $X_b$, as shown in Fig. 9-2(c) and (d), respectively.

Next, we find that the vertical deflection at $b$ resulting from the uniform load [Fig. 9-2(c)] is given by

$$\Delta_b' = \frac{wl^4}{8EI}$$

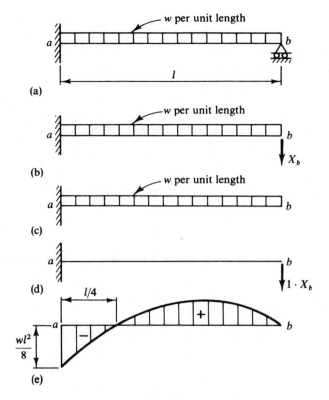

**Fig. 9-2**

and that the vertical deflection at $b$ because of a unit load applied at $b$ in place of $X_b$ [Fig. 9-2(d)] is given by

$$\delta_{bb} = \frac{l^3}{3EI}$$

Note that $\Delta_b'$ and the flexibility coefficient $\delta_{bb}$ may be found by any method described in Chapter 8.

Applying compatibility equation

$$\Delta_b = \Delta_b' + \delta_{bb}X_b = 0$$

we obtain

$$\frac{wl^4}{8EI} + \left(\frac{l^3}{3EI}\right)X_b = 0$$

from which

$$X_b = -\frac{3wl}{8}$$

The minus sign indicates an upward reaction.

With reaction at $b$ determined, we find that the beam reduces to a statically determinate one. We can readily obtain reaction components at $a$ from the equilibrium equations:

$$\sum F_y = 0 \qquad V_a = wl - \tfrac{3}{8}wl = \tfrac{5}{8}wl \qquad \text{(upward)}$$

$$\sum M_a = 0 \qquad M_a = \tfrac{1}{2}wl^2 - \tfrac{3}{8}wl^2 = \tfrac{1}{8}wl^2 \qquad \text{(counterclockwise)}$$

The moment diagram for the beam is shown in Fig. 9-2(e).

*Solution 2* The beam in Fig. 9-2(a) can be rendered statically determinate by removing the fixed support and replacing it with a hinged support. In addition to the original uniform loading, a redundant moment $M_a$ is then applied to the primary structure, a simple beam, as shown in Fig. 9-3(a). The unknown $M_a$ can be solved by the condition of compatibility that the rotation at end $a$ must be zero.

The rotation at end $a$ for the primary structure due to the uniform loading alone [Fig. 9-3(b)] is given by

$$\theta'_a = \frac{wl^3}{24EI}$$

and that due to a unit couple applied at end $a$ [Fig. 9-3(c)] is given by

$$\delta_{aa} = \frac{l}{3EI}$$

Using the compatibility equation

$$\theta_a = \theta'_a + \delta_{aa}M_a = \frac{wl^3}{24EI} + \frac{M_a l}{3EI} = 0$$

we solve for

$$M_a = -\tfrac{1}{8}wl^2$$

The minus sign indicates a counterclockwise moment. After $M_a$ is determined, the rest of the analysis can be carried out without difficulty.

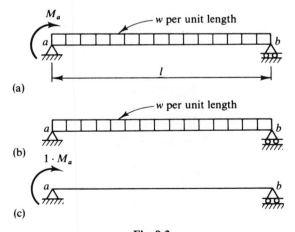

(a)

(b)

(c)

**Fig. 9-3**

*Solution 3* From the previous solutions we come to recognize that we are free to select redundants in analyzing a statically indeterminate structure, the only restriction being that the redundants should be so selected that a *stable* cut structure remains. Figure 9-4 will serve as an illustration. Let us cut the beam at midspan section $c$ and introduce in its place a hinge, so that the beam is stable and determinate. A pair of redundant couples, called $M_c$, together with the original loading are then applied to the primary structure, as shown in Fig. 9-4(a).

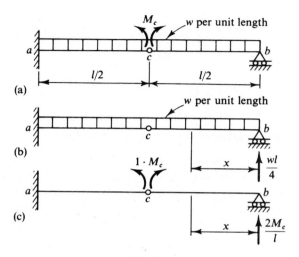

**Fig. 9-4**

The redundant $M_c$ is solved by the condition of compatibility that the rotation of the left side relative to the right side at section $c$ must be zero.

Using the method of virtual work we evaluate the relative rotation at $c$ due to the external loading alone [Fig. 9-4(b)] as

$$\theta'_c = \int_0^l \frac{Mm \; dx}{EI} = \int_0^l \frac{[(wlx/4) - (wx^2/2)](2x/l) \; dx}{EI} = -\frac{wl^3}{12EI}$$

and that due to a pair of unit couples acting at $c$ [Fig. 9-4(c)] as

$$\delta_{cc} = \int_0^l \frac{m^2 \; dx}{EI} = \int_0^l \frac{(2x/l)^2 \; dx}{EI} = \frac{4l}{3EI}$$

Setting the total relative angular displacement at $c$ equal to zero, we have

$$-\frac{wl^3}{12EI} + M_c\left(\frac{4l}{3EI}\right) = 0$$

from which
$$M_c = +\frac{wl^2}{16}$$

After $M_c$ is determined, the rest of the analysis can easily be carried out.

**Example 9-2.** Suppose that the support at $b$ of Example 9-1 is elastic and the spring flexibility is $f$ (displacement per unit force), as shown in Fig. 9-5. Determine reaction at $b$ (the spring force), denoted by $X_b$.

**Fig. 9-5**

Assume downward $X_b$ (i.e., tension in the spring) as positive. The compatibility is

$$\Delta'_b + \delta_{bb}X_b + fX_b = 0$$

This equation can be explained by putting it in the form

$$\Delta'_b - \delta_{bb}(-X_b) = f(-X_b)$$

Since $(-X_b)$ represents the compression in the spring, the equation indicates that the downward deflection at $b$ caused by the beam load minus that caused by upward reaction should be equal to the spring contraction.

By substituting $\Delta'_b = wl^4/8EI$, $\delta_{bb} = l^3/3EI$ in the equation above, we obtain

$$\frac{wl^4}{8EI} + \frac{X_b l^3}{3EI} + fX_b = 0$$

from which

$$X_b = -\frac{3}{8}wl\left[\frac{1}{1 + (3fEI/l^3)}\right]$$

The minus sign indicates an upward reaction.

For an nonyielding support, $f = 0$, the equation above gives

$$X_b = -\tfrac{3}{8}wl$$

as found previously.

If a beam is provided with $n$ redundant elastic supports having spring flexibilities $f_1, f_2, \ldots, f_n$, respectively, then the general compatibility equation is

$$
\begin{Bmatrix} \Delta'_1 \\ \Delta'_2 \\ \cdot \\ \cdot \\ \Delta'_n \end{Bmatrix} +
\begin{bmatrix} \delta_{11} + f_1 & \delta_{12} & \cdots & \delta_{1n} \\ \delta_{21} & \delta_{22} + f_2 & \cdots & \delta_{2n} \\ \cdot & & & \\ \cdot & & & \\ \delta_{n1} & \delta_{n2} & \cdots & \delta_{nn} + f_n \end{bmatrix}
\begin{Bmatrix} X_1 \\ X_2 \\ \cdot \\ \cdot \\ X_n \end{Bmatrix} =
\begin{Bmatrix} 0 \\ 0 \\ \cdot \\ \cdot \\ 0 \end{Bmatrix} \qquad (9\text{-}12)
$$

**Example 9-3.** Find the reactions for the beam with two sections shown in Fig. 9-6(a).

In this problem it may be convenient to select the vertical reaction at support $b$ as redundant. The beam is then considered as a simple beam subject to the

180

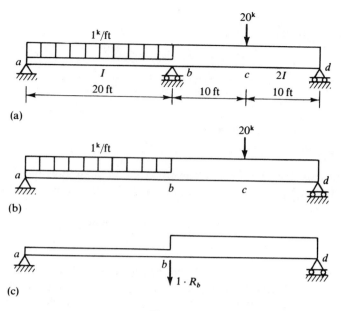

Fig. 9-6

origina! loading and the redundant $R_b$ as shown in Fig. 9-6(b) and (c), respectively.

The compatibility requires

$$\Delta_b = \Delta_b' + \delta_{bb}R_b = 0$$

Using the method of virtual work, we have

$$\int \frac{Mm_b \, dx}{EI} + R_b \int \frac{(m_b)^2 \, dx}{EI} = 0$$

from which

$$R_b = -\frac{\int Mm_b \, dx/EI}{\int (m_b)^2 \, dx/EI}$$

where $M$ = bending moment at any section of the primary beam caused by the original loading [Fig. 9-6(b)]

$m_b$ = bending moment at the same section of the primary beam caused by a unit load in place of the redundant $R_b$ [Fig. 9-6(c)]

The solution is completely shown in Table 9-1.

$$R_b = -\frac{\int_0^{20} \frac{(20x - x^2/2)(x/2) \, dx}{EI} + \int_0^{10} \frac{(20x)(x/2) \, dx}{2EI} + \int_{10}^{20} \frac{(200)(x/2) \, dx}{2EI}}{\int_0^{20} \frac{(x/2)^2 \, dx}{EI} + \int_0^{10} \frac{(x/2)^2 \, dx}{2EI} + \int_{10}^{20} \frac{(x/2)^2 \, dx}{2EI}}$$

$$= -25.84 \text{ kips}$$

181

TABLE 9-1

| Section | Origin | Limit (ft) | $M$ (ft-kips) | $m_b$ (ft-kips) | $I$ |
|---------|--------|-----------|---------------|-----------------|-----|
| $ab$ | $a$ | 0 to 20 | $20x - \dfrac{(1)(x)^2}{2}$ | $\dfrac{x}{2}$ | $I$ |
| $dc$ | $d$ | 0 to 10 | $20x$ | $\dfrac{x}{2}$ | $2I$ |
| $cb$ | $d$ | 10 to 20 | $20x - 20(x - 10)$ or $200$ | $\dfrac{x}{2}$ | $2I$ |

The negative sign indicates an upward reaction at support $b$.

After $R_b$ is obtained, we can readily find the reactions at the other two supports by statics. That is,

$$R_a = R_b = 20 - (\tfrac{1}{2})(25.84) = 7.08 \text{ kips}$$

acting upward.

The end moments for a fixed-end beam, called *fixed-end moments*, are important in the methods of slope deflection and of moment distribution, which are discussed in later chapters. The following examples are attempts to solve fixed-end moments due to common types of loading by the method of consistent deformations.

**Example 9-4.** The fixed-end beam of uniform cross section subjected to a single concentrated load shown in Fig. 9-7(a) is statically indeterminate to the second degree since the horizontal force does not exist. End moments $M_A$ and $M_B$ are selected as redundants. The original beam is then considered as equivalent to a simple beam (not shown) under the combined action of a concentrated force $P$ and redundant moments $M_A$ and $M_B$. It is convenient to apply the conjugate-beam method to determine $M_A$ and $M_B$ based on the condition that the slope and deflection at either end of the fixed-end beam must be zero. In other words, there will be no support reactions for the conjugate beam, and the positive and negative $M/EI$ diagrams (elastic loads) given in Fig. 9-7(b) must form a balanced system. Thus, from $\sum F_y = 0$,

$$\frac{Pab}{2EI} - \frac{M_A l}{2EI} - \frac{M_B l}{2EI} = 0$$

or
$$M_A + M_B = \frac{Pab}{l} \tag{9-13}$$

From $\sum M_B = 0$,

$$\left(\frac{Pab}{2EI}\right)\left(\frac{l + b}{3}\right) - \left(\frac{M_A l}{2EI}\right)\left(\frac{2l}{3}\right) - \left(\frac{M_B l}{2EI}\right)\left(\frac{l}{3}\right) = 0$$

or
$$2M_A + M_B = \frac{Pab}{l} + \frac{Pab^2}{l^2} \tag{9-14}$$

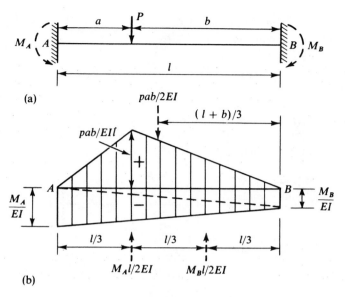

(a)

(b)

**Fig. 9-7**

Solving Eqs. 9-13 and 9-14 simultaneously, we obtain

$$M_A = \frac{Pab^2}{l^2} \qquad M_B = \frac{Pa^2b}{l^2} \tag{9-15}$$

**Example 9-5.** Find the end moments of a fixed-end beam of constant $EI$ caused by a uniform load, as shown in Fig. 9-8(a).

Because of symmetry, the beam is statically indeterminate to the first degree,

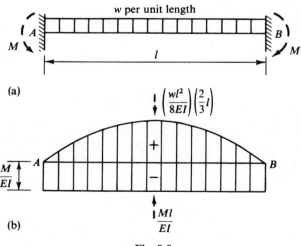

(a)

(b)

**Fig. 9-8**

183

since $M_A = M_B = M$, as indicated in Fig. 9-8(a). By the method of conjugate beam [Fig. 9-8(b)], $\sum F_y = 0$,

$$\left(\frac{wl^2}{8EI}\right)\left(\frac{2l}{3}\right) - \frac{Ml}{EI} = 0$$

from which
$$M = \tfrac{1}{12} wl^2 \qquad (9\text{-}16)$$

**Example 9-6.** If the fixed-end beam is loaded with an external couple $M$ as shown in Fig. 9-9(a), the deflected elastic shape will be somewhat like that shown by the dotted line, which gives the sense of the end moments as indicated.

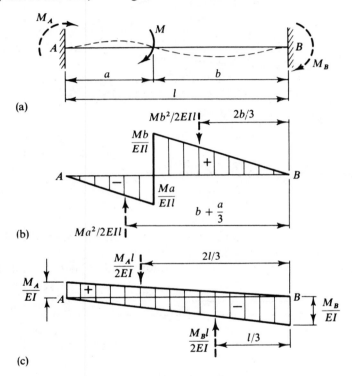

**Fig. 9-9**

As before, end moments $M_A$ and $M_B$ are chosen as redundants. The elastic loads based on the moment diagrams divided by $EI$ plotted for external moment $M$ and redundants $M_A$ and $M_B$, as given in Fig. 9-9(b) and (c), must be in equilibrium themselves. From $\sum F_y = 0$,

$$\frac{M_A l}{2EI} + \frac{Mb^2}{2EIl} - \frac{M_B l}{2EI} - \frac{Ma^2}{2EIl} = 0$$

or
$$M_A - M_B = \frac{M(a^2 - b^2)}{l^2} \qquad (9\text{-}17)$$

From $\sum M_B = 0$,

$$\left(\frac{M_A l}{2EI}\right)\left(\frac{2l}{3}\right) + \left(\frac{Mb^2}{2EIl}\right)\left(\frac{2b}{3}\right) - \left(\frac{M_B l}{2EI}\right)\left(\frac{l}{3}\right) - \left(\frac{Ma^2}{2EIl}\right)\left(b + \frac{a}{3}\right) = 0$$

or

$$2M_A - M_B = \frac{M[a^2 + 2b(a - b)]}{l^2} \tag{9-18}$$

Solving Eqs. 9-17 and 9-18 simultaneously, we obtain

$$M_A = \frac{Mb}{l^2}(2a - b) \qquad M_B = \frac{Ma}{l^2}(2b - a) \tag{9-19}$$

Note that $M_A$ and $M_B$ bear the same sense as the externally applied $M$, as indicated in Fig. 9-9(a), if $a > l/3$ and $b > l/3$.

## 9-3. ANALYSIS OF STATICALLY INDETERMINATE RIGID FRAMES BY THE METHOD OF CONSISTENT DEFORMATIONS

The general procedure illustrated in Sec. 9-2 in solving statically indeterminate beams can be applied equally well to the analysis of statically indeterminate rigid frames, as in the following example.

**Example 9-7.** For the loaded rigid frame shown in Fig. 9-10(a), find the reaction components at the fixed end $a$, and plot the moment diagram for the entire frame. Assume the same $EI$ for all members.

To do this, we start by removing support $a$ and introducing in its place three redundant reaction components $X_1$, $X_2$, and $X_3$, as shown in Fig. 9-10(b). These can be taken as the superposition of four basic cases, as shown in Fig. 9-10(c), (d), (e), and ($f$), respectively. Since end $a$ is fixed, compatibility requires that

$$\begin{Bmatrix} \Delta_1' \\ \Delta_2' \\ \Delta_3' \end{Bmatrix} + \begin{bmatrix} \delta_{11} & \delta_{12} & \delta_{13} \\ \delta_{21} & \delta_{22} & \delta_{23} \\ \delta_{31} & \delta_{32} & \delta_{33} \end{bmatrix} \begin{Bmatrix} X_1 \\ X_2 \\ X_3 \end{Bmatrix} = \begin{Bmatrix} 0 \\ 0 \\ 0 \end{Bmatrix} \tag{9-20}$$

Taking advantages of Examples 8-4 and 8-5, we note that

$$\begin{Bmatrix} \Delta_1' \\ \Delta_2' \\ \Delta_3' \end{Bmatrix} = \frac{1}{EI} \begin{Bmatrix} 5,000 \\ -7,500 \\ -800 \end{Bmatrix}$$

and

$$\begin{bmatrix} \delta_{11} & \delta_{12} & \delta_{13} \\ \delta_{21} & \delta_{22} & \delta_{23} \\ \delta_{31} & \delta_{32} & \delta_{33} \end{bmatrix} = \frac{1}{EI} \begin{bmatrix} 1,667 & -1,000 & -200 \\ -1,000 & 1,333 & 150 \\ -200 & 150 & 30 \end{bmatrix}$$

Fig. 9-10

Substituting these values in Eq. 9-20, we obtain

$$\begin{Bmatrix} 5{,}000 \\ -7{,}500 \\ -800 \end{Bmatrix} + \begin{bmatrix} 1{,}667 & -1{,}000 & -200 \\ -1{,}000 & 1{,}333 & 150 \\ -200 & 150 & 30 \end{bmatrix} \begin{Bmatrix} X_1 \\ X_2 \\ X_3 \end{Bmatrix} = \begin{Bmatrix} 0 \\ 0 \\ 0 \end{Bmatrix} \qquad (9\text{-}21)$$

Solving, we obtain

$$\begin{Bmatrix} X_1 \\ X_2 \\ X_3 \end{Bmatrix} = \begin{Bmatrix} 1 \\ 6 \\ 3.33 \text{ ft} \end{Bmatrix} \text{ kips}$$

Note that the solution of this problem could be simplified by setting $X_2 =$ 6 kips in Eq. 9-20, since we know this value beforehand because of the symmetry of the loaded frame.

The final results are shown in Fig. 9-10(g); the moment diagram for the whole frame is shown in Fig. 9-10(h). A sketch of the elastic deformation of the frame due to bending distortion is shown by the dashed line in Fig. 9-10(i). Note that in this case there is one point of inflection in each column and two points of inflection in the beam.

By referring to Example 9-7 we see that by using the method of consistent deformations in analyzing a rigid frame, we encounter tedious calculations of the flexibility coefficients. The work, if done by hand, will become intolerable if the problem involves as many redundants as a rigid frame usually does. As a matter of fact, the method of consistent deformations is seldom used for analysis of rigid frames by hand calculation, since a solution can be much more easily obtained by the method of slope deflection or of moment distribution. However, with the development of high-speed *electronic computers*, this method has regained considerable strength in the scope of structural analysis, as will be seen in Chapter 15.

## 9-4. ANALYSIS OF STATICALLY INDETERMINATE
### TRUSSES BY THE METHOD OF CONSISTENT
### DEFORMATIONS

The indeterminateness of a truss may be due to redundant supports or redundant bars or both. If it results from redundant supports, the procedure for attack is the same as that described for a continuous beam. If the superfluous element is a bar, the bar is considered to be cut at a section and replaced by two equal and opposite axial redundant forces representing the internal action for that bar. The condition equation is such that the relative axial displacement between the two sides at the cut section caused by the combined effect of the original loading and the redundants should be zero.

**Example 9-8.** Analyze the continuous truss in Fig. 9-11(a). Assume that $E = 30,000$ kips/in.$^2$ and $L(\text{ft})/A(\text{in.}^2) = 1$ for all members.

In this problem it is convenient to select the central support as the redundant element. We begin by removing support $c$ and introducing in its place a redundant reaction $X_c$, as shown in Fig. 9-11(b). The primary structure is then a simply supported truss subjected to an external load of 64 kips at joint $b$ and a redundant $X_c$. The effects can be separated, respectively, as shown in Fig. 9-11(c) and (d).

Since support $c$ is on a rigid foundation, the compatibility equation can be expressed by

$$\Delta_c = \Delta_c' + \delta_{cc}X_c = 0$$

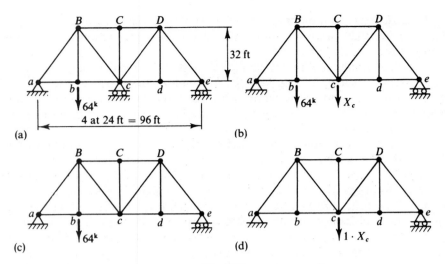

**Fig. 9-11**

Using virtual work gives

$$\sum \frac{S'u_cL}{AE} + X_c \sum \frac{u_c^2 L}{AE} = 0$$

from which

$$X_c = -\frac{\sum (S'u_cL/AE)}{\sum (u_c^2 L/AE)}$$

where $S'$ = internal force in any member of the primary truss due to the original loading [Fig. 9-11(c)]

$u_c$ = internal force in the same member of the primary truss due to a unit force at $c$ [Fig. 9-11(d)]

The solution is shown completely in Table 9-2.

$$X_c = -\frac{122}{3.25} = -37.5 \text{ kips}$$

The negative sign indicates an upward reaction at support $c$.

After $X_c$ is determined, we can readily obtain each bar force $S$ from

$$S = S' + u_c X_c$$

as given in the last column of Table 9-2.

**Example 9-9.** Analyze the truss in Fig. 9-12(a). Assume that $E = 30,000$ kips/in.$^2$ and $L(\text{ft})/A(\text{in.}^2) = 1$ for all members.

The truss in Fig. 9-12(a) has two redundant elements, one in the reaction component and the other in the bar. Let us select the horizontal component of reaction at the right end hinge and the internal force in bar $Cd$ as redundants. We then have a primary truss loaded as shown in Fig. 9-12(b) in which the original hinged support $e$ is replaced by a roller acted on by a redundant hori-

# TABLE 9-2

| Member | $\dfrac{L}{A}$ (ft/in.²) | $S'$ (kips) | $u_c$ | $\dfrac{S' u_c L}{A}$ (ft-kips/in.²) | $\dfrac{u_c^2 L}{A}$ (ft/in.²) | $S = S' + u_c X_c$ (kips) |
|--------|--------|--------|--------|--------|--------|--------|
| ab | 1 | +36 | +3/8 | +13.5 | +9/64 | $36 - 14.1 = +21.9$ |
| bc | 1 | +36 | +3/8 | +13.5 | +9/64 | $36 - 14.1 = +21.9$ |
| cd | 1 | +12 | +3/8 | +4.5 | +9/64 | $12 - 14.1 = -2.1$ |
| de | 1 | +12 | +3/8 | +4.5 | +9/64 | $12 - 14.1 = -2.1$ |
| BC | 1 | −24 | −3/4 | +18 | +36/64 | $-24 + 28.2 = +4.2$ |
| CD | 1 | −24 | −3/4 | +18 | +36/64 | $-24 + 28.2 = +4.2$ |
| aB | 1 | −60 | −5/8 | +37.5 | +25/64 | $-60 + 23.4 = -36.6$ |
| Bb | 1 | +64 | 0 | 0 | 0 | $+64 + 0 = +64$ |
| Bc | 1 | −20 | +5/8 | −12.5 | +25/64 | $-20 - 23.4 = -43.4$ |
| Cc | 1 | 0 | 0 | 0 | 0 | 0 |
| cD | 1 | +20 | +5/8 | +12.5 | +25/64 | $+20 - 23.4 = -3.4$ |
| Dd | 1 | 0 | 0 | 0 | 0 | 0 |
| De | 1 | −20 | −5/8 | +12.5 | +25/64 | $-20 + 23.4 = +3.4$ |
| | | | $\Sigma$ | +122 | +3.25 | |

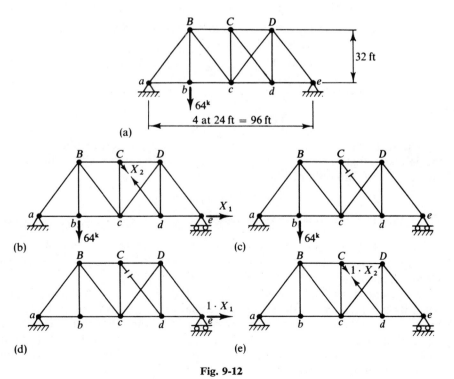

Fig. 9-12

189

zontal reaction $X_1$ and the bar $Cd$ is cut and a pair of redundant forces $X_2$ applied to it. This may again be replaced by the three basic cases shown in Fig. 9-12(c)–(e). Since both the horizontal movement at spuport $e$ and the relative axial displacement between the cut ends of bar $Cd$ are zero, we have

$$\begin{Bmatrix} \Delta_1 \\ \Delta_2 \end{Bmatrix} = \begin{Bmatrix} \Delta'_1 \\ \Delta'_2 \end{Bmatrix} + \begin{bmatrix} \delta_{11} & \delta_{12} \\ \delta_{21} & \delta_{22} \end{bmatrix} \begin{Bmatrix} X_1 \\ X_2 \end{Bmatrix} = \begin{Bmatrix} 0 \\ 0 \end{Bmatrix} \tag{9-22}$$

or

$$\begin{Bmatrix} \sum \dfrac{S'u_1L}{AE} \\[2mm] \sum \dfrac{S'u_2L}{AE} \end{Bmatrix} + \begin{bmatrix} \sum \dfrac{u_1^2L}{AE} & \sum \dfrac{u_1u_2L}{AE} \\[2mm] \sum \dfrac{u_1u_2L}{AE} & \sum \dfrac{u_2^2L}{AE} \end{bmatrix} \begin{Bmatrix} X_1 \\ X_2 \end{Bmatrix} = \begin{Bmatrix} 0 \\ 0 \end{Bmatrix} \tag{9-23}$$

Note that

$S'$ = internal force in any bar of the primary truss due to the original loading [Fig. 9-12(c)]

$u_1$ = internal force in the same bar of the primary truss due to a unit horizontal force acting at $e$ [Fig. 9-12(d)]

$u_2$ = internal force in the same bar of the primary truss due to a pair of unit axial forces acting at the cut ends of bar $Cd$ [Fig. 9-12(e)]

Using the values summed up in Table 9-3, we reduce Eq. 9-23 to

$$\begin{Bmatrix} 96 \\ 27.2 \end{Bmatrix} + \begin{bmatrix} 4 & -\tfrac{3}{5} \\ -\tfrac{3}{5} & 4 \end{bmatrix} \begin{Bmatrix} X_1 \\ X_2 \end{Bmatrix} = \begin{Bmatrix} 0 \\ 0 \end{Bmatrix} \tag{9-24}$$

Solving, we obtain

$$\begin{Bmatrix} X_1 \\ X_2 \end{Bmatrix} = \begin{Bmatrix} -25.6 \\ -10.6 \end{Bmatrix} \text{ kips}$$

The negative signs indicate that the horizontal reaction at hinge $e$ acts to the left and that the axial force in member $Cd$ is compressive. The rest of member forces are obtained by

$$S = S' + u_1X_1 + u_2X_2$$

The complete solution is shown in Table 9-3.

**Example 9-10.** Analyze the truss in Fig. 9-13(a) subject to a rise of 50°F at the top chords $BC$ and $CD$. Assume $\alpha = 0.0000065$ in./in./1°F; $E = 30,000$ kips/in.$^2$; and $L(\text{ft})/A(\text{in.}^2) = 1$ for all members.

The truss is statically indeterminate to the first degree. Cut bar $Cd$ and select its bar force $X_1$ as the redundant as shown in Fig. 9-13(b). The primary truss is then a simply supported truss subjected to the temperature rise in the top chords and the redundant axial force $X_1$. Since the relative axial displacement between the cut ends due to the combined effect of temperature rise and $X_1$ must be zero, we have

$$\Delta_1 = \Delta'_1 + \delta_{11}X_1 = 0$$

TABLE 9-3

| Member | $\dfrac{L}{A}$ (ft/in.²) | $S'$ (kips) | $u_1$ | $u_2$ | $\dfrac{S'u_1 L}{A}$ (ft-kips/in.²) | $\dfrac{S'u_2 L}{A}$ (ft-kips/in.²) | $\dfrac{u_1^2 L}{A}$ (ft/in.²) | $\dfrac{u_2^2 L}{A}$ (ft/in.²) | $\dfrac{u_1 u_2 L}{A}$ (ft/in.²) | $S = S' + u_1 X_1 + u_2 X_2$ (kips) |
|---|---|---|---|---|---|---|---|---|---|---|
| $ab$ | 1 | +36 | +1 | 0 | +36 | 0 | +1 | 0 | 0 | $36 - 25.6 + 0 = +10.4$ |
| $bc$ | 1 | +36 | +1 | 0 | +36 | 0 | +1 | 0 | 0 | $36 - 25.6 + 0 = +10.4$ |
| $cd$ | 1 | +12 | +1 | -3/5 | +12 | -7.2 | +1 | +9/25 | -3/5 | $12 - 25.6 + 6.4 = -7.0$ |
| $de$ | 1 | +12 | +1 | 0 | +12 | 0 | +1 | 0 | 0 | $12 - 25.6 + 0 = -13.6$ |
| $BC$ | 1 | -24 | 0 | 0 | 0 | 0 | 0 | 0 | 0 | $-24 + 0 + 0 = -24$ |
| $CD$ | 1 | -24 | 0 | -3/5 | 0 | +14.4 | 0 | +9/25 | 0 | $-24 + 0 + 6.4 = -17.6$ |
| $aB$ | 1 | -60 | 0 | 0 | 0 | 0 | 0 | 0 | 0 | $-60 + 0 + 0 = -60$ |
| $Bb$ | 1 | +64 | 0 | 0 | 0 | 0 | 0 | 0 | 0 | $+64 + 0 + 0 = +64$ |
| $Bc$ | 1 | -20 | 0 | 0 | 0 | 0 | 0 | 0 | 0 | $-20 + 0 + 0 = -20$ |
| $Cc$ | 1 | 0 | 0 | -4/5 | 0 | 0 | 0 | +16/25 | 0 | $0 + 0 + 8.5 = +8.5$ |
| $Cd$ | 1 | 0 | 0 | +1 | 0 | 0 | 0 | +1 | 0 | $0 + 0 - 10.6 = -10.6$ |
| $cD$ | 1 | +20 | 0 | +1 | 0 | +20 | 0 | +1 | 0 | $20 + 0 - 10.6 = +9.4$ |
| $Dd$ | 1 | 0 | 0 | -4/5 | 0 | 0 | 0 | +16/25 | 0 | $0 + 0 + 8.5 = +8.5$ |
| $De$ | 1 | -20 | 0 | 0 | 0 | 0 | 0 | 0 | 0 | $-20 + 0 + 0 = -20$ |
| | | | | $\Sigma$ | +96 | +27.2 | +4 | +4 | -3/5 | |

191

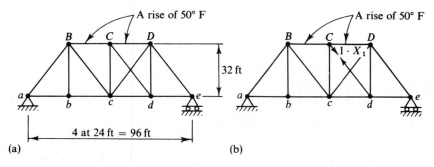

Fig. 9-13

or
$$\sum u_1(\alpha t^\circ L) + X_1 \sum \frac{u_1^2 L}{AE} = 0$$

where $\Delta_1'$ = relative displacement between the cut ends of the primary truss due to the temperature rise = $\sum u_1(\alpha t^\circ L)$ (see Eq. 8-20)

$u_1$ = internal force in any member of the primary truss due to a pair of unit axial forces acting at the ends of the cut section

The solution is shown in Table 9-4.

$$\frac{4X_1}{30,000} - 0.00468 = 0$$

$$X_1 = 35.1 \text{ kips} \qquad \text{(tension)}$$

Although these illustrations are aimed at statically indeterminate trusses

TABLE 9-4

| Member | $\dfrac{L}{A}$ (ft/in.²) | $u_1$ | $\alpha t^\circ L$ (ft) | $u_1 \alpha t^\circ L$ (ft) | $\dfrac{u_1^2 L}{A}$ (ft/in.²) | $S = u_1 X_1$ (kips) |
|--------|------|-------|---------|----------|------|------|
| ab | 1 | 0 | 0 | 0 | 0 | 0 |
| bc | 1 | 0 | 0 | 0 | 0 | 0 |
| cd | 1 | $-3/5$ | 0 | 0 | $+9/25$ | $-21.1$ |
| de | 1 | 0 | 0 | 0 | 0 | 0 |
| BC | 1 | 0 | $+0.0078$ | 0 | 0 | 0 |
| CD | 1 | $-3/5$ | $+0.0078$ | $-0.00468$ | $+9/25$ | $-21.1$ |
| aB | 1 | 0 | 0 | 0 | 0 | 0 |
| Bb | 1 | 0 | 0 | 0 | 0 | 0 |
| Bc | 1 | 0 | 0 | 0 | 0 | 0 |
| Cc | 1 | $-4/5$ | 0 | 0 | $+16/25$ | $-28.1$ |
| Cd | 1 | $+1$ | 0 | 0 | $+1$ | $+35.1$ |
| cD | 1 | $+1$ | 0 | 0 | $+1$ | $+35.1$ |
| Dd | 1 | $-4/5$ | 0 | 0 | $+16/25$ | $-28.1$ |
| De | 1 | 0 | 0 | 0 | 0 | 0 |
| | | | $\Sigma$ | $-0.00468$ | $+4$ | |

with one or two redundants, the procedure described can be extended to trusses with many degrees of redundancy.

## PROBLEMS

**9-1.** Analyze the beam in Fig. 9-14 by the method of consistent deformations. (a) Use the reaction at center support $b$ as redundant. (b) Use the internal moment at $b$ as redundant. Assume constant $EI$.

Fig. 9-14

**9-2.** Determine the reaction at $b$ in Fig. 9-15 by the method of consistent deformations. Assume constant $EI$.

Fig. 9-15

**9-3.** Find the reaction at $b$ in Fig. 9-16 by the method of consistent deformations. Assume constant $E$.

Fig. 9-16

**9-4.** For the system shown in Fig. 9-17, determine, by the method of consistent deformations, the reaction at support $e$. The flexibility of the spring $f = 0.2$ cm/kN of force; the bending rigidity of the beam $EI = 30,000$ kN $\cdot$ m².

**Fig. 9-17**

**9-5.** Find the fixed-end moments for the beams in Fig. 9-18 by the method of consistent deformations. Assume constant *EI*.

(a)                                    (b)

**Fig. 9-18**

**9-6.** Use the method of consistent deformations to determine the horizontal reaction at support *c* of the rigid frame shown in Fig. 9-19. Assume that $E = 20,000$ kN/cm² and $I = 20,000$ cm⁴.

**Fig. 9-19**

**9-7.** Analyze the rigid frame shown in Fig. 9-20 by the method of consistent deformations. Use the reaction components at support *a* as the redundants. All members have the same value of *EI*.

194

**Fig. 9-20**

**9-8.** Repeat Prob. 9-7 using the moment forces at joints $b$, $c$, and $d$ as the redundants.

**9-9.** Analyze the truss in Fig. 9-21 by the method of consistent deformations. Assume that $E = 20,000$ kN/cm² and $A = 25$ cm² for all members.

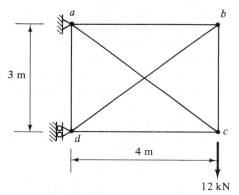

**Fig. 9-21**

**9-10.** Analyze each of the trusses in Fig. 9-22 by the method of consistent deformations. Assume that $E = 30,000$ kips/in.² and $L(ft)/A(in.^2) = 2$ for all members.

**Fig. 9-22**

**9-11.** Analyze the truss in Fig. 9-22(a) (without the external load) subject to a rise in temperature of 50°F for member $BC$. Assume that $\alpha = 0.0000065$ in./in./1°F.

**195**

# 10

## ANALYSIS OF STATICALLY

## INDETERMINATE STRUCTURES BY

## THE METHOD OF LEAST WORK

### 10-1. GENERAL

The method of consistent deformations, discussed in Chapter 9, involves superposition equations for the elastic deformations of the primary structure at the points of application of the redundants $X_1, X_2, \ldots, X_n$, the primary structure being stable and determinate and subjected to external actions, together with $n$ redundant forces. The expressions that the displacement at each of $n$ redundants equals zero for a loaded structure with nonyielding supports may be set up by the use of Castigliano's theorem as

$$
\begin{Bmatrix} \Delta_1 \\ \Delta_2 \\ \cdot \\ \cdot \\ \cdot \\ \Delta_n \end{Bmatrix} = \begin{Bmatrix} \dfrac{\partial W}{\partial X_1} \\ \dfrac{\partial W}{\partial X_2} \\ \cdot \\ \cdot \\ \cdot \\ \dfrac{\partial W}{\partial X_n} \end{Bmatrix} = \begin{Bmatrix} 0 \\ 0 \\ \cdot \\ \cdot \\ \cdot \\ 0 \end{Bmatrix}
\qquad (10\text{-}1)
$$

where $W$ is the *total strain energy* of the primary structure and is therefore a function of the external loads and the unknown redundant forces $X_1, X_2, \ldots, X_n$. There are as many simultaneous equations as the number of unknown redundants involved. Equation 10-1,

$$
\frac{\partial W}{\partial X_1} = \frac{\partial W}{\partial X_2} = \cdots = \frac{\partial W}{\partial X_n} = 0
$$

is known as Castigliano's compatibility equation and it may be stated as follows: *The redundants must have such value that the total strain energy of the structure is a minimum consistent with equilibrium.* For this reason it is sometimes referred to as the *theorem of least work*.

It is apparent from the foregoing proof that the method of least work and the method of consistent deformations are, in reality, identical. The choice between the two methods is a matter of personal preference. Because of the simpler setup, the method of least work is more widely used by structural engineers. However, this method is limited to the computation of the internal forces produced only by external loads on a structure mounted on unyielding supports. It cannot be used to determine stresses caused by temperature change, support movements, fabrication errors, and so on, and thus is not so general as the method of consistent deformations.

## 10-2. ANALYSIS OF STATICALLY INDETERMINATE BEAMS BY THE METHOD OF LEAST WORK

The procedure involves expressing the total strain energy in the primary structure in terms of the original loads and the one or more redundant elements and placing each of the derivatives of this expression with respect to the unknown redundants equal to zero, thus solving for the unknowns.

As before, we consider bending moment to be the only significant factor contributing to the internal energy of the beams or rigid frames. Therefore, the total strain energy can be expressed by

$$W = \int \frac{M^2 \, dx}{2EI}$$

Setting the derivative of this expression with respect to any redundant $X_i$ equal to zero gives the equation of least work as

$$\int \frac{M(\partial M/\partial X_i) \, dx}{EI} = 0$$

Therefore, for a statically indeterminate beam (or rigid frame) with $n$ redundants, we can write a set of $n$ simultaneous least-work equations:

$$\begin{Bmatrix} \dfrac{\partial W}{\partial X_1} \\ \dfrac{\partial W}{\partial X_2} \\ \cdot \\ \cdot \\ \cdot \\ \dfrac{\partial W}{\partial X_n} \end{Bmatrix} = \begin{Bmatrix} \displaystyle\int \dfrac{M(\partial M/\partial X_1) \, dx}{EI} \\ \displaystyle\int \dfrac{M(\partial M/\partial X_2) \, dx}{EI} \\ \cdot \\ \cdot \\ \cdot \\ \displaystyle\int \dfrac{M(\partial M/\partial X_n) \, dx}{EI} \end{Bmatrix} = \begin{Bmatrix} 0 \\ 0 \\ \cdot \\ \cdot \\ \cdot \\ 0 \end{Bmatrix} \qquad (10\text{-}2)$$

to solve all the unknown redundants.

**Example 10-1.** The beam in Fig. 10-1(a) was solved by consistent deformations in Example 9-1. Let us re-solve it by the method of least work.

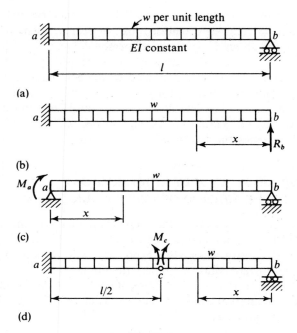

**Fig. 10-1**

*Solution 1* Select the reaction at $b$, called $R_b$, as redundant, and take the end $b$ as the origin [see Fig. 10-1(b)]. Thus,

$$M = R_b x - \frac{wx^2}{2} \qquad \text{and} \qquad \frac{\partial M}{\partial R_b} = x$$

Substituting these in the expression

$$\frac{\partial W}{\partial R_b} = \int_0^l \frac{M(\partial M/\partial R_b)\, dx}{EI} = 0$$

we obtain

$$\int_0^l \frac{[R_b x - (wx^2/2)]x\, dx}{EI} = 0$$

or

$$\frac{R_b l^3}{3} - \frac{wl^4}{8} = 0$$

from which

$$R_b = \frac{3wl}{8}$$

acting in the direction shown.

*Solution 2* Let us choose the end moment at $a$, called $M_a$, as redundant and take the origin at $a$ as shown in Fig. 10-1(c).

$$M = M_a + \left(\frac{wl}{2} - \frac{M_a}{l}\right)x - \frac{wx^2}{2} \qquad \text{and} \qquad \frac{\partial M}{\partial M_a} = 1 - \frac{x}{l}$$

Setting

$$\frac{\partial W}{\partial M_a} = \int_0^l \frac{M(\partial M/\partial M_a)\,dx}{EI}$$

$$= \int_0^l \frac{\{M_a + [(wl/2) - (M_a/l)]x - (wx^2/2)\}\{1 - (x/l)\}\,dx}{EI} = 0$$

we solve for
$$M_a = -\frac{1}{8}wl^2$$

The negative sign indicates a counterclockwise moment.

*Solution* 3 Suppose that we select the internal moment at midspan section $c$, called $M_c$, as the redundant. We will then have the loaded primary structure shown in Fig. 10-1(d).

By statics, the reaction at end $b$ is $(wl/4 + 2M_c/l)$ acting upward. Using $b$ as the origin, we have

$$M = \left(\frac{wl}{4} + \frac{2M_c}{l}\right)x - \frac{wx^2}{2} \quad \text{and} \quad \frac{\partial M}{\partial M_c} = \frac{2x}{l}$$

Setting
$$\frac{\partial W}{\partial M_c} = \frac{1}{EI}\int_0^l \left[\left(\frac{wl}{4} + \frac{2M_c}{l}\right)x - \frac{wx^2}{2}\right]\frac{2x}{l}\,dx = 0$$

we obtain
$$M_c = \frac{wl^2}{16}$$

acting in the assumed direction.

**Example 10-2.** The reaction at support $b$ in Fig. 10-2(a) was solved by the method of consistent deformations in Example 9-3 and will now be re-solved by the method of least work.

If we choose the reaction at support $b$, called $R_b$, as redundant, we will have the loaded primary structure shown in Fig. 10-2(b).

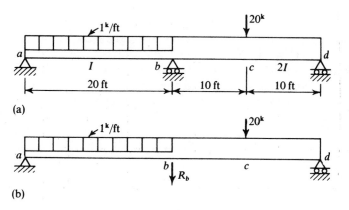

(a)

(b)

**Fig. 10-2**

The solution for the equation

$$\frac{\partial W}{\partial R_b} = \int \frac{M(\partial M/\partial R_b)\,dx}{EI} = 0$$

is contained in Table 10-1.

$$\int_0^{20} \frac{\{[20 + (R_b/2)]x - (x^2/2)\}x/2\,dx}{EI} + \int_0^{10} \frac{\{[20 + (R_b/2)]x\}x/2\,dx}{2EI}$$

$$+ \int_{10}^{20} \frac{\{(R_b x/2) + 200\}x/2\,dx}{2EI} = 0$$

$$R_b = -25.84 \text{ kips}$$

The negative sign indicates an upward reaction at support $b$.

TABLE 10-1

| Section | Origin | Limit (ft) | $M$ (ft-kips) | $\frac{\partial M}{\partial R_b}$ (ft) | $I$ |
|---------|--------|-----------|---------------|------------------------------------------|-----|
| $ab$ | $a$ | 0 to 20 | $\left(20 + \frac{R_b}{2}\right)x - \frac{(1)x^2}{2}$ | $\frac{x}{2}$ | $I$ |
| $dc$ | $d$ | 0 to 10 | $\left(20 + \frac{R_b}{2}\right)x$ | $\frac{x}{2}$ | $2I$ |
| $cb$ | $d$ | 10 to 20 | $\left(20 + \frac{R_b}{2}\right)x - 20(x - 10)$<br>or $\frac{R_b x}{2} + 200$ | $\frac{x}{2}$ | $2I$ |

$$\int_0^{20} \frac{\{[20 + (R_b/2)]x - (x^2/2)\}x/2\,dx}{EI} + \int_0^{10} \frac{\{[20 + (R_b/2)]x\}x/2\,dx}{2EI}$$

$$+ \int_{10}^{20} \frac{\{(R_b x/2) + 200\}x/2\,dx}{2EI} = 0$$

$$R_b = -25.84 \text{ kips}$$

To determine, by the method of least work, the *fixed-end moments* of beam $AB$ due to general loading, as shown in Fig. 10-3(a), we select the left-end reaction components $M_A$ and $V_A$ as redundants, as shown in Fig. 10-3(b). The primary structure is a cantilever subjected to the original loads on the span together with the redundant forces $M_A$ and $V_A$ at the left end. Applying the method of least work, we obtain

$$\frac{\partial W}{\partial M_A} = \int_0^l \frac{M(\partial M/\partial M_A)\,dx}{EI} = 0 \tag{10-3}$$

$$\frac{\partial W}{\partial V_A} = \int_0^l \frac{M(\partial M/\partial V_A)\,dx}{EI} = 0 \tag{10-4}$$

Since the bending moment at any section of the primary structure is given by

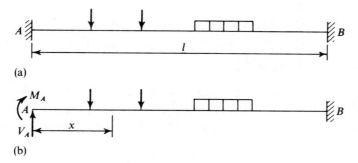

(a)

(b)

**Fig. 10-3**

$$M = M' + M_A + V_A x$$

where $M'$ indicates the bending moment at the same section of the primary structure resulting from the original loads on the span, we have

$$\frac{\partial M}{\partial M_A} = 1 \quad \text{and} \quad \frac{\partial M}{\partial V_A} = x$$

Substituting these in Eqs. 10-3 and 10-4 results in the following two equations:

$$\int_0^l \frac{M\, dx}{EI} = 0 \tag{10-5}$$

$$\int_0^l \frac{Mx\, dx}{EI} = 0 \tag{10-6}$$

to solve for redundants $M_A$ and $V_A$.

For a beam of uniform section with constant $EI$, Eqs. 10-5 and 10-6 reduce to

$$\int_0^l M\, dx = 0 \tag{10-7}$$

$$\int_0^l Mx\, dx = 0 \tag{10-8}$$

**Example 10-3.** The fixed-end moments of the beam shown in Fig. 10-4 were solved by the method of consistent deformations in Example 9-4 and will be re-solved by using Eqs. 10-7 and 10-8.

**Fig. 10-4**

Taking the origin at $A$, we note that

$$M = M_A + V_A x \qquad\qquad 0 \leq x \leq a$$

$$M = M_A + V_A x - P(x - a) \qquad a \leq x \leq l$$

Applying Eqs. 10-7 and 10-8 gives

$$\int_0^l M\,dx = \int_0^a (M_A + V_A x)\,dx + \int_a^l [M_A + V_A x - P(x - a)]\,dx$$

$$= \int_0^l (M_A + V_A x)\,dx + \int_a^l [-P(x - a)]\,dx = 0$$

or
$$M_A l + \frac{V_A l^2}{2} - \frac{P b^2}{2} = 0 \tag{10-9}$$

and $\displaystyle \int_0^l Mx\,dx = \int_0^a (M_A + V_A x)x\,dx + \int_a^l [M_A + V_A x - P(x - a)]x\,dx$

$$= \int_0^l (M_A + V_A x)x\,dx + \int_a^l [-P(x - a)]x\,dx = 0$$

or
$$\frac{M_A l^2}{2} + \frac{V_A l^3}{3} - \frac{P b^2(a + 2l)}{6} = 0 \tag{10-10}$$

Solving Eqs. 10-9 and 10-10 simultaneously, we obtain

$$M_A = -\frac{Pab^2}{l^2} \qquad V_A = \frac{Pb^2(l + 2a)}{l^3}$$

Similarly,
$$M_B = -\frac{Pa^2 b}{l^2} \qquad V_B = \frac{Pa^2(l + 2b)}{l^3}$$

## 10-3. ANALYSIS OF STATICALLY INDETERMINATE
## RIGID FRAMES BY THE METHOD OF LEAST WORK

The procedure in analyzing a statically indeterminate rigid frame by the method of least work is similar to that described in Sec. 10-2 for statically indeterminate beams. The following examples serve to illustrate the application of this method.

**Example 10-4.** Consider the loaded frame shown in Fig. 10-5(a), which was analyzed by the method of consistent deformations (Example 9-7). Let us now analyze it by least work.

Let us take the reaction components $M_a$, $H_a$, and $V_a$ at end $a$ as redundants, as shown in Fig. 10-5(b). Because of symmetry, we may replace $V_a$ with 6 kips so that there will be only two redundants $M_a$ and $H_a$ left, as indicated. The solutions for the equations of least work,

$$\frac{\partial W}{\partial M_a} = 0 \qquad \text{or} \qquad \int_F \frac{M(\partial M/\partial M_a)\,dx}{EI} = 0$$

$$\frac{\partial W}{\partial H_a} = 0 \qquad \text{or} \qquad \int_F \frac{M(\partial M/\partial H_a)\,dx}{EI} = 0$$

are contained in Table 10-2.

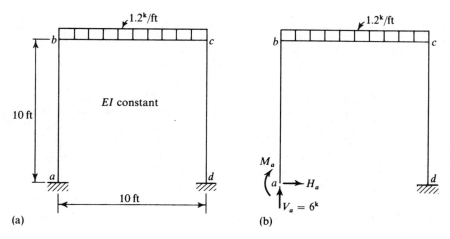

Fig. 10-5

TABLE 10-2

| Member | Origin | Limit (ft) | $M$ (ft-kips) | $\dfrac{\partial M}{\partial M_a}$ | $\dfrac{\partial M}{\partial H_a}$ (ft) |
|--------|--------|-----------|---------------|-----------------|-----------------|
| $ab$ | $a$ | 0 to 10 | $M_a - H_a x$ | 1 | $-x$ |
| $bc$ | $b$ | 0 to 10 | $M_a - 10H_a + 6x - \dfrac{(1.2)x^2}{2}$ | 1 | $-10$ |
| $cd$ | $c$ | 0 to 10 | $M_a + H_a(x - 10)$ | 1 | $x - 10$ |

$$\int_F M(\partial M/\partial M_a)\,dx$$

$$= \int_0^{10} (M_a - H_a x)(1)\,dx + \int_0^{10} \left[ M_a - 10H_a + 6x - \frac{(1.2)x^2}{2} \right](1)\,dx$$

$$+ \int_0^{10} [M_a + H_a(x - 10)](1)\,dx$$

$$= (10M_a - 50H_a) + (10M_a - 100H_a + 100) + (10M_a - 50H_a)$$

$$= 0$$

or $$3M_a - 20H_a + 10 = 0 \qquad\qquad (10\text{-}11)$$

and

$$\int_F M(\partial M/\partial H_a)\,dx$$

$$= \int_0^{10} (M_a - H_a x)(-x)\,dx + \int_0^{10} \left[ M_a - 10H_a + 6x - \frac{(1.2)x^2}{2} \right](-10)\,dx$$

$$+ \int_0^{10} [M_a + H_a(x - 10)](x - 10)\,dx$$

$$= \left( -50M_a + \frac{1{,}000}{3}H_a \right) + (-100M_a + 1{,}000H_a - 1{,}000)$$

$$+ \left( -50M_a + \frac{1{,}000}{3}H_a \right) = 0$$

203

or                                $-3M_a + 25H_a - 15 = 0$                    (10-12)

Solving Eqs. 10-11 and 10-12 simultaneously, we obtain

$$H_a = 1.0 \text{ kip} \qquad M_a = 3.33 \text{ ft-kips}$$

**Example 10-5.** Analyze the frame in Fig. 10-5(a) by taking the internal shear, thrust, and moment in the midspan section of the beam as redundants.

Because of symmetry, the shear must be zero in the midspan section $e$ of the beam, and only thrust and bending moment are left as redundants, as shown in Fig. 10-6(a). The solution can be simplified by considering only half of the frame, as shown in Fig. 10-6(b) and Table 10-3.

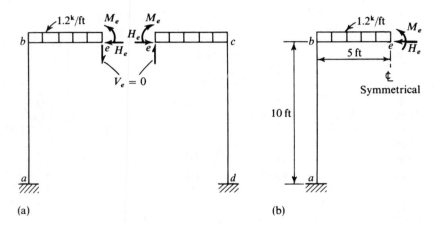

**Fig. 10-6**

**TABLE 10-3**

| Member | Origin | Limit (ft) | $M$ (ft-kips) | $\dfrac{\partial M}{\partial M_e}$ | $\dfrac{\partial M}{\partial H_e}$ (ft) |
|--------|--------|-----------|---------------|------------------------------------|------------------------------------------|
| $eb$ | $e$ | 0 to 5 | $M_e - \dfrac{(1.2)x^2}{2}$ | 1 | 0 |
| $ba$ | $b$ | 0 to 10 | $M_e + H_e x - 15$ | 1 | $x$ |

Applying

$$\frac{\partial W}{\partial M_e} = 0 \qquad \text{or} \qquad \int_F \frac{M(\partial M/\partial M_e)\, dx}{EI} = 0$$

we have

$$\frac{2}{EI}\left[\int_0^5 \left(M_e - \frac{(1.2)x^2}{2}\right)(1)\, dx + \int_0^{10} (M_e + H_e x - 15)(1)\, dx\right] = 0$$

or                        $3M_e + 10H_e - 35 = 0$                    (10-13)

Applying

$$\frac{\partial W}{\partial H_e} = 0 \qquad \text{or} \qquad \int_F \frac{M(\partial M/\partial H_e)\, dx}{EI} = 0$$

we have

$$\frac{2}{EI}\left[\int_0^{10} (M_e + H_e x - 15)x\, dx\right] = 0$$

or

$$3M_e + 20H_e - 45 = 0 \tag{10-14}$$

Solving Eqs. 10-13 and 10-14 simultaneously gives

$$H_e = 1.0 \text{ kip} \qquad M_e = 8.33 \text{ ft-kips}$$

from which we obtain

$$H_a = 1.0 \text{ kip} \qquad M_a = 3.33 \text{ ft-kips}$$

as previously found.

In comparison with Example 10-4 we see that, by proper selection of the redundants through taking advantage of the symmetry of the structure, we minimize considerably the numerical computations.

For a highly indeterminate rigid frame, such as the one shown in Fig. 10-7(a), the procedure of the analysis remains the same. The frame is statically indeterminate to the 24th degree. We may cut it back to three determinate structures and substitute the redundants $X_1, X_2, \ldots, X_{24}$ at the cut sections as shown in Fig. 10-7(b).

From least work, we have 24 equations to solve for all the redundants simultaneously, namely,

$$\begin{Bmatrix} \dfrac{\partial W}{\partial X_1} \\[2mm] \dfrac{\partial W}{\partial X_2} \\[2mm] \cdot \\ \cdot \\ \cdot \\ \dfrac{\partial W}{\partial X_{24}} \end{Bmatrix} = \begin{Bmatrix} 0 \\[2mm] 0 \\[2mm] \cdot \\ \cdot \\ \cdot \\ 0 \end{Bmatrix}$$

where $W$ is the total strain energy of the frame due to the external loads and redundant forces. The principle is neat and elegant, whereas the numerical calculations involved in the equations above are so cumbersome that it is almost impossible for a structural engineer to reach an exact solution for the system with only a slide rule or desk calculating machine. Formerly, to handle a practical problem like this, a grossly simplified model of the actual structure was often used. However, with the advent of the digital computer, this unsatisfactory situation underwent a drastic change, and the solving of simultaneous equations can now be performed in a matter of minutes.

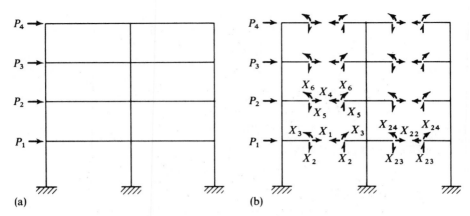

(a)  (b)

**Fig. 10-7**

## 10-4. ANALYSIS OF STATICALLY INDETERMINATE TRUSSES BY THE METHOD OF LEAST WORK

Statically indeterminate trusses can be analyzed by the method of least work in a similar manner. Referring to Eq. 8-10, we find the total strain energy (internal work) of a truss is expressed by

$$W = \Sigma \frac{S^2 L}{2AE}$$

Setting the derivative of this expression with respect to any redundant $X_i$ equal to zero gives the equation of least work as

$$\Sigma \frac{S(\partial S/\partial X_i)L}{AE} = 0$$

Thus, for a statically indeterminate truss with $n$ redundant elements, we have a set of $n$ simultaneous least-work equations available for their solution,

$$\begin{Bmatrix} \Sigma \dfrac{S(\partial S/\partial X_1)L}{AE} \\ \Sigma \dfrac{S(\partial S/\partial X_2)L}{AE} \\ \cdot \\ \cdot \\ \cdot \\ \Sigma \dfrac{S(\partial S/\partial X_n)L}{AE} \end{Bmatrix} = \begin{Bmatrix} 0 \\ 0 \\ \cdot \\ \cdot \\ \cdot \\ 0 \end{Bmatrix} \qquad (10\text{-}15)$$

The application of this procedure is illustrated in the following example.

**Example 10-6.** Analyze the truss in Fig. 10-8(a). Assume that $E = 30{,}000$ kips/in.$^2$ and $L(\text{ft})/A(\text{in.}^2) = 1$ for all members.

206

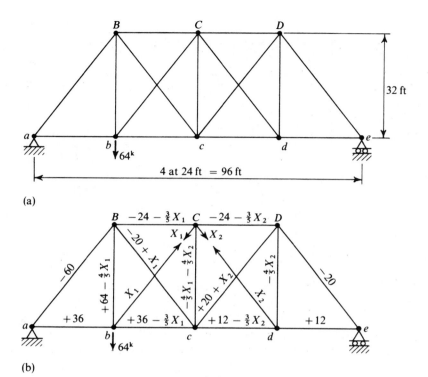

(a)

(b)

**Fig. 10-8**

The truss is statically indeterminate to the second degree. We may take bars $bC$ and $Cd$ as redundant members. As shown in Fig. 10-8(b), these bars are cut and replaced by redundant axial forces $X_1$ and $X_2$, respectively. The internal force for each bar is then computed in terms of the external load and redundant forces as indicated. The unknowns $X_1$ and $X_2$ are then solved by the simultaneous equations

$$\sum \frac{S(\partial S/\partial X_1)L}{AE} = 0 \quad \text{and} \quad \sum \frac{S(\partial S/\partial X_2)L}{AE} = 0$$

as prepared in Table 10-4. Setting

$$-78.4 + 4X_1 + 0.64X_2 = 0 \tag{10-16}$$

$$27.2 + 0.64X_1 + 4X_2 = 0 \tag{10-17}$$

and solving Eqs. 10-16 and 10-17 simultaneously, we obtain

$$X_1 = +21.2 \text{ kips} \qquad X_2 = -10.2 \text{ kips}$$

The answer for each of the bar forces is given in the last column of Table 10-4. Note that this procedure can be extended to trusses with many redundants.

TABLE 10-4

| Member | $\frac{L}{A}$ (ft/in.²) | $S$ (kips) | $\frac{\partial S}{\partial X_1}$ | $\frac{\partial S}{\partial X_2}$ | $\frac{S(\partial S/\partial X_1)L}{A}$ (ft-kips/in.²) | $\frac{S(\partial S/\partial X_2)L}{A}$ (ft-kips/in.²) | Answer (kips) |
|---|---|---|---|---|---|---|---|
| $ab$ | 1 | $36 + 0 + 0$ | 0 | 0 | 0 | 0 | $+36$ |
| $bc$ | 1 | $36 - (3/5)X_1 + 0$ | $-3/5$ | 0 | $-21.6 + (9/25)X_1 + 0$ | 0 | $+23.3$ |
| $cd$ | 1 | $12 + 0 - (3/5)X_2$ | 0 | $-3/5$ | 0 | $-7.2 + 0 + (9/25)X_2$ | $+18.2$ |
| $de$ | 1 | $12 + 0 + 0$ | 0 | 0 | 0 | 0 | $+12$ |
| $BC$ | 1 | $-24 - (3/5)X_1 + 0$ | $-3/5$ | 0 | $14.4 + (9/25)X_1 + 0$ | 0 | $-36.7$ |
| $CD$ | 1 | $-24 + 0 - (3/5)X_2$ | 0 | $-3/5$ | 0 | $14.4 + 0 + (9/25)X_2$ | $-17.8$ |
| $aB$ | 1 | $-60 + 0 + 0$ | 0 | 0 | 0 | 0 | $-60$ |
| $Bb$ | 1 | $64 - (4/5)X_1 + 0$ | $-4/5$ | 0 | $-51.2 + (16/25)X_1 + 0$ | 0 | $+47$ |
| $Bc$ | 1 | $-20 + X_1 + 0$ | $+1$ | 0 | $-20 + X_1 + 0$ | 0 | $+1.2$ |
| $bC$ | 1 | $0 + X_1 + 0$ | $+1$ | 0 | $0 + X_1 + 0$ | 0 | $+21.2$ |
| $Cc$ | 1 | $0 - (4/5)X_1 - (4/5)X_2$ | $-4/5$ | $-4/5$ | $0 + (16/25)X_1 + (16/25)X_2$ | $0 + (16/25)X_1 + (16/25)X_2$ | $-8.8$ |
| $Cd$ | 1 | $0 + 0 + X_2$ | 0 | $+1$ | 0 | $0 + 0 + X_2$ | $-10.2$ |
| $cD$ | 1 | $20 + 0 + X_2$ | 0 | $+1$ | 0 | $20 + 0 + X_2$ | $+9.8$ |
| $Dd$ | 1 | $0 + 0 - (4/5)X_2$ | 0 | $-4/5$ | 0 | $0 + 0 + (16/25)X_2$ | $+8.2$ |
| $De$ | 1 | $-20 + 0 + 0$ | 0 | 0 | 0 | 0 | $-20$ |
| | | | | $\Sigma$ | $-78.4 + 4X_1 + 0.64X_2$ | $27.2 + 0.64X_1 + 4X_2$ | |

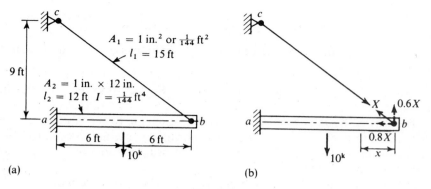

**Fig. 10-9**

## 10-5. ANALYSIS OF STATICALLY INDETERMINATE COMPOSITE STRUCTURES BY THE METHOD OF LEAST WORK

Structures made up of some members which are two-force members carrying only axial forces and others which are not are called composite structures. They are conveniently analyzed by the method of least work, as illustrated in the following examples.

**Example 10-7.** Figure 10-9(a) shows a cantilever beam whose other end is supported by a rod. Find the force in the rod. $E = 30,000$ kips/in.$^2$.

The structure is statically indeterminate to the first degree. Select the force in the tie rod as the redundant $X$ as shown in Fig. 10-9(b). Then the internal work in the rod is

$$\frac{X^2 l_1}{2 A_1 E}$$

and the internal work in the beam is equal to

$$\int_0^6 \frac{(0.6Xx)^2 \, dx}{2EI} + \int_6^{12} \frac{[0.6Xx - 10(x-6)]^2 \, dx}{2EI} + \frac{(-0.8X)^2 l_2}{2 A_2 E}$$

Applying $\partial W / \partial X = 0$ gives

$$\frac{X l_1}{A_1 E} + \int_0^6 \frac{(0.6Xx)(0.6x) \, dx}{EI} + \int_6^{12} \frac{[0.6Xx - 10(x-6)][0.6x] \, dx}{EI}$$
$$+ \frac{(-0.8X)(-0.8)l_2}{A_2 E} = 0$$

or

$$\frac{15X}{1/144} + \frac{207.4X - 3,024 + 1,944}{1/144} + \frac{(0.64X)(12)}{12/144} = 0$$

After simplifying, we find that

$$15X + 207.4X - 1,080 + 0.64X = 0$$

**209**

which yields

$$X = 4.84 \text{ kips} \quad \text{(tension)}$$

The effect of the axial force of beam on the strain energy is small and can be neglected.

**Example 10-8.** Analyze the trussed beam in Fig. 10-10(a). $E = 30,000$ kips/in.$^2$.

The trussed beam shown is statically indeterminate to the first degree. This problem is most conveniently solved by choosing the internal force in rod $bd$ as the redundant [see Fig. 10-10(b)] and by expressing the internal strain energy in terms of this unknown.

Referring to Fig. 10-10(b) and applying $\partial W / \partial X = 0$, we obtain

$$\frac{(X)(1)(9)}{5/144} + 2\left[\frac{(\frac{5}{6}X)(\frac{5}{6})(15)}{4/144}\right] + \frac{(\frac{2}{3}X)(\frac{2}{3})(24)}{12/144}$$

$$+ 2\int_0^{12} \frac{[(12 - X/2)x - (1)(x)^2/2][-x/2]\,dx}{1/144} = 0$$

or $\qquad\qquad 1.8X + 5.2X + 0.89X - 4,320 + 288X = 0$

from which $\qquad\qquad X = 14.6 \text{ kips} \quad \text{(compression)}$

After the compressive force in strut $bd$ is determined, the rest of the analysis can be carried out without difficulty.

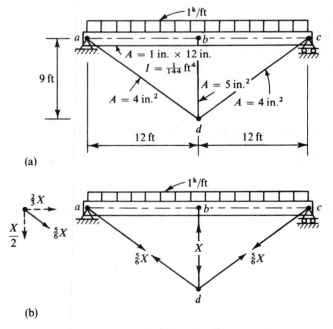

(a)

(b)

**Fig. 10-10**

**Example 10-9.** Analyze the truss shown in Fig. 10-11(a) resulting from the application of a uniform load of 15 kN/m over member *ab*. Assume a constant *E* for the structure.

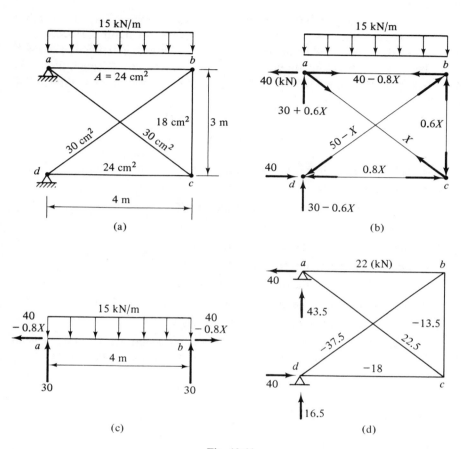

**Fig. 10-11**

Let us select the internal force in member *ac*, called *X*, as the redundant. After the horizontal reactions at supports *a* and *d* are determined by equilibrium, we can express the rest of member forces and vertical reactions at *a* and *d* in terms of *X* as shown in Fig. 10-11(b).

Next, we use the compatibility equation

$$\frac{\partial W}{\partial X} = \Sigma \frac{S \frac{\partial S}{\partial X} L}{AE} = 0$$

or     $$\Sigma S \frac{\partial S}{\partial X} = 0 \qquad \left(\frac{L}{AE} = \text{constant for all members}\right)$$

to obtain    $(X)(1) + (50 - X)(-1) + (0.8X)(0.8)$
$+ (40 - 0.8X)(-0.8) + (0.6X)(0.6) = 0$

solving for    $X = 22.5$ kN

Note that if a transverse load is applied to a truss member, then the member acts as a beam, and the system becomes composite in nature. However, in analyzing this system, the bending effect need not enter the expression $\partial W/\partial X = 0$, because the bending moment in member $ab$ is not a function of $X$, as indicated in the free-body diagram of $ab$ [see Fig. 10-11(c)]. The final result is shown in Fig. 10-11(d).

## PROBLEMS

**10-1.** Solve Prob. 9-1 by the method of least work.

**10-2.** Solve Prob. 9-2 by the method of least work.

**10-3.** Solve Prob. 9-3 by the method of least work.

**10-4.** Solve Prob. 9-5 by the method of least work.

**10-5.** Solve Prob. 9-6 by the method of least work.

**10-6.** Solve Prob. 9-7 by the method of least work.

**10-7.** Solve Prob. 9-8 by the method of least work.

**10-8.** Solve Prob. 9-9 by the method of least work.

**10-9.** Solve Prob. 9-10 by the method of least work.

**10-10.** Find the internal force for the tie rod $ac$ of the composite structure shown in Fig. 10-12, and sketch the moment diagram for member $ab$. $E = 30,000$ kips/in.$^2$.

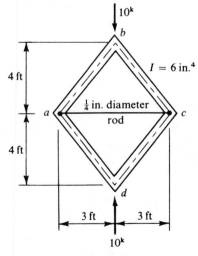

**Fig. 10-12**

**10-11.** Find the internal forces for all rods of the composite structure shown in Fig. 10-13, and sketch the moment diagram for beam *ab*. $E = 30,000$ kips/in.²

**Fig. 10-13**

**10-12.** For the truss and load shown in Fig. 10-14, find the axial forces in all members. $E = 20,000$ kN/cm².

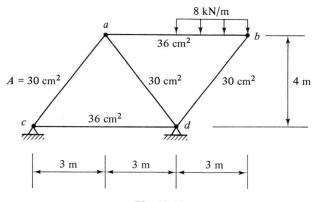

**Fig. 10-14**

# 11

## INFLUENCE LINES FOR STATICALLY

## INDETERMINATE STRUCTURES

### 11-1. MAXWELL'S LAW OF RECIPROCAL DELECTIONS

Referring to Fig. 11-1, we note that the law simply states that

$$\Delta_{21} = \Delta_{12} \qquad (11\text{-}1)$$

where $\Delta_{21}$ = deflection at point 2 due to the load $P$ applied at point 1
$\Delta_{12}$ = deflection at point 1 along the original line of action of $P$ due to the same load applied at point 2 along the original deflection $\Delta_{21}$

(a)                                                    (b)

**Fig. 11-1**

To prove this statement, both deflections are evaluated by the method of virtual work. Thus, from Fig. 11-1(a) we have

$$\Delta_{21} = \int_0^l \frac{M_1 m_2 \, dx}{EI}$$

where $M_1$ = moment at any section due to load $P$ applied at point 1
$m_2$ = moment at the same section due to a unit load applied at point 2 along the desired deflection

Similarly, from Fig. 11-1(b) we obtain that

$$\Delta_{12} = \int_0^l \frac{M_2 m_1 \, dx}{EI}$$

but $\qquad\qquad M_1 = Pm_1 \qquad$ and $\qquad M_2 = Pm_2$

It is readily seen that

$$\Delta_{21} = \int_0^l \frac{(Pm_1)m_2 \, dx}{EI} = \int_0^l \frac{(Pm_2)m_1 \, dx}{EI} = \int_0^l \frac{M_2 m_1 \, dx}{EI} = \Delta_{12}$$

The special case is that $P = 1$, for which we can write

$$\delta_{21} = \delta_{12} \tag{11-2}$$

where $\delta_{21}$ = deflection at point 2 resulting from a unit load applied at point 1
$\delta_{12}$ = deflection at point 1 along the original line of action due to a unit load applied at point 2 along the original deflection $\delta_{21}$

We have hitherto demonstrated the law in regard to applied forces and their corresponding linear deflections. However, the reciprocity extends also to rotational displacement. For the case of two unit couples applied separately to any two points of a structure, the law is: *The rotational deflection at point 2 on a structure caused by a unit couple at point 1 is equal to the rotational deflection at point 1 due to a unit couple at point 2.*

By the reason of virtual work we also observe that the *rotational deflection at point 2 due to a unit force at point 1 is equal in magnitude to the linear deflection at point 1 along the original force due to a unit couple at point 2.*

Maxwell's law is perfectly general and is applicable to any type of structure as long as the material of the structure is elastic and follows Hooke's law.

## 11-2. BETTI'S LAW

Maxwell's theorem of reciprocity is a special case of *Betti's law*, which may be expressed briefly by

$$W_{12} = W_{21} \tag{11-3}$$

where $W_{12}$ = external work done by a force system $P_1$ already on the structure during a distortion caused by a force system $P_2$ applied later
$W_{21}$ = external work done by $P_2$ during a distortion caused by $P_1$ when $P_1$ is the second in order of application

To prove this theorem, let us consider the beam in Fig. 11-2(a), where $P_1$ (representing the force system 1) is applied first. The elastic curve caused by $P_1$ is indicated by the dashed line. During this operation the work done is

$$\tfrac{1}{2} P_1(P_1\delta_{11})$$

$P_2$ (representing the force system 2) is applied next, which causes further displacement of the beam [see Fig. 11-2(a)]. The additional work done is

$$P_1(P_2\delta_{12}) + \tfrac{1}{2}P_2(P_2\delta_{22})$$

(a)                                                    (b)

**Fig. 11-2**

The total work done is

$$\tfrac{1}{2}P_1(P_1\delta_{11}) + P_1(P_2\delta_{12}) + \tfrac{1}{2}P_2(P_2\delta_{22}) \tag{11-4}$$

If we reverse the order of application of two systems of forces [see Fig. 11-2(b)] the total work, obtained similarly, can be expressed by

$$\tfrac{1}{2}P_2(P_2\delta_{22}) + P_2(P_1\delta_{21}) + \tfrac{1}{2}P_1(P_1\delta_{11}) \tag{11-5}$$

Equating Eq. 11-4 to Eq. 11-5 gives

$$P_1(P_2\delta_{12}) = P_2(P_1\delta_{21}) \tag{11-6}$$

as asserted by Eq. 11-3.

Equation 11-6 also indicates that

$$\delta_{12} = \delta_{21}$$

which is the *law of reciprocal deflections*.

Note that Betti's law depends on the principle of independence of effect of loads, which is true as long as the displacements are small and also independent. The law is applicable to any type of elastic structure with unyielding supports and at constant temperature. It holds for generalized forces and deflections, that is, for couples and rotations as well as for forces and linear displacements.

## 11-3. INFLUENCE LINES AS DEFLECTED STRUCTURES: THE MÜLLER-BRESLAU PRINCIPLE

Suppose that we want the influence line for the reaction at support $b$ of the indeterminate beam $abc$ shown in Fig. 11-3(a). The influence ordinate at any point $i$, distance $x$ from the left end, is obtained by placing a unit load at that point and computing the reaction at support $b$. The procedure for finding this reaction is contained in the following steps.

1. Remove the support at $b$ and introduce in its place a redundant reaction called $R_b$.

2. Consider beam $ac$ as the primary structure subjected to the combined effects of the unit force at $i$ and $R_b$ [see Fig. 11-3(b)].

3. Use the condition of compatibility that the total deflection at point $b$ must be zero,

$$\Delta_b = R_b\delta_{bb} - \delta_{bi} = 0$$

See Fig. 11-3(c) for $\delta_{bi}$ and Fig. 11-3(d) for $\delta_{bb}$.

$$R_b = \frac{\delta_{bi}}{\delta_{bb}}$$

4. Use the reciprocity,

$$\delta_{bi} = \delta_{ib}$$

to obtain

$$R_b = \frac{\delta_{ib}}{\delta_{bb}} \tag{11-7}$$

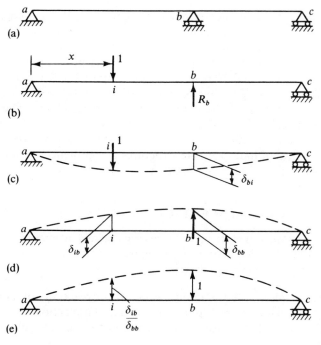

**Fig. 11-3**

Note that in Eq. 11-7 the numerator $\delta_{ib}$ represents the ordinate of the deflection curve of the primary beam $ac$ caused by applying a unit force at $b$. The denominator $\delta_{bb}$ is only a special case of $\delta_{ib}$ (i.e., $\delta_{bb} = \delta_{ib}$ if $i = b$), as shown in Fig. 11-3(d). Each ordinate of the curve of Fig. 11-3(d) divided by $\delta_{bb}$ will give the corresponding influence ordinate for $R_b$ (see Eq. 11-7), as shown in Fig. 11-3(e).

Referring to Fig. 11-3(e), we note that at $b$, $\delta_{bb}/\delta_{bb} = 1$. Hence, the influence line for $R_b$ is nothing more than the deflected structure resulting from removal of the support at $b$ and introduction in its place of a unit deflection along the line of reaction.

We have hitherto used the reaction of a support as illustration. This, then, is the Müller–Breslau principle and may be stated as: *The ordinates of the influence line for any element* (reaction, axial force, shear, or moment) *of a structure equal those of the deflection curve obtained by removing the restraint corresponding to that element from the structure and introducing in its place a unit load divided by the deflection at the point of application of the unit load.*

This may be rephrased as: *The deflected structure resulting from a unit displacement corresponding to the action for which the influence line is desired gives the influence line for that action.*

It may be noted that when applying the Müller–Breslau principle, if the ac-

217

tual structure is statically indeterminate to more than the first degree, the primary structure which remains after the removal of a redundant will still be statically indeterminate. This does not change our general procedure. However, the statically indeterminate primary structure must first be analyzed by one of the methods in Chapters 9 and 10 before we can compute the necessary deflections.

The Müller–Breslau principle is applicable to any type of structure—beam, truss, or rigid frame; statically determinate or indeterminate. The reader will find that the virtual-work method (Chapter 6) for obtaining influence lines for statically determinate beams and the Müller–Breslau principle are applied identically. In case of indeterminate structures this principle is limited to structures of elastic material obeying Hooke's law.

### 11-4. SKETCH OF INFLUENCE LINES

The Müller–Breslau principle provides a very convenient method for sketching qualitative influence lines for indeterminate structures and is the basis for certain experimental model analyses.

1. In the simplest case the influence line for a reaction component can be sketched by removing the restraint and allowing the reaction to move through a *unit displacement*. The deflected structure will then be the influence line for the reaction.

Thus, the dashed line in Fig. 11-4(a) shows the influence line for the vertical reaction at support *a* of the three-span continuous beam. In Fig. 11-4(b) the dashed line indicates the influence line for the fixed-end moment at support *a* of the fixed-end beam *ab*. The curve is obtained by replacing the fixed support at *a* with a hinge support and by introducing a unit rotation. In Fig. 11-4(c) is shown the construction of the influence line for the horizontal reaction component at the fixed support *d* of a portal frame. Note that the fixed support

**Fig. 11-4**

at *d* is replaced by a roller and slide acted on by a horizontal force so as to produce a unit horizontal displacement.

Note the following:

a. The vertical deflections of the structure will be influence line ordinates for the vertical loads on the structure.

b. The horizontal deflections of the structure will be influence line ordinates for the horizontal loads on the structure.

c. The rotation of the tangents of the structure will be influence line ordinates for the moment load on the structure.

2. The moment influence line for a section of a beam or rigid frame can be drawn by cutting the section and allowing a pair of equal and opposite moments to produce a *unit relative rotation* (but no relative translation) for the two sides of the section considered. The deflected structure will then be the influence line for the moment. Thus, the influence line for the moment at the midspan section of a three-span continuous beam is the dashed line in Fig. 11-5.

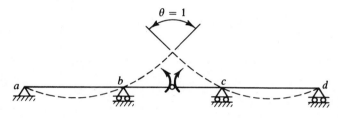

**Fig. 11-5**

3. The shear influence line for a section of a beam or rigid frame can be drawn by cutting the section and applying a pair of equal and opposite shearing forces to produce a *unit relative transverse displacement* (but no relative rotation) for the two sides of the section considered. The deflected structure will then be the influence line for the shear. The influence line for the shear at the midspan section of the fixed-end beam *ab* is shown in Fig. 11-6 by the dashed lines.

**Fig. 11-6**

4. To obtain the influence line for the axial force in a bar, we cut the bar and apply a pair of equal and opposite axial forces so as to cause a *unit relative axial displacement* for the two cut ends. The deflected structure will give the desired influence line.

Figure 11-7 serves as a simple illustration of obtaining the influence line for the bar force in *bC* of the indeterminate truss. The vertical components of the

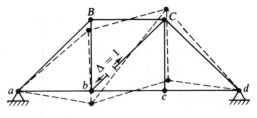

Fig. 11-7

panel point deflections are thus the influence ordinates for the vertical panel loads.

For highly indeterminate continuous beams or rigid frames the technique of sketching qualitative influence lines, based on the Müller–Breslau principle, is extremely useful in determining the loading patterns for design. Figure 11-8 shows typical influence lines for a five-span continuous beam.

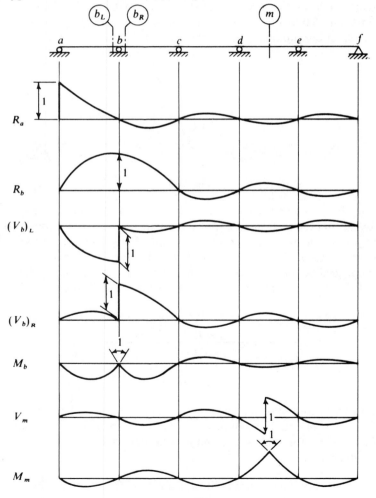

Fig. 11-8

The sketches in Fig. 11-8 indicate that if a maximum $R_a$ is desired, then spans *ab*, *cd*, and *ef* should be loaded; if the maximum values of reaction, of shear, and of bending moment at *b* are desired, then spans *ab*, *bc*, and *de* should be loaded.

Figure 11-9(a) shows the influence line for the positive moment at the mid-span section of *A3-B3* of the frame shown. The uniform loading pattern for obtaining the maximum positive moment of this section is shown in Fig. 11-9(b).

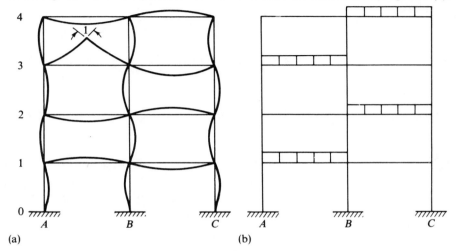

(a)                                                    (b)

**Fig. 11-9**

## 11-5. NUMERICAL EXAMPLES

**Example 11-1.** Construct by the Müller–Breslau principle the influence line for the reaction at support *b* of Fig. 11-10(a).

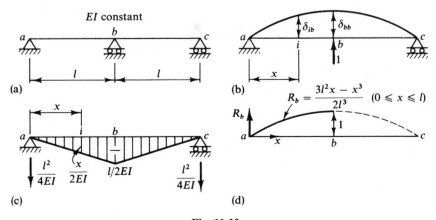

**Fig. 11-10**

We begin by removing the support at $b$ and placing a unit load along the line of reaction, as shown in Fig. 11-10(b). The ordinates of the resulting deflection curve, in Fig. 11-10(b), divided by $\delta_{bb}$ give the corresponding influence ordinates for the reaction at $b$, called $R_b$.

Probably the easiest method for computing the ordinate of the curve in Fig. 11-10(b) at any point $i$ a distance $x$ from the left end, is the conjugate-beam method, shown in Fig. 11-10(c).

Thus,

$$\delta_{ib} = \left(\frac{l^2}{4EI}\right)(x) - \left(\frac{x}{2EI}\right)\left(\frac{x}{2}\right)\left(\frac{x}{3}\right) = \frac{3l^2x - x^3}{12EI} \qquad (0 \le x \le l)$$

Substituting $x = l$ in the expression above gives

$$\delta_{bb} = \frac{l^3}{6EI}$$

The influence ordinate for any point $i$ $(0 \le x \le l)$ is governed by the equation

$$R_b = \frac{\delta_{ib}}{\delta_{bb}} = \frac{3l^2x - x^3}{2l^3}$$

as shown in Fig. 11-10(d). Because of symmetry about the middle support, we can accomplish the other half, as shown by the dashed line in Fig. 11-10(d).

**Example 11-2.** The influence lines for other elements may easily be deduced from the influence line of $R_b$, previously obtained. For instance, let us find the influence line for the moment at $b$, called $M_b$, for the beam in Fig. 11-10(a) by using the results of Example 11-1.

We observe that the influence line for $M_b$, which will be symmetrical about the middle support, may be obtained by solving the reaction at $a$ and then taking the moment about $b$. For a unit load placed in the left span, as shown in Fig. 11-11(a), by taking

(a)　　　　　　　　　　　　　　　　　　　(b)

$$R_b = \frac{3l^2x - x^3}{2l^3} \qquad (0 \le x \le l)$$

**Fig. 11-11**

$$\sum M_c = 0 \qquad 2R_a l - (1)(2l - x) + \left(\frac{3l^2x - x^3}{2l^3}\right)(l) = 0$$

we obtain $$R_a = \frac{4l^3 - 5l^2x + x^3}{4l^3} \qquad (0 \le x \le l)$$

from which

$$M_b = R_a l - (1)(l - x) = -\frac{x(l^2 - x^2)}{4l^2} \qquad (0 \le x \le l)$$

This is plotted in Fig. 11-11(b); the other half is indicated by the dashed line.

**Example 11-3.** Without reference to Example 11-1 or 11-2, compute the ordinates at 2-ft intervals of the influence line for the moment at the midspan section $d$ of $ab$ for the beam shown in Fig. 11-12(a).

The ability to resist moment at section $d$ is first destroyed by inserting a pin.

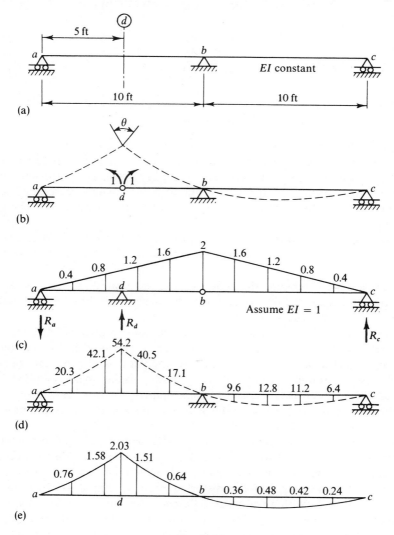

(a)

(b)

(c)

(d)

(e)

**Fig. 11-12**

Unit couples are applied on each side of the pin to produce certain relative rotation, denoted by $\theta$, between the two sides, as shown in Fig. 11-12(b). The conjugate beam and elastic load are then obtained, as shown in Fig. 11-12(c). Note that we assume $EI = 1$ ($EI$ will be canceled out in the final stage of calculation); hence, the elastic load of Fig. 11-12(c) is the moment diagram based on Fig. 11-12(b).

Referring to Fig. 11-12(c), we may solve reactions $R_a$, $R_c$, and $R_d$ by the equilibrium equations and the condition equation $M_b = 0$. Thus,

$$\sum M_a = 0 \qquad 20R_c + 5R_d - \frac{(2)(20)}{2}(10) = 0 \qquad (11\text{-}8)$$

$$\sum M_c = 0 \qquad 15R_d - 20R_a - \frac{(2)(20)}{2}(10) = 0 \qquad (11\text{-}9)$$

$$M_b = 0 \qquad 10R_c - \frac{(2)(10)}{2}\left(\frac{10}{3}\right) = 0 \qquad (11\text{-}10)$$

Solving Eqs. 11-8, 11-9, and 11-10 simultaneously, we obtain

$$R_a = 10 \qquad R_c = 3.33 \qquad R_d = 26.67$$

Note that $R_d$ is the shear difference between the left and right sides at section $d$ of the conjugate beam and thus equals the relative rotation $\theta$ between corresponding portions of the distorted beam in Fig. 11-12(b). The various moments of the conjugate beam at 2-ft intervals and at point $d$ which correspond to the deflections of the distorted beam in Fig. 11-12(b) are computed as shown in Fig. 11-12(d).

These values divided by 26.67 (so as to make $\theta = 1$) will give the ordinates of the influence line for the moment at section $d$, as shown in Fig. 11-12(e).

**Example 11-4.** Compute the influence ordinates at 2-ft intervals for the shear at the midspan section $d$ of $ab$ for the beam shown in Fig. 11-13(a) without reference to Example 11-1 or 11-2.

We start by removing the shearing resistance at section $d$ without impairing the capacity for resistance to moment. This can be accomplished by cutting the beam and inserting a slide device. Next, we apply a pair of equal and opposite unit forces to produce certain relative vertical displacement, denoted by $s$, between the two cut ends without causing relative rotation, as shown in Fig. 11-13(b). Also indicated in Fig. 11-13(b) are the induced reactions and moments at the cut ends required by equilibrium.

The conjugate beam together with the elastic loads is shown in Fig. 11-13(c). Attention should be paid to the elastic load $M_d$ acting at $d$. This is necessary since the relative vertical displacement (without relative rotation) between the two cut ends at section $d$ of the distorted beam in Fig. 11-13(b) requires a moment difference (without a shear difference) for the two sides at $d$ of the conjugate beam shown in Fig. 11-13(c). This can be fulfilled only by applying a moment at $d$ for the conjugate beam. Referring to Fig. 11-13(c), we solve

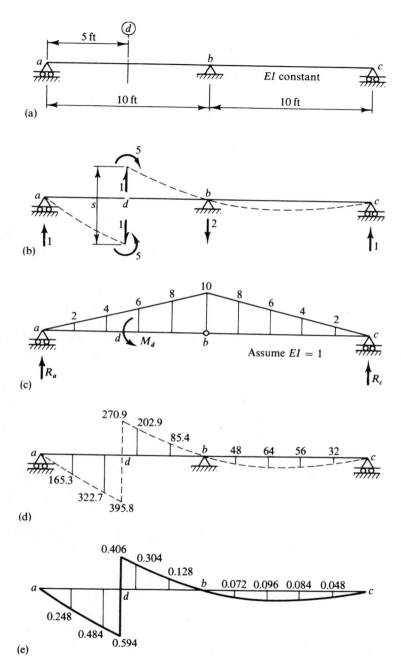

5 ft

d

a          b          c

EI constant

10 ft          10 ft

(a)

5

1

a     d          b

s

1

1          2          1

5

(b)

2     4     6     8     10     8     6     4     2

a          d     M_d          b          c

R_a          Assume EI = 1          R_c

(c)

270.9     202.9     85.4

a          d          b     48     64     56     32     c

165.3

322.7

395.8

(d)

0.406     0.304     0.128

a          d          b     0.072  0.096  0.084  0.048     c

0.248

0.484

0.594

(e)

Fig. 11-13

225

reactions $R_a$, $R_c$, and $M_d$ by the equilibrium equations and the condition equation $M_b = 0$. Thus,

$$\sum M_a = 0 \qquad 20R_c + M_d - \frac{(10)(20)}{2}(10) = 0 \qquad (11\text{-}11)$$

$$\sum M_c = 0 \qquad 20R_a - M_d - \frac{(10)(20)}{2}(10) = 0 \qquad (11\text{-}12)$$

$$M_b = 0 \qquad 10R_c - \frac{(10)(10)}{2}\left(\frac{10}{3}\right) = 0 \qquad (11\text{-}13)$$

Solving Eqs. 11-11, 11-12, and 11-13 simultaneously, we obtain

$$R_a = 83.33 \qquad R_c = 16.67 \qquad M_d = 666.67$$

Note that $M_d$ is the moment difference between the left and right sides at section $d$ of the conjugate beam and, therefore, equals the relative deflection $s$ between the corresponding portions of the distorted beam shown in Fig. 11-13(b). The various moments of the conjugate beam at 2-ft intervals and at $d$ are then computed, as shown in Fig. 11-13(d).

These values divided by $M_d = 666.67$ (so as to make $s = 1$) will give the ordinates of the shear influence line for section $d$, as shown in Fig. 11-13(e).

**Example 11-5.** Construct the influence line for the reaction at support $c$ of the truss shown in Fig. 11-14(a). Assume that $E = $ a constant and $L(\text{ft})/A(\text{in.}^2) = 1$ for all members.

We remove the support at $c$ and introduce in its place a unit force, as shown in Fig. 11-14(b). The vertical deflections of all the lower-chord panel points of the distorted truss are then computed as given in Table 11-1, in which $u_b$ is the bar force in any member due to a unit load acting upward at joint $b$, and

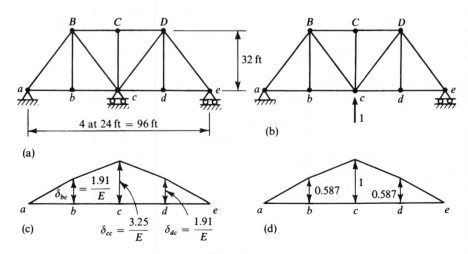

**Fig. 11-14**

TABLE 11-1

| Member | $\dfrac{L}{A}$ | $u_b$ | $u_c$ | $\dfrac{u_b u_c L}{A}$ | $\dfrac{u_c^2 L}{A}$ |
|--------|------|-------|-------|----------------|----------------|
| ab | 1 | $-9/16$ | $-3/8$ | $+27/128$ | $+9/64$ |
| bc | 1 | $-9/16$ | $-3/8$ | $+27/128$ | $+9/64$ |
| cd | 1 | $-3/16$ | $-3/8$ | $+9/128$ | $+9/64$ |
| de | 1 | $-3/16$ | $-3/8$ | $+9/128$ | $+9/64$ |
| BC | 1 | $+3/8$ | $+3/4$ | $+36/128$ | $+36/64$ |
| CD | 1 | $+3/8$ | $+3/4$ | $+36/128$ | $+36/64$ |
| aB | 1 | $+15/16$ | $+5/8$ | $+75/128$ | $+25/64$ |
| Bb | 1 | $-1$ | $0$ | $0$ | $0$ |
| Bc | 1 | $+5/16$ | $-5/8$ | $-25/128$ | $+25/64$ |
| Cc | 1 | $0$ | $0$ | $0$ | $0$ |
| cD | 1 | $-5/16$ | $-5/8$ | $+25/128$ | $+25/64$ |
| Dd | 1 | $0$ | $0$ | $0$ | $0$ |
| De | 1 | $+5/16$ | $+5/8$ | $+25/128$ | $+25/64$ |
|  |  |  | $\Sigma$ | $+1.91$ | $+3.25$ |

$u_c$ is the bar force in the same member due to a unit load acting upward at joint $c$. The results are shown in Fig. 11-14(c). Note that in this particular case $\delta_{bc} = \delta_{dc}$.

$$\delta_{bc} = \Sigma \frac{u_b u_c L}{AE} = \frac{1.91}{E}$$

$$\delta_{cc} = \Sigma \frac{u_c^2 L}{AE} = \frac{3.25}{E}$$

The ordinates of the deflection curve in Fig. 11-14(c) divided by the value of $\delta_{cc}$ will give the influence ordinates for the reaction at support $c$, as shown in Fig. 11-14(d).

**Example 11-6.** Find the influence line for the bar force in member $BC$ of the truss in Fig. 11-14(a).

We can accomplish this easily if we use the results of the preceding example. As a unit load is placed at panel point $b$ [see Fig. 11-15(a)], we have

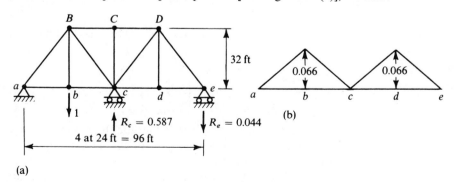

(a)

(b)

**Fig. 11-15**

227

$$R_c = 0.587 \quad \text{(up)}$$

From $\sum M_a = 0$, $\qquad R_e = 0.044 \quad \text{(down)}$

By taking the section and applying $\sum M_c = 0$, we obtain the bar force in $BC$ as

$$S_{BC} = \frac{(0.044)(48)}{32} = +0.066 \quad \text{(tension)}$$

Because of symmetry, $S_{BC} = +0.066$ when the unit load is placed at panel point $d$. Note also that $S_{BC} = 0$ when the unit load is at points $a$, $c$, and $e$. We thus complete the influence line for $S_{BC}$, as shown in Fig. 11-15(b).

## 11-6. SOLVING INFLUENCE ORDINATES FOR STATICALLY INDETERMINATE STRUCTURES BY CONSISTENT DEFORMATIONS

For highly indeterminate structures the influence ordinates for various functions may be found by using a computer for a number of equations based on consistent deformations. Let us consider the four-span continuous beam shown in Fig. 11-16(a). Find the influence line for the reaction at support $b$, called $R_b$.

We start by removing support $b$ and applying to it a force $X_b$ along the line of reaction so as to produce a unit displacement at $b$. Then, by the Müller–Breslau principle the elastic line of the distorted beam, shown by the dashed line in Fig. 11-16(b), will be the influence line for $R_b$. Also indicated in Fig. 11-16(b) are the induced reactions at supports $c$ and $d$, called $X_c$ and $X_d$, respectively, and the ordinate of the curve at any point $i$, called $\Delta_i$, a distance $x$ from the left end.

To obtain the value of $\Delta_i$ by the method of consistent deformations, we regard the indeterminate beam in Fig. 11-16(b) as a simple beam $ae$ (primary structure) subjected to forces $X_b$, $X_c$, and $X_d$, the effects of which can be separated by the principle of superposition into three basic cases, as shown in Fig. 11-16(c), (d), and (e), respectively. Thus,

$$\Delta_i = \delta_{ib}X_b + \delta_{ic}X_c + \delta_{id}X_d$$

in which the unknowns $X_b$, $X_c$, and $X_d$ can be solved by the compatibility condition

$$\begin{Bmatrix} \Delta_b \\ \Delta_c \\ \Delta_d \end{Bmatrix} = \begin{bmatrix} \delta_{bb} & \delta_{bc} & \delta_{bd} \\ \delta_{cb} & \delta_{cc} & \delta_{cd} \\ \delta_{db} & \delta_{dc} & \delta_{dd} \end{bmatrix} \begin{Bmatrix} X_b \\ X_c \\ X_d \end{Bmatrix} = \begin{Bmatrix} 1 \\ 0 \\ 0 \end{Bmatrix} \qquad (11\text{-}14)$$

In a similar manner, we can find the influence line for the reaction at each of the other supports. This done, the influence lines for the moment and shear at various points can be deduced from them.

Without the aid of a calculating machine, the Müller–Breslau procedures for quantitative influence lines are often tedious, especially if an indeterminate

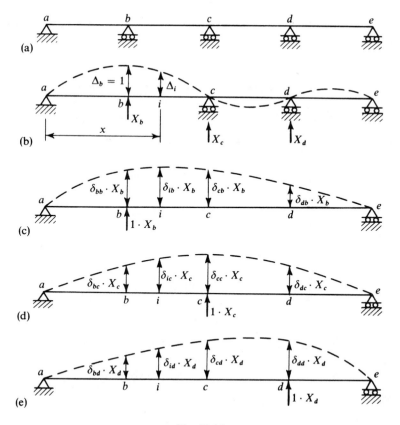

**Fig. 11-16**

structure remains after establishing the initial displacement conditions along the given redundant. A much faster way by hand calculation is the moment-distribution procedure, which we discuss in Sec. 14-6.

## PROBLEMS

**11-1.** Compute the ordinates at 2-m intervals of the influence line for the reaction at $a$ of the beam shown in Fig. 11-17.

**Fig. 11-17**

**11-2.** Compute the influence ordinates at 2-m intervals for the end moment at *a* of the beam shown in Fig. 11-18.

**Fig. 11-18**

**11-3.** Refer to the beam in Fig. 11-19. Compute the influence ordinates at 2-m intervals for the following elements: (a) the reaction at support *a*, (b) the moment at *b*, and (c) the shear at the midspan section of *ab*.

**Fig. 11-19**

**11-4.** Sketch the influence lines for $R_a$, $R_c$, $M_c$, $V_c$ (left), $V_c$ (right), and $M$ and $V$ for the midspan section of *bc* of the beam shown in Fig. 11-20.

**Fig. 11-20**

**11-5.** (a) Assume $L/A$ to be constant for all bars. Prepare an influence line for the bar force in *cd* of the truss shown in Fig. 11-21. (b) Use the results of (a) to prepare the influence line for the bar force in *BC*.

**Fig. 11-21**

**11-6.** Sketch the influence lines for the shear and moment in the midspan section of member *ab* of the rigid frame shown in Fig. 11-22.

**11-7.** Sketch the influence line for moment at end *b* of member *ab* in Fig. 11-22.

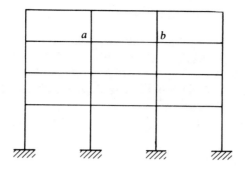

**Fig. 11-22**

# 12

## ANALYSIS OF STATICALLY
## INDETERMINATE BEAMS AND RIGID
## FRAMES BY THE SLOPE-DEFLECTION METHOD

### 12-1. GENERAL

Throughout the previous chapters the methods of analyzing statically indeterminate structures have used forces (reactions or internal forces) as the basic unknowns. These are often referred to as the force or action methods. Displacements, however, may be used equally well as unknowns. Methods using displacements as the basic unknowns are called displacement methods. One of the important displacement methods is the *method of slope deflection*, based on determining the rotations and deflections of various joints from which the end moments for each member are found.

The slope-deflection method may be used in analyzing all types of statically indeterminate beams and rigid frames composed of prismatic or nonprismatic members. However, in this chapter we discuss exclusively beams and rigid frames made of prismatic members.

### 12-2. BASIC SLOPE-DEFLECTION EQUATIONS

The basis of the slope-deflection method lies in the *slope-deflection equations*, which express the end moments of each member in terms of the end distortions of that member.

Consider member *ab* shown in Fig. 12-1, which is isolated from a loaded statically indeterminate beam or rigid frame (not shown). The member is deformed (see the dashed line) with end rotations $\theta_a$ and $\theta_b$ and relative deflection $\Delta$ between the ends. Obviously, the induced end moments at *a* and *b*, called $M_{ab}$ and $M_{ba}$, are related to the elastic distortions at both ends as well as to the load on span *ab*, if any. Thus,

$$M_{ab} = f(\theta_a, \theta_b, \Delta, \text{load on span}) \tag{12-1}$$

$$M_{ba} = g(\theta_a, \theta_b, \Delta, \text{load on span}) \tag{12-2}$$

where $f$ and $g$ are symbols for functions.

**Fig. 12-1**

To find the expressions of Eqs. 12-1 and 12-2, let us first establish the following sign convention for slope deflection:

1. The moment acting on the end of a member (not joint) is positive when clockwise.
2. The rotation at the end of a member is positive when the tangent to the deformed curve at the end rotates clockwise from its original position.
3. The relative deflection between ends of a member is positive when it corresponds to a clockwise rotation of the member (the straight line joining the ends of the elastic curve).

All signs of end distortions and moments shown in Fig. 12-1 are positive. The sign conventions established here are purely arbitrary and could be replaced by any other convenient system; but once these conventions have been adopted, we will restrict ourselves to this system.

Next, let us refer to Fig. 12-1 and observe that the end moments $M_{ab}$ and $M_{ba}$ may be considered as the algebraic sum of four separate effects:

1. The moment due to end rotation $\theta_a$ while the other end $b$ is fixed
2. The moment due to end rotation $\theta_b$ while end $a$ is fixed
3. The moment due to a relative deflection $\Delta$ between the ends of the member without altering the existing slopes of tangents at the ends
4. The moment caused by placing the actual loads on the span without altering the existing end distortions

In each of these cases the corresponding end moments can be evaluated as follows:

1. Consider member $ab$ supported as shown in Fig. 12-2(a). The dashed lines indicate the deformed shape. Note that end $a$ is rotated through an angle $\theta_a$, whereas end $b$ is fixed ($\theta_b = 0$); there is no relative end displacement between $a$ and $b$ ($\Delta = 0$). The corresponding end moments at $a$ and $b$, denoted by $M'_{ab}$ and $M'_{ba}$, respectively, can best be found by the method of conjugate beam, which is shown in Fig. 12-2(b) with the moment diagram divided by $EI$ as its elastic loads and $\theta_a$ as its reaction so that the positive shear in the conjugate

233

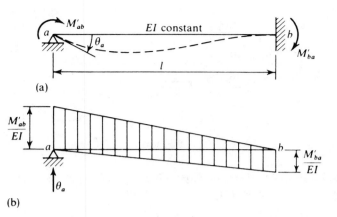

(a)

(b)

Fig. 12-2

beam gives the desired positive slope in actual beam. From equilibrium conditions:

$$\sum M_a = 0 \qquad \left(\frac{M'_{ab}l}{2EI}\right)\left(\frac{l}{3}\right) - \left(\frac{M'_{ba}l}{2EI}\right)\left(\frac{2l}{3}\right) = 0 \qquad (12\text{-}3)$$

$$\sum M_b = 0 \qquad (\theta_a l) - \left(\frac{M'_{ab}l}{2EI}\right)\left(\frac{2l}{3}\right) + \left(\frac{M'_{ba}l}{2EI}\right)\left(\frac{l}{3}\right) = 0 \qquad (12\text{-}4)$$

From Eq. 12-3,

$$M'_{ba} = \tfrac{1}{2}M'_{ab} \qquad (12\text{-}5)$$

Substituting Eq. 12-5 in Eq. 12-4 gives

$$M'_{ab} = \frac{4EI\theta_a}{l} \qquad (12\text{-}6)$$

Thus,

$$M'_{ba} = \frac{2EI\theta_a}{l} \qquad (12\text{-}7)$$

2. Consider member $ab$ supported and deformed as shown in Fig. 12-3 where end $b$ is rotated through an angle $\theta_b$ and end $a$ is fixed. The corresponding end moment at $b$, called $M''_{ba}$, and the moment at $a$, called $M''_{ab}$, are obtained similarly:

$$M''_{ab} = \tfrac{1}{2}M''_{ba} \qquad (12\text{-}8)$$

$$M''_{ab} = \frac{2EI\theta_b}{l} \qquad (12\text{-}9)$$

Fig. 12-3

$$M''_{ba} = \frac{4EI\theta_b}{l} \tag{12-10}$$

3. To find the moments developed at the ends as the result of a pure relative deflection $\Delta$ between the two ends without causing end rotations, let us consider the fixed-end beam in Fig. 12-4(a). Because of the symmetry of the deformation with reference to the midpoint of the member [see the dashed lines in Fig. 12-4(a)], the two end moments must be equal. Thus, if we let $M'''_{ab}$ and $M'''_{ba}$ be the end moments at $a$ and $b$, respectively, we have

$$M'''_{ab} = M'''_{ba} = -M$$

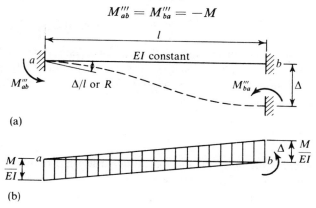

(a)

(b)

**Fig. 12-4**

The negative sign indicates that $M'''_{ab}$ and $M'''_{ba}$ are counterclockwise. The value of $M$ may be found by the method of conjugate beam, as shown in Fig. 12-4(b). Note that, besides the distributed elastic loads of the $M/EI$ diagram, a couple acts at end $b$ equal to $\Delta$ corresponding to the deflection at $b$ of the base structure. From $\sum M = 0$, we have

$$\left(\frac{Ml}{2EI}\right)\left(\frac{l}{3}\right) - \Delta = 0$$

or

$$M = \frac{6EI\Delta}{l^2}$$

Thus, the moments developed at the ends of a member due to a pure relative end displacement are given by

$$M'''_{ab} = M'''_{ba} = -\frac{6EI\Delta}{l^2} \tag{12-11}$$

4. Finally, the moments induced at the ends of a member without causing end distortions when the external loads are placed on the span are nothing more than the fixed-end moments, usually denoted by $M^F_{ab}$ and $M^F_{ba}$.

Summing up the four elements listed above, we have

$$M_{ab} = M'_{ab} + M''_{ab} + M'''_{ab} \pm M^F_{ab}$$
$$M_{ba} = M'_{ba} + M''_{ba} + M'''_{ba} \pm M^F_{ba}$$

Using Eqs. 12-6, 12-7, and 12-9 to 12-11, we find that

$$M_{ab} = \frac{4EI\theta_a}{l} + \frac{2EI\theta_b}{l} - \frac{6EI\Delta}{l^2} \pm M_{ab}^F$$

$$M_{ba} = \frac{2EI\theta_a}{l} + \frac{4EI\theta_b}{l} - \frac{6EI\Delta}{l^2} \pm M_{ba}^F$$

Rearranging gives

$$M_{ab} = 2E\frac{I}{l}\left(2\theta_a + \theta_b - 3\frac{\Delta}{l}\right) \pm M_{ab}^F \tag{12-12}$$

$$M_{ba} = 2E\frac{I}{l}\left(2\theta_b + \theta_a - 3\frac{\Delta}{l}\right) \pm M_{ba}^F \tag{12-13}$$

which are the basic equations of slope deflection for a general deformed member of uniform cross section. The equations express end moment $M_{ab}$ and $M_{ba}$ in terms of the end slopes $(\theta_a, \theta_b)$, the relative deflection between the two ends ($\Delta$), and the loading on the span $ab$.

If we let

$$\frac{I}{l} = K \qquad \frac{\Delta}{l} = R$$

$K$ being the *stiffness factor of the member* and $R$ the rigid-body *rotation of the member* [see Fig. 12-4(a)], the equations become

$$M_{ab} = 2EK(2\theta_a + \theta_b - 3R) \pm M_{ab}^F \tag{12-14}$$

$$M_{ba} = 2EK(2\theta_b + \theta_a - 3R) \pm M_{ba}^F \tag{12-15}$$

The signs and values of $M_{ab}^F$ and $M_{ba}^F$ depend on the loading condition on span $ab$. If the member $ab$ carries no load itself, then $M_{ab}^F = M_{ba}^F = 0$. See Secs. 9-2 and 10-2 for the evaluation of the fixed-end moments. For reference, the fixed-end moments for a straight member of constant $EI$ due to the common types of loading are given in Table 12-1.


## 12-3. PROCEDURE OF ANALYSIS BY
## THE SLOPE-DEFLECTION METHOD

The slope-deflection method consists of writing a series of slope-deflection equations expressing the end moments for all members in terms of the slope (rotation) and the deflection (relative translation) of various joints, or of quantities proportional to them, and of solving these unknown displacements by a number of equilibrium equations, which these end moments must satisfy. Once the displacements have been determined, the end moments for each member may be figured back. The solution thus obtained is unique, since it satisfies the equilibrium equations and end conditions (compatibility conditions) embedded in the slope-deflection equations.

To illustrate, let us consider the frame in Fig. 12-5(a). First we draw the free-body diagrams for all members as shown in Fig. 12-5(b), where the un-

TABLE 12-1

| $M_{ab}^F$ | Loading Case | $M_{ba}^F$ |
|---|---|---|
| $-\dfrac{Pl}{8}$ | | $+\dfrac{Pl}{8}$ |
| $-\dfrac{Pcd^2}{l^2}$ | | $+\dfrac{Pc^2d}{l^2}$ |
| $-\alpha(1-\alpha)Pl$ | | $+\alpha(1-\alpha)Pl$ |
| $-\dfrac{wl^2}{12}$ | $w$ per unit length downward | $+\dfrac{wl^2}{12}$ |
| $-\dfrac{wl^2}{20}$ | | $+\dfrac{wl^2}{30}$ |
| $-\dfrac{11wl^2}{192}$ | | $+\dfrac{5wl^2}{192}$ |
| $-\dfrac{wl^2}{15}$ | Parabolic loading | $+\dfrac{wl^2}{15}$ |
| $+\dfrac{Md(2c-d)}{l^2}$ | | $+\dfrac{Mc(2d-c)}{l^2}$ |

known end moments for each member are assumed positive (i.e., acting clockwise according to our sign convention).

Next, we observe that ends $a$ and $d$ of the frame are fixed and will undergo no rotation ($\theta_a = \theta_d = 0$) or linear displacement. Joint $b$, owing to the restriction of length $ab$ (we neglect the small change in length in $ab$ due to the axial forces) and the support $a$, cannot move otherwise but rotates about $a$. However, since the deformations of the frame are extremely small as compared to the length, we may replace arc length with tangent length without appreciable error. With joint $b$, and therefore $c$, moving a horizontal distance $\Delta$ to the right [see Fig. 12-5(a)], there is a relative deflection $\Delta$ between joints $a$ and $b$

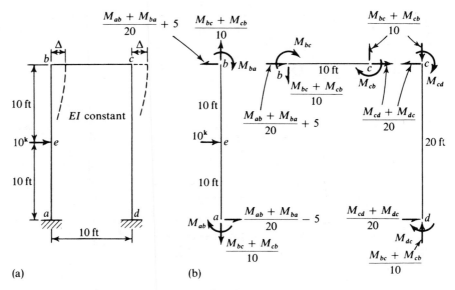

**Fig. 12-5**

and also joints $c$ and $d$. There is no relative deflection between joints $b$ and $c$ if the small lengthening or shortening in $ab$ or $cd$, caused primarily by axial forces, are neglected. There are some joint rotations at $b$ and $c$. Attention should be paid to the fact that, when a rigid frame is deformed, each rigid joint is considered to rotate as a whole. For instance, members $ba$ and $bc$ rotate the same angle $\theta_b$ at joint $b$. Similarly, members $cb$ and $cd$ rotate the same angle $\theta_c$ at joint $c$.

To determine the end moments for each member shown in Fig. 12-5(b), we write a series of slope-deflection equations as follows:

$$M_{ab} = 2EK_{ab}(\theta_b - 3R) - M_{ab}^F = 2E\left(\frac{I}{20}\right)(\theta_b - 3R) - 25$$

$$M_{ba} = 2EK_{ab}(2\theta_b - 3R) + M_{ba}^F = 2E\left(\frac{I}{20}\right)(2\theta_b - 3R) + 25$$

$$M_{bc} = 2EK_{bc}(2\theta_b + \theta_c) = 2E\left(\frac{I}{10}\right)(2\theta_b + \theta_c)$$

$$M_{cb} = 2EK_{bc}(2\theta_c + \theta_b) = 2E\left(\frac{I}{10}\right)(2\theta_c + \theta_b) \qquad (12\text{-}16)$$

$$M_{cd} = 2EK_{cd}(2\theta_c - 3R) = 2E\left(\frac{I}{20}\right)(2\theta_c - 3R)$$

$$M_{dc} = 2EK_{cd}(\theta_c - 3R) = 2E\left(\frac{I}{20}\right)(\theta_c - 3R)$$

Involved in these expressions are three unknowns: $\theta_b$, $\theta_c$, and $R$ (or $\Delta/20$). These can be solved by the three equations of statics the end moments must satisfy.

238

By taking joints $b$ and $c$ as free bodies, we obtain immediately two equilibrium equations:

$$\Sigma M_{\text{joint } b} = 0 \quad \text{or} \quad M_{ba} + M_{bc} = 0 \tag{12-17}$$

$$\Sigma M_{\text{joint } c} = 0 \quad \text{or} \quad M_{cb} + M_{cd} = 0 \tag{12-18}$$

Usually, we have as many joint equilibrium equations as the number of joint displacements unknowns involved. However, with the member rotation $R$ unknown, a third equation may be more conveniently secured from the equilibrium of the structure. Referring to Fig. 12-5, by taking the whole frame as free body, we see that the horizontal shear in ends $a$ and $d$ must balance the horizontal external force acting on the frame. Thus,

$$10 + \left(\frac{M_{ab} + M_{ba}}{20} - 5\right) + \left(\frac{M_{cd} + M_{dc}}{20}\right) = 0$$

or

$$M_{ab} + M_{ba} + M_{cd} + M_{dc} + 100 = 0 \tag{12-19}$$

Before we try to substitute the expressions of Eq. 12-16 in Eqs. 12-17, 12-18, and 12-19, we should note that if our purpose is to determine the end moments but not to obtain the exact values of the slope and deflection of each joint, then we may substitute the relative values for the coefficients $2EI/l$, usually some simple integers, in the slope-deflection equations in order to facilitate the calculation. This can be done because such a substitution will only magnify the values of $\theta$ and $\Delta$ but will not affect the final result of the end moments. Thus, if we set

$$2E\frac{I}{20} = 1$$

accordingly,

$$2E\left(\frac{I}{10}\right) = 2$$

and the moment expressions of Eq. 12-16 become

$$\begin{aligned}
M_{ab} &= \theta_b - 3R - 25 \\
M_{ba} &= 2\theta_b - 3R + 25 \\
M_{bc} &= 2(2\theta_b + \theta_c) \\
M_{cb} &= 2(2\theta_c + \theta_b) \\
M_{cd} &= 2\theta_c - 3R \\
M_{dc} &= \theta_c - 3R
\end{aligned} \tag{12-20}$$

Substituting these in Eqs. 12-17, 12-18, and 12-19 yields

$$6\theta_b + 2\theta_c - 3R + 25 = 0 \tag{12-21}$$

$$2\theta_b + 6\theta_c - 3R = 0 \tag{12-22}$$

$$3\theta_b + 3\theta_c - 12R + 100 = 0 \tag{12-23}$$

Solving Eqs. 12-21, 12-22, and 12-23 simultaneously, we obtain

$$\theta_b = -1.20 \qquad \theta_c = 5.05 \qquad R = 9.30$$

Note that the values thus obtained are only the relative values of the slope and the deflection for the various joints. They must be divided by the factor $2EI/20$ to give the absolute values of the slope and deflection.

To determine the end moments for each member of the frame, substitute $\theta_b = -1.20$, $\theta_c = 5.05$, $R = 9.30$ in Eq. 12-20 and obtain

$$M_{ab} = -54.10 \text{ ft-kips} \qquad \text{(counterclockwise)}$$
$$M_{ba} = -5.30 \text{ ft-kips} \qquad \text{(counterclockwise)}$$
$$M_{bc} = +5.30 \text{ ft-kips} \qquad \text{(clockwise)}$$
$$M_{cb} = +17.80 \text{ ft-kips} \qquad \text{(clockwise)}$$
$$M_{cd} = -17.80 \text{ ft-kips} \qquad \text{(counterclockwise)}$$
$$M_{dc} = -22.80 \text{ ft-kips} \qquad \text{(counterclockwise)}$$

The answer diagram for the end actions for each member of the frame is shown in Fig. 12-6(a), wihch is based on Fig. 12-5(b).

**Fig. 12-6**

The moment diagram for the frame is drawn as in Fig. 12-6(b). The moment is plotted on the compressive side of each member. In this particular case each member has one point of inflection corresponding to the point of zero moment.

Finally, we can sketch the elastic curve of the deformed structure, as shown in Fig. 12-6(c), by using the values (or relative values) of the joint rotations and deflections together with the bending moment diagram. Note particularly the following:

1. The elastic curve of the deformed frame bends according to the bending moment diagram.

2. Both joints $b$ and $c$ deflect to the right the same horizontal distance.

3. Joint $b$ rotates counterclockwise while joint $c$ rotates clockwise.

4. Since joints $b$ and $c$ are rigid, the tangents to the elastic curves $ba$ and $bc$ at $b$ and the tangents to the elastic curves $cb$ and $cd$ at $c$ should be perpendicular to each other so as to maintain the original formation at the joints of the unloaded frame.

## 12-4. ANALYSIS OF STATICALLY INDETERMINATE BEAMS BY THE SLOPE-DEFLECTION METHOD

The application of the slope-deflection method in solving statically indeterminate beams will be illustrated in the following examples.

**Example 12-1.** Figure 12-7(a) shows a continuous, two-section beam, all the supports of which are immovable. We wish to draw the shear and bending moment diagrams for the beam. The solution for the end moments and shears is contained in the following steps:

1. Since $2EK_{ab} = 2E(2I)/16 = EI/4$ and $2EK_{bc} = 2EI/12 = EI/6$, if we let $2EK_{ab} = 3$, then $2EK_{bc} = 2$ relatively. The relative values of $2EK$ are shown circled in Fig. 12-7(a).

2. By inspection $\theta_a = 0$ (fixed end at $a$) and $R_{ab} = R_{bc} = 0$ (immovable supports at $a$, $b$, $c$).

3. Calculate the fixed-end moments as follows:

$$M_{ab}^F = -\frac{(7.5)(16)}{8} = -15 \text{ ft-kips}$$

$$M_{ba}^F = +\frac{(7.5)(16)}{8} = +15 \text{ ft-kips}$$

$$M_{bc}^F = -\frac{(1)(12)^2}{12} = -12 \text{ ft-kips}$$

$$M_{cb}^F = +\frac{(1)(12)^2}{12} = +12 \text{ ft-kips}$$

(a)

(b)

(c)

**Fig. 12-7**

4. Write the slope-deflection equations using the relative $2EK$ values.

$$M_{ab} = (3)(\theta_b) - 15 \tag{12-24}$$

$$M_{ba} = (3)(2\theta_b) + 15 = 6\theta_b + 15 \tag{12-25}$$

$$M_{bc} = (2)(2\theta_b + \theta_c) - 12 = 4\theta_b + 2\theta_c - 12 \tag{12-26}$$

$$M_{cb} = (2)(2\theta_c + \theta_b) + 12 = 2\theta_b + 4\theta_c + 12 \tag{12-27}$$

5. Involved in the equations above are two unknowns, $\theta_b$ and $\theta_c$, that can be solved by two joint conditions:

$$\sum M_{\text{joint } b} = M_{ba} + M_{bc} = 0 \tag{12-28}$$

$$M_{cb} = (5)(3) = 15 \tag{12-29}$$

Substituting Eqs. 12-25 and 12-26 in Eq. 12-28, and Eq. 12-27 in Eq. 12-29 yields

$$10\theta_b + 2\theta_c + 3 = 0 \tag{12-30}$$

and

$$2\theta_b + 4\theta_c - 3 = 0 \tag{12-31}$$

Solving yields $\qquad \theta_b = -\tfrac{1}{2} \qquad \theta_c = 1$

6. Substituting the above values in step 4, we obtain

$$M_{ab} = -16.5 \text{ ft-kips}$$

$$M_{ba} = +12.0 \text{ ft-kips}$$

$$M_{bc} = -12.0 \text{ ft-kips}$$

$$M_{cb} = +15.0 \text{ ft-kips}$$

242

7. Having determined the end moments for each member, we can find the end shears and, therefore, the reactions.

$$R_a = V_{ab} = \frac{7.5}{2} + \frac{16.5 - 12}{16} = 4.03 \text{ kips} \qquad \text{(up)}$$

$$V_{ba} = \frac{-7.5}{2} + \frac{16.5 - 12}{16} = -3.47 \text{ kips} \qquad \text{(up)}$$

$$V_{bc} = \frac{12}{2} - \frac{15 - 12}{12} = 5.75 \text{ kips} \qquad \text{(up)}$$

$$R_b = 3.47 + 5.75 = 9.22 \text{ kips} \qquad \text{(up)}$$

$$V_{cb} = -\frac{12}{2} - \frac{15 - 12}{12} = -6.25 \text{ kips} \qquad \text{(up)}$$

$$V_{cd} = 5 \text{ kips} \qquad \text{(up)}$$

$$R_c = 6.25 + 5 = 11.25 \text{ kips} \qquad \text{(up)}$$

8. The shear and moment diagrams are now drawn, as shown in Fig. 12-7(b) and (c), respectively.

**Example 12-2.** For the system and load shown in Fig. 12-8, find a general expression for the spring force. Given $E = 20{,}000 \text{ kN/cm}^2$, $I = 5{,}000 \text{ cm}^4$, $k = 5 \text{ kN/cm}$, $w = 4 \text{ kN/m}$, and $l = 3$ m, determine the value of the spring force.

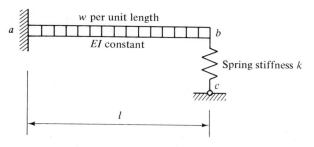

**Fig. 12-8**

Assuming that the contraction of spring at $b$ is $\Delta$, we write the moment equations as

$$M_{ab} = 2E\frac{I}{l}\left(\theta_b - 3\frac{\Delta}{l}\right) - \frac{wl^2}{12} \qquad (12\text{-}32)$$

$$M_{ba} = 2E\frac{I}{l}\left(2\theta_b - 3\frac{\Delta}{l}\right) + \frac{wl^2}{12} \qquad (12\text{-}33)$$

Using $M_{ba} = 0$, we have

$$\theta_b = \frac{3\Delta}{2l} - \frac{wl^3}{48EI} \qquad (12\text{-}34)$$

Substituting Eq. 12-34 in Eq. 12-32 gives

$$M_{ab} = -\left(\frac{3EI\Delta}{l^2} + \frac{wl^2}{8}\right) \tag{12-35}$$

Let the spring force be $X$; hence, $X = k\Delta$ or $\Delta = X/k$. Thus,

$$M_{ab} = -\left(\frac{3EIX}{kl^2} + \frac{wl^2}{8}\right) \tag{12-36}$$

Since the compressive force in the spring is equal to the end shear at $b$,

$$X = \frac{wl}{2} + \frac{M_{ab}}{l}$$

Using Eq. 12-36, we obtain

$$X = \frac{3}{8}wl\left(\frac{1}{1 + \dfrac{3EI}{kl^3}}\right) \tag{12-37}$$

A substitution of $E = 20,000$ kN/cm², $I = 5,000$ cm⁴, $k = 5$ kN/cm, $l = 3$ m $= 300$ cm, and $wl = (4)(3) = 12$ kN in Eq. 12-37 yields

$$X = \left(\frac{3}{8}\right)(12)\left(\frac{1}{1 + \dfrac{(3)(10)^8}{(5)(3)^3(10)^6}}\right) = 1.2 \text{ kN}$$

Some special cases can be derived by specifying the value of spring stiffness. They are given in Fig. 12-9. For instance, $k = 0$ means no axial resistance (free end) (i.e., $X = 0$); $k = \infty$ means an unyielding support, $X = \frac{3}{8} wl$. If the spring is replaced by an elastic link with length $L$, cross-sectional area $A$, and modulus of elasticity $E$, then we use $k = AE/L$ in Eq. 12-37.

$k = 0$ $\qquad\qquad$ $k = \infty$ $\qquad\qquad$ $k = \dfrac{AE}{L}$

**Fig. 12-9**

## 12-5. ANALYSIS OF STATICALLY INDETERMINATE
## RIGID FRAMES WITHOUT JOINT TRANSLATION
## BY THE SLOPE-DEFLECTION METHOD

Some rigid frames, such as those shown in Fig. 12-10(a)–(d), are so constructed that translations of joints are prevented. Others, although capable of joint translation in construction, will undergo no joint translation because of the symmetry of the structure and the loading about a certain axis, such as those shown in Fig. 12-10(e)–(g).

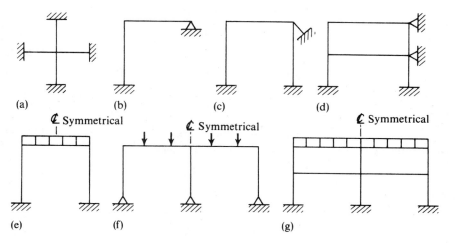

(a)  (b)  (c)  (d)

(e)  (f)  (g)

**Fig. 12-10**

In both cases

$$R = 0$$

in the equations of slope deflection so that the analysis is considerably simplified, as we shall see in the following examples.

**Example 12-3.** The end moments for the frame shown in Fig. 12-11 were solved by the methods of consistent deformation (Example 9-7) and least work (Examples 10-4 and 10-5) and will be re-solved by slope deflection.

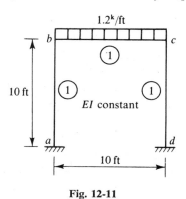

**Fig. 12-11**

The analysis is contained in the following steps:

1. The relative $2EK$ values for all members are shown encircled.
2. $\theta_a = \theta_d = R = 0$, and because of symmetry, $\theta_c = -\theta_b$.
3. $M_{bc}^F = -M_{cb}^F = -(1.2)(10)^2/12 = -10$ ft-kips.

245

4. Equations of slope deflection are then given by

$$M_{ab} = -M_{dc} = (1)(\theta_b) \tag{12-38}$$

$$M_{ba} = -M_{cd} = (1)(2\theta_b) \tag{12-39}$$

$$M_{bc} = -M_{cb} = (1)(2\theta_b - \theta_b) - 10 \tag{12-40}$$

5. There is only one unknown, $\theta_b$, involved in this analysis, and it can be solved by

$$\sum M_{\text{joint } b} = M_{ba} + M_{bc} = 0 \tag{12-41}$$

Substituting Eqs. 12-39 and 12-40 in Eq. 12-41 gives

$$3\theta_b - 10 = 0$$

from which

$$\theta_b = 3.33$$

6. Going back to step 4, we obtain

$$M_{ab} = -M_{dc} = 3.33 \text{ ft-kips}$$

$$M_{ba} = -M_{cd} = 6.67 \text{ ft-kips}$$

$$M_{bc} = -M_{cb} = -6.67 \text{ ft-kips}$$

**Example 12-4.** Analyze the frame in Fig. 12-12 if the support at $a$ yields 0.0016 radian clockwise. Assume that $EI = 10{,}000$ kips-ft².

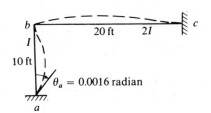

a

**Fig. 12-12**

$$M_{ab} = \frac{2EI}{10}(2\theta_a + \theta_b) = \frac{EI}{5}[(2)(0.0016) + \theta_b]$$

$$M_{ba} = \frac{2EI}{10}(2\theta_b + \theta_a) = \frac{EI}{5}(2\theta_b + 0.0016)$$

$$M_{bc} = \frac{2E(2I)}{20}(2\theta_b + \theta_c) = \frac{EI}{5}(2\theta_b) \qquad (\theta_c = 0)$$

$$M_{cb} = \frac{2E(2I)}{20}(2\theta_c + \theta_b) = \frac{EI}{5}(\theta_b)$$

The unknown $\theta_b$ is solved by

$$\sum M_{\text{joint } b} = M_{ba} + M_{bc} = 0$$

or

$$4\theta_b + 0.0016 = 0$$

from which

$$\theta_b = -0.0004$$

With $\theta_b$ determined, all end moments can be figured as

$$M_{ab} = \frac{10,000}{5}(0.0032 - 0.0004) = 5.6 \text{ ft-kips}$$

$$M_{ba} = \frac{10,000}{5}(-0.0008 + 0.0016) = 1.6 \text{ ft-kips}$$

$$M_{bc} = \frac{10,000}{5}(-0.0008) = -1.6 \text{ ft-kips}$$

$$M_{cb} = \frac{10,000}{5}(-0.0004) = -0.8 \text{ ft-kips}$$

The deformed structure is indicated by the dashed lines in Fig. 12-12.

## 12-6. ANALYSIS OF STATICALLY INDETERMINATE RIGID FRAMES WITH ONE DEGREE OF FREEDOM OF JOINT TRANSLATION BY THE SLOPE-DEFLECTION METHOD

Figure 12-13 shows several examples of rigid frames with one degree of freedom of joint translation. In each of these examples, if the translation of one joint is given or assumed, the translation of all other joints can be deduced from it.

For instance, suppose that, in the frame in Fig. 12-13(a), joint $a$ moves to

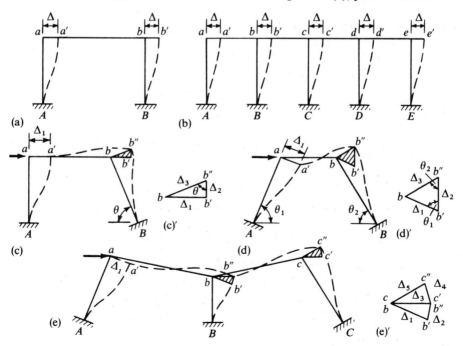

**Fig. 12-13**

$a'$, a distance $\Delta$. Since we neglect any slight change in the length of a member due to axial forces, and since the rotations of members are small, joint $a$ moves essentially perpendicular to member $Aa$, and in this case, horizontally. Similarly, joint $b$ at the top of column $Bb$ must move horizontally. Furthermore, as $b$ is the end of member $ab$ and the change of axial length of $ab$ is neglected, the horizontal movement of $b$ [see $bb'$ in Fig. 12-13(a)] must also be $\Delta$ (i.e., $aa' = bb' = \Delta$).

In the same manner, we reason that if joint $a$ in Fig. 12-13(b) moves a horizontal distance $\Delta$ to $a'$, the tops of all other columns must have the same horizontal displacement.

Let us now consider the case of Fig. 12-13(c) in which the frame is acted on by a lateral force at the top of column $Aa$. Joint $a$ cannot move other than horizontally, say a distance $\Delta_1$ to $a'$. To find the final location of joint $b$, we imagine that the frame is temporarily disconnected at $b$. Point $b$, being the end of member $ab$, will move to $b'$ the displacement $\Delta_1$, if free of other effects except that due to the movement of $a$. However, $b$ is also the end of member $Bb$; the final position of $b$, called $b''$, must be determined by two arcs that restrict the motion of $b$—one from $a'$ with a radius equal to $ab$ (or $a'b'$) and the other from $B$ with a radius equal to $Bb$. Since the deformations of the frame are very small in proportion to the length, it is permissible to substitute the tangents for the arcs, as shown in Fig. 12-13(c). The displacement diagram is shown separately in Fig. 12-13(c)$'$, for which we note that $\Delta_1$ is the relative displacement between ends of member $Aa$, $\Delta_2$ that between the ends of $ab$, and $\Delta_3$ that between the ends of $Bb$. The relationship between $\Delta_1$, $\Delta_2$, and $\Delta_3$ can be expressed by sine law:

$$\frac{\Delta_1}{\sin \theta} = \frac{\Delta_2}{\sin (90° - \theta)} = \frac{\Delta_3}{\sin 90°}$$

from which

$$\Delta_1 = \Delta_2 \tan \theta = \Delta_3 \sin \theta$$

The joint displacements of Fig. 12-13(d) are similar to that described for Fig. 12-13(c), noting that joint $a$ should move perpendicularly to member $Aa$. From the displacement diagram shown in Fig. 12-13(d)$'$ we see that

$$\frac{\Delta_1}{\sin \theta_2} = \frac{\Delta_2}{\sin (\theta_1 + \theta_2)} = \frac{\Delta_3}{\sin \theta_1}$$

The procedure described above is now extended to a two-span frame such as the one shown in Fig. 12-13(e) together with the joint-displacement diagram in Fig. 12-13(e)$'$. It may be extended to any number of spans. With the relative deflection between the ends for each member clarified, it becomes a simple matter to apply the slope-deflection equations.

**Example 12-5.** Find the end moments for each member of the portal frame shown in Fig. 12-14(a) resulting from a lateral force $P$ acting on the top of the column. Assume constant $EI$ throughout the entire frame.

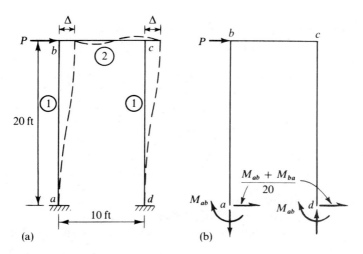

Fig. 12-14

Because of the lateral force $P$ acting at $b$, the frame deflects to the right. Both joints $b$ and $c$ move the same horizontal distance $\Delta$, as indicated. There are also rotations $\theta_b$ at joint $b$ and $\theta_c$ at joint $c$. Now, since an equal and opposite force $P$ acting at $c$ would completely balance the original force at $b$, and would thus return the structure to the original position (except some small change of length in $bc$), $\theta_b$ must be equal to $\theta_c$. Thus,

$$M_{ab} = M_{dc} = (1)(\theta_b - 3R) \qquad (\theta_a = \theta_d = 0)$$
$$M_{ba} = M_{cd} = (1)(2\theta_b - 3R)$$
$$M_{bc} = M_{cb} = (2)(2\theta_b + \theta_b) = 6\theta_b$$

This special case in which the end moments and joint rotations of one side of the center-line axis of the structure are the same as those of other side is termed *antisymmetry*, in contrast to the case of *symmetry*, in which the values of the end moments and joint rotations of one side of the center-line axis of the structure are equal but opposite to those of the other side, according to the sign convention of slope deflection.

The unknowns, $\theta_b$ and $R$, are then solved by two equilibrium equations, one for joint $b$ and the other for the entire frame.

$$\sum M_{\text{joint } b} = M_{ba} + M_{bc} = 0$$

or

$$8\theta_b - 3R = 0 \qquad (12\text{-}42)$$

By isolating the frame from the supports [Fig. 12-14(b)],

$$\sum F_x = (2)\frac{M_{ab} + M_{ba}}{20} + P = 0$$

or

$$3\theta_b - 6R + 10P = 0 \qquad (12\text{-}43)$$

249

Solving Eqs. 12-42 and 12-43 simultaneously, we obtain

$$\theta_b = \frac{10}{13}P \qquad 3R = \frac{80}{13}P$$

Substituting these in equations for end moments, we obtain

$$M_{ab} = M_{dc} = -\frac{70}{13}P$$

$$M_{ba} = M_{cd} = -\frac{60}{13}P$$

$$M_{bc} = M_{cb} = +\frac{60}{13}P$$

**Example 12-6.** Draw the bending moment diagram for the frame shown in Fig. 12-15(a). The relative values of $2EK$ are encircled.

To find the end moments, we begin by sketching the relative displacement diagram, as shown in Fig. 12-15(b). Thus,

$$R_{ab} = \frac{\Delta}{l_{ab}} = \frac{\Delta}{15}$$

$$R_{bc} = -\frac{\frac{3}{4}\Delta}{l_{bc}} = -\frac{\frac{3}{4}\Delta}{15} = -\frac{\Delta}{20}$$

$$R_{cd} = \frac{\frac{5}{4}\Delta}{l_{cd}} = \frac{\frac{5}{4}\Delta}{25} = \frac{\Delta}{20}$$

**Fig. 12-15**

If we let $R_{ab} = R$, then $R_{bc} = -3R/4$ and $R_{cd} = 3R/4$. The expressions for the various end moments are now written

$$M_{ab} = (1)(\theta_b - 3R) \qquad (\theta_a = 0)$$

$$M_{ba} = (1)(2\theta_b - 3R)$$

$$M_{bc} = (2)[2\theta_b + \theta_c + (3)(\tfrac{3}{4}R)]$$

$$M_{cb} = (2)[2\theta_c + \theta_b + (3)(\tfrac{3}{4}R)]$$

$$M_{cd} = (2)[2\theta_c - (3)(\tfrac{3}{4}R)] \qquad (\theta_d = 0)$$

$$M_{dc} = (2)[\theta_c - (3)(\tfrac{3}{4}R)]$$

Two of the three condition equations required to evaluate the three independent unknowns $\theta_b$, $\theta_c$, and $R$ are from $\sum M = 0$ for joints $b$ and $c$. Thus,

$$M_{ba} + M_{bc} = 0$$

$$6\theta_b + 2\theta_c + 1.5R = 0 \qquad \text{(12-44)}$$

$$M_{cb} + M_{cd} = 0$$

$$\theta_b + 4\theta_c = 0 \qquad \text{(12-45)}$$

The third condition equation can best be found by expressing $\sum M_o = 0$ for the entire frame, $o$ being the center of moment chosen at the intersection of the two legs [see Fig. 12-15(c)], since this eliminates the axial forces from the equation. Thus,

$$M_{ab} + M_{dc} - (100)(20) - \left(\frac{M_{ab} + M_{ba}}{15}\right)(35) - \left(\frac{M_{cd} + M_{dc}}{25}\right)(50) = 0$$

$$-6\theta_b - 10\theta_c + 24.5R - 2{,}000 = 0 \quad \text{(12-46)}$$

Solving Eqs. 12-44, 12-45, and 12-46 simultaneously, we obtain

$$\theta_b = -21.4 \qquad \theta_c = 5.36 \qquad R = 78.6$$

Substituting these values in the moment expressions yields

$$M_{ab} = -257.2 \text{ ft-kips} \qquad M_{ba} = -278.6 \text{ ft-kips}$$

$$M_{bc} = 278.6 \text{ ft-kips} \qquad M_{cb} = 332.2 \text{ ft-kips}$$

$$M_{cd} = -332.2 \text{ ft-kips} \qquad M_{dc} = -343.0 \text{ ft-kips}$$

as shown in Fig. 12-15(d).

## 12-7. ANALYSIS OF STATICALLY INDETERMINATE RIGID FRAMES WITH TWO DEGREES OF FREEDOM OF JOINT TRANSLATION BY THE SLOPE-DEFLECTION METHOD

The number of degrees of freedom for joint translation in a rigid frame equals the number of independent joint translations that can be given to the frame. Figure 12-16 shows several examples of rigid frames with two degrees of freedom of joint translation.

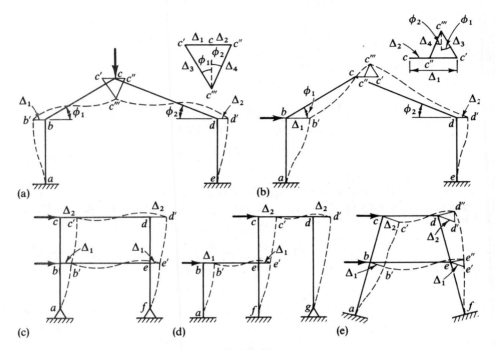

**Fig. 12-16**

In Fig. 12-16(a) is shown an unsymmetrical gable bent subjected to a vertical load at the top. Under this pressure joint $b$ will move a distance $\Delta_1$ to $b'$ and joint $d$ a distance $\Delta_2$ to $d'$, as indicated. To locate the position of $c$, let us imagine that joint $c$ is temporarily disconnected. Joint $c$, being the end of member $bc$, will move to $c'$ ($cc' = bb' = \Delta_1$). Joint $c$, being the end of member $cd$, will move to $c''$ ($c'' = dd' = \Delta_2$). The final position of $c$ is $c'''$, which is the intersection of the line perpendicular to $bc$ drawn at $c'$ and the line perpendicular to $cd$ drawn at $c''$, as indicated in Fig. 12-16(a). From the displacement diagram $\Delta_3$ (the relative displacement between ends $b$ and $c$) and $\Delta_4$ (that between $c$ and $d$) is related to $\Delta_1$ (that between $a$ and $b$) and $\Delta_2$ (that between $d$ and $e$) by the sine law,

$$\frac{\Delta_1 + \Delta_2}{\sin(\phi_1 + \phi_2)} = \frac{\Delta_3}{\sin(90° - \phi_2)} = \frac{\Delta_4}{\sin(90° - \phi_1)}$$

or

$$\frac{\Delta_1 + \Delta_2}{\sin(\phi_1 + \phi_2)} = \frac{\Delta_3}{\cos\phi_2} = \frac{\Delta_4}{\cos\phi_1}$$

so that $\Delta_3$ and $\Delta_4$ can be deduced from $\Delta_1$ and $\Delta_2$.

The case shown in Fig. 12-16(b) is similar to that of Fig. 12-16(a) except that we assume the tops of the two legs move in the same direction as the result of a lateral force applied at $b$. The relationships of joint displacements are expressed by

$$\frac{\Delta_1 - \Delta_2}{\sin(\phi_1 + \phi_2)} = \frac{\Delta_3}{\sin(90° - \phi_2)} = \frac{\Delta_4}{\sin(90° - \phi_1)}$$

252

or
$$\frac{\Delta_1 - \Delta_2}{\sin(\phi_1 + \phi_2)} = \frac{\Delta_3}{\cos \phi_2} = \frac{\Delta_4}{\cos \phi_1}$$

Figure 12-16(c) and (e) show the joint displacements in two-story frames, and Fig. 12-16(d) shows the joint displacements in a two-stage frame. In each of these cases, the arrangement of the structure is such that the joint translations of the first floor are not required to be in any fixed relationship to those of the second floor (i.e., the translations of joints $b$ and $e$ do not have fixed ratio to the translations of joints $c$ and $d$). The procedure for finding joint displacements on each of the floors is the same as that discussed in Sec. 12-6 for one-story frames. With the relative end displacement for each member consistently determined, it becomes rather easy to analyze the frame by the method of slope deflection.

**Example 12-7.** Find all the end moments for the gable bent shown in Fig. 12-17(a). Assume constant $EI$ throughout the entire frame so that the relative $2EK$ values are the encircled numbers.

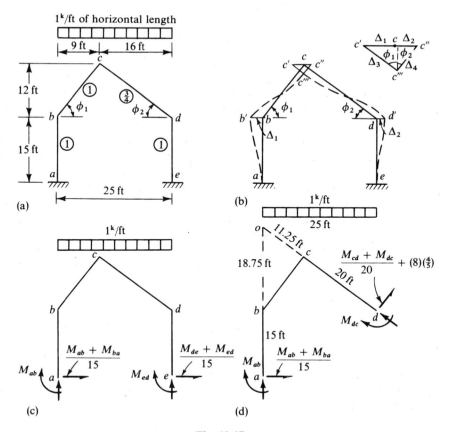

**Fig. 12-17**

We start by sketching the joint displacements of the frame as shown in Fig. 12-17(b), for which we note that

$$\Delta_3 = \frac{(\Delta_1 + \Delta_2) \sin (90° - \phi_2)}{\sin (\phi_1 + \phi_2)} = \frac{(\Delta_1 + \Delta_2) \cos \phi_2}{\sin [180° - (\phi_1 + \phi_2)]} = \frac{4}{5}(\Delta_1 + \Delta_2)$$

since, if we refer to Fig. 12-17(a), $\cos \phi_2 = \frac{4}{5}$ and $180° - (\phi_1 + \phi_2) = \sphericalangle c = 90°$ in this case. Similarly,

$$\Delta_4 = \frac{(\Delta_1 + \Delta_2) \cos \phi_1}{\sin [180° - (\phi_1 + \phi_2)]} = \frac{3}{5}(\Delta_1 + \Delta_2)$$

Thus, the rotation of each member can be expressed in terms of $\Delta_1$ and $\Delta_2$.

$$R_{ab} = -\frac{\Delta_1}{15} \qquad\qquad R_{bc} = \left(\frac{1}{15}\right)\left(\frac{4}{5}\right)(\Delta_1 + \Delta_2)$$

$$R_{cd} = -\left(\frac{1}{20}\right)\left(\frac{3}{5}\right)(\Delta_1 + \Delta_2) \qquad R_{de} = \frac{\Delta_2}{15}$$

If we let $\Delta_1/15 = R_1$ and $\Delta_2/15 = R_2$, we have

$$R_{ab} = -R_1 \qquad\qquad R_{bc} = 0.8(R_1 + R_2)$$
$$R_{cd} = -0.45(R_1 + R_2) \qquad R_{de} = R_2$$

The fixed-end moments are found to be

$$M_{bc}^F = -M_{cb}^F = -\frac{(1)(9)^2}{12} = -6.750 \text{ ft-kips}$$

$$M_{cd}^F = -M_{dc}^F = -\frac{(1)(16)^2}{12} = -21.333 \text{ ft-kips}$$

The expressions for various end moments are then written

$$M_{ab} = (1)(\theta_b + 3R_1) \qquad (\theta_a = 0)$$
$$M_{ba} = (1)(2\theta_b + 3R_1)$$
$$M_{bc} = (1)[2\theta_b + \theta_c - 2.4(R_1 + R_2)] - 6.750$$
$$M_{cb} = (1)[2\theta_c + \theta_b - 2.4(R_1 + R_2)] + 6.750$$
$$M_{cd} = (\tfrac{3}{4})[2\theta_c + \theta_d + 1.35(R_1 + R_2)] - 21.333$$
$$M_{dc} = (\tfrac{3}{4})[2\theta_d + \theta_c + 1.35(R_1 + R_2)] + 21.333$$
$$M_{de} = (1)(2\theta_d - 3R_2) \qquad (\theta_e = 0)$$
$$M_{ed} = (1)(\theta_d - 3R_2)$$

Five statical equations are needed to solve the independent unknowns $\theta_b$, $\theta_c$, $\theta_d$, $R_1$, and $R_2$—three from $\sum M = 0$ for joints $b$, $c$, and $d$; one from shear balance ($\sum F_x = 0$) for the entire frame [see Fig. 12-17(c)]; and one from $\sum M_o = 0$ for the portion of the frame shown in Fig. 12-17(d). Thus,

$$\sum M_{\text{joint } b} = M_{ba} + M_{bc} = 0$$
$$4\theta_b + \theta_c + 0.6R_1 - 2.4R_2 - 6.750 = 0 \qquad (12\text{-}47)$$

$$\sum M_{\text{joint } c} = M_{cb} + M_{cd} = 0$$

$$\theta_b + 3.50\theta_c + 0.750\theta_d - 1.387(R_1 + R_2) - 14.583 = 0 \qquad (12\text{-}48)$$

$$\sum M_{\text{joint } d} = M_{dc} + M_{de} = 0$$

$$0.750\theta_c + 3.50\theta_d + 1.013R_1 - 1.987R_2 + 21.333 = 0 \qquad (12\text{-}49)$$

$\sum F_x = 0$ for the frame of Fig. 12-17(c):

$$\frac{M_{ab} + M_{ba}}{15} + \frac{M_{de} + M_{ed}}{15} = 0$$

$$\theta_b + \theta_d + 2R_1 - 2R_2 = 0 \qquad (12\text{-}50)$$

$\sum M_o = 0$ for the portion of the frame [Fig. 12-17(d)]:

$$M_{ab} - \left(\frac{M_{ab} + M_{ba}}{15}\right)(33.75) - \left(\frac{M_{cd} + M_{dc}}{20}\right)(31.25) - (8)\left(\frac{4}{5}\right)(31.25)$$

$$+ M_{dc} + \frac{(1)(25)^2}{2} = 0$$

$$(\theta_b + 3R_1) - 2.25(3\theta_b + 6R_1) - 1.5625[2.25\theta_c + 2.25\theta_d + 2.025(R_1 + R_2)]$$
$$- 200 + [0.750\theta_c + 1.50\theta_d + 1.013(R_1 + R_2) + 21.333] + 312.5 = 0$$

or

$$5.750\theta_b + 2.766\theta_c + 2.016\theta_d + 12.651R_1 + 2.151R_2 - 133.833 = 0 \quad (12\text{-}51)$$

Solving Eqs. 12-47, 12-48, 12-49, 12-50, and 12-51 simultaneously, we obtain

$$\theta_b = 0.341 \qquad \theta_c = 11.039 \qquad \theta_d = -8.342$$
$$R_1 = 8.543 \qquad R_2 = 4.561$$

Substituting these in the expressions for the various end moments, we obtain

$$M_{ab} = 26 \text{ ft-kips}$$
$$M_{ba} = -M_{bc} = 26.3 \text{ ft-kips}$$
$$M_{cb} = -M_{cd} = -2.3 \text{ ft-kips}$$
$$M_{dc} = -M_{de} = 30.3 \text{ ft-kips}$$
$$M_{ed} = -22 \text{ ft-kips}$$

**Example 12-8.** Analyze the frame in Fig. 12-18(a). Assume that all members are of uniform cross section so that the relative values of $2EK$ are the circled numbers. Because of the action of the lateral forces, the frame will deflect to the right. Assume that joint $b$ and joint $e$, on the first-floor level, move a horizontal distance $\Delta_1$; joint $c$ and joint $d$, on the second-floor level, move a horizontal distance $\Delta_2$, as shown in Fig. 12-18(b). Antisymmetry exists in this case. Thus,

$$\theta_b = \theta_e \qquad \theta_c = \theta_d \qquad R_{ab} = R_{ef} = \frac{\Delta_1}{20} \qquad R_{bc} = R_{de} = \frac{\Delta_2 - \Delta_1}{20}$$

Now since

$$\theta_a = \theta_f = R_{be} = R_{cd} = 0$$

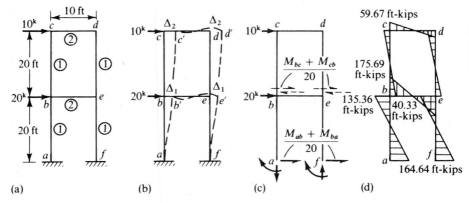

(a)  (b)  (c)  (d)

**Fig. 12-18**

the problem involves a total of four unknowns, $\theta_b$, $\theta_c$, $R_{ab}$, and $R_{bc}$, which are to be solved by four equations of statics, two from $\sum M = 0$ for joints $b$ and $c$ and the other two from shear balance for the frame.

The equations expressing end moments are then written

$$M_{ab} = M_{fe} = (1)(\theta_b - 3R_{ab}) \qquad (\theta_a = 0)$$
$$M_{ba} = M_{ef} = (1)(2\theta_b - 3R_{ab})$$
$$M_{bc} = M_{ed} = (1)(2\theta_b + \theta_c - 3R_{bc})$$
$$M_{be} = M_{eb} = (2)(2\theta_b + \theta_e) = 6\theta_b \qquad (\theta_b = \theta_e)$$
$$M_{cb} = M_{de} = (1)(2\theta_c + \theta_b - 3R_{bc})$$
$$M_{cd} = M_{dc} = (2)(2\theta_c + \theta_d) = 6\theta_c \qquad (\theta_c = \theta_d)$$

The equations from joint equilibrium in moment are then established:

$$M_{ba} + M_{bc} + M_{be} = 0$$
$$10\theta_b + \theta_c - 3R_{ab} - 3R_{bc} = 0 \tag{12-52}$$
$$M_{cb} + M_{cd} = 0$$
$$\theta_b + 8\theta_c - 3R_{bc} = 0 \tag{12-53}$$

The third equation is from shear balance for the entire frame isolated from the supports [see Fig. 12-18(c)]:

$$(2)\left(\frac{M_{ab} + M_{ba}}{20}\right) + 20 + 10 = 0$$
$$\theta_b - 2R_{ab} + 100 = 0 \tag{12-54}$$

The fourth equation results from considering the free body cut out by a horizontal section just above $be$; the shear in the two legs [see the dashed lines in Fig. 12-18(c)] must balance the lateral force of 10 kips. Thus,

$$(2)\left(\frac{M_{bc} + M_{cb}}{20}\right) + 10 = 0$$
$$3\theta_b + 3\theta_c - 6R_{bc} + 100 = 0 \tag{12-55}$$

Solving Eqs. 12-52, 12-53, 12-54, and 12-55 simultaneously, we obtain

$$\theta_b = 29.282 \qquad \theta_c = 9.945$$
$$R_{ab} = 64.641 \qquad R_{cd} = 36.282$$

Substituting these in the moment equations, we arrive at

$$M_{ab} = M_{fe} = -164.64 \text{ ft-kips}$$
$$M_{ba} = M_{ef} = -135.36 \text{ ft-kips}$$
$$M_{bc} = M_{ed} = -40.33 \text{ ft-kips}$$
$$M_{be} = M_{eb} = +175.69 \text{ ft-kips}$$
$$M_{cb} = M_{de} = -59.67 \text{ ft-kips}$$
$$M_{cd} = M_{dc} = +59.67 \text{ ft-kips}$$

as plotted in Fig. 12-18(d).

## 12-8. ANALYSIS OF STATICALLY INDETERMINATE RIGID FRAMES WITH SEVERAL DEGREES OF FREEDOM OF JOINT TRANSLATION BY THE SLOPE-DEFLECTION METHOD

The slope-deflection procedure described in the preceding sections may be extended to the analysis of frames with more than two degrees of freedom of joint translation. Solving simultaneous slope-deflection equations in terms of joint rotations and translations, although time consuming, is frequently simpler and more easily applied than any method previously discussed. Consider the four-story, two-bay building frame shown in Fig. 12-19. It would require the solution of 24 simultaneous equations in terms of the total internal work of the frame by the method of least work (see Sec. 10-3), compared with 16 equations needed by the method of slope deflection: 12 equations expressing $\sum M = 0$

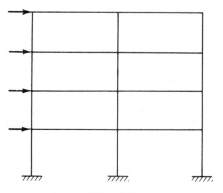

**Fig. 12-19**

for each of the 12 joints having rotations and four equations expressing the shear balance for each floor level of the four stories.

On the other hand, in the analysis of certain types of frames, such as the one shown in Fig. 12-20, the slope-deflection method becomes difficult to apply, since it would require the solution of 13 simultaneous equations—six equations expressing $\sum M = 0$ for joints $b, c, d, e, f,$ and $g$; one equation expressing $\sum F_x = 0$ for the entire frame; and six equations expressing $\sum M = 0$ about the six moment centers $O_1, O_2, O_3, O_4, O_5,$ and $O_6$ for the portions of frame indicated for each moment center in Fig. 12-20. However, if we use the method of least work, we must solve only three simultaneous equations in terms of the total internal work of the structure. Many engineering short cuts, such as the *elastic-center method* and *column-analogy method*, used in solving this type of structure are the direct outcome of least work or consistent deformation. Because of their limited use, they are not treated in this book.

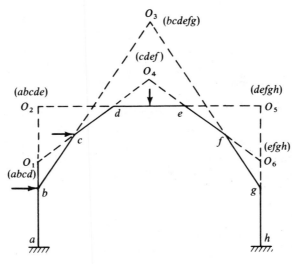

**Fig. 12-20**

## 12-9. MATRIX FORMULATION OF
### SLOPE-DEFLECTION PROCEDURE

In slope-deflection method, the joint displacements appear as the basic unknown in the simultaneous equations, which express the joint equilibrium of forces in terms of joint rotations and translations. The matrix formulation of these equilibrium equations will readily throw light on the analysis, revealing the essence of matrix stiffness procedure. This can best be illustrated by going over the problem in Fig. 12-5.

We start by Eq. 12-16 and, with some modifications, rewrite the slope-deflection equations as

$$M_{ab} = 2E\frac{I_{ab}}{l_{ab}}\left(\theta_b - \frac{3\Delta}{l_{ab}}\right) - 25$$

$$M_{ba} = 2E\frac{I_{ab}}{l_{ab}}\left(2\theta_b - \frac{3\Delta}{l_{ab}}\right) + 25$$

$$M_{bc} = 2E\frac{I_{bc}}{l_{bc}}(2\theta_b + \theta_c)$$

$$M_{cb} = 2E\frac{I_{bc}}{l_{bc}}(2\theta_c + \theta_b) \tag{12-56}$$

$$M_{cd} = 2E\frac{I_{cd}}{l_{cd}}\left(2\theta_c - \frac{3\Delta}{l_{cd}}\right)$$

$$M_{dc} = 2E\frac{I_{cd}}{l_{cd}}\left(\theta_c - \frac{3\Delta}{l_{cd}}\right)$$

We also rewrite the joint equilibrium equations as

$$M_{ba} + M_{bc} = 0$$

$$M_{cb} + M_{cd} = 0 \tag{12-57}$$

$$\frac{M_{ab} + M_{ba}}{l_{ab}} + \frac{M_{cd} + M_{dc}}{l_{cd}} + 5 = 0$$

Substituting the various expressions of Eq. 12-56 in Eq. 12-57, we have

$$\left(\frac{4EI_{ab}}{l_{ab}} + \frac{4EI_{bc}}{l_{bc}}\right)\theta_b + \frac{2EI_{bc}}{l_{bc}}\theta_c - \frac{6EI_{ab}}{l_{ab}^2}\Delta + 25 = 0$$

$$\frac{2EI_{bc}}{l_{bc}}\theta_b + \left(\frac{4EI_{bc}}{l_{bc}} + \frac{4EI_{cd}}{l_{cd}}\right)\theta_c - \frac{6EI_{cd}}{l_{cd}^2}\Delta = 0 \tag{12-58}$$

$$\frac{6EI_{ab}}{l_{ab}^2}\theta_b + \frac{6EI_{cd}}{l_{cd}^2}\theta_c - \left(\frac{12EI_{ab}}{l_{ab}^3} + \frac{12EI_{cd}}{l_{cd}^3}\right)\Delta + 5 = 0$$

Equation 12-58 in matrix form is

FJA      $\theta_b = 1$      $\theta_c = 1$      $\Delta = 1$      JL

$$\begin{Bmatrix} 25 \\ 0 \\ -5 \end{Bmatrix} + \begin{bmatrix} \dfrac{4EI_{ab}}{l_{ab}} + \dfrac{4EI_{bc}}{l_{bc}} & \dfrac{2EI_{bc}}{l_{bc}} & -\dfrac{6EI_{ab}}{l_{ab}^2} \\[2mm] \dfrac{2EI_{bc}}{l_{bc}} & \dfrac{4EI_{bc}}{l_{bc}} + \dfrac{4EI_{cd}}{l_{cd}} & -\dfrac{6EI_{cd}}{l_{cd}^2} \\[2mm] -\dfrac{6EI_{ab}}{l_{ab}^2} & -\dfrac{6EI_{cd}}{l_{cd}^2} & \dfrac{12EI_{ab}}{l_{ab}^3} + \dfrac{12EI_{cd}}{l_{cd}^3} \end{bmatrix} \begin{Bmatrix} \theta_b \\ \theta_c \\ \Delta \end{Bmatrix} = \begin{Bmatrix} 0 \\ 0 \\ 0 \end{Bmatrix}$$

$$\tag{12-59}$$

Equation 12-59 can be put in a generalized form as

$$F' + SD = F \tag{12-60}$$

We come to notice that the column matrix $F'$ on the left-hand side of Eq. 12-59 or 12-60 represents the fixed-joint action (FJA) under applied member loads; while the square matrix $S$ is a stiffness matrix, each column of which gives the various joint actions required to produce a certain unit joint displacement. Note that the stiffness matrix is symmetrical. The column matrix $D$ represents joint displacements. The column matrix $F$ on the right-hand side of the same equation contains the actual joint loads (JL) corresponding to the joint displacements. In our case there are no loads actually acting on joints $b$ and $c$, so that the elements in JL matrix are zeros.

It may be interesting to compare Eq. 12-60 with Eq. 9-11 and find out the duality between the displacement method (stiffness method) and force method (flexibility method).

The analysis therefore can be looked upon as contained in the following major steps:

1. Analyze a restrained structure with all joints fixed and subjected to member loads only [Fig. 12-21(a)]. With reference to Fig. 12-21(b), we see that

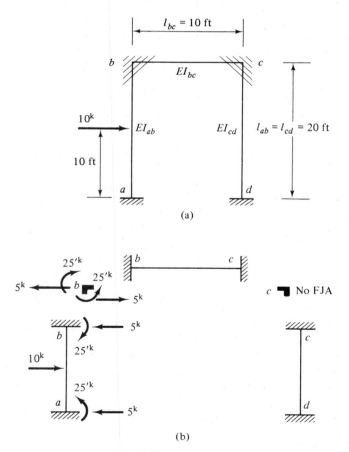

(a)

(b)

Fig. 12-21

locking the joint $b$ against rotation requires an artificial moment force 25 ft-kips acting clockwise in order to balance the internal moment exerted by member $ab$ at that joint. Likewise, preventing the lateral translation of joint $b$ requires a lateral force of 5 kips acting to the left to balance the internal shear exerted by member $ab$ at joint $b$. To be simple, the fixed-joint action is obtained from the algebraic sum of fixed-end actions of the related members.

2. Apply joint displacements individually and successively to joints $b$ and $c$ (free joints) so that the altered structure is restored to the actual displaced configuration. Figure 12-22 shows the separate cases of unit joint displacement from which we obtained the joint forces corresponding to these displacements. For instance, turning a unit rotation of joint $b$ would require a moment force of $(4EI_{ab}/l_{ab} + 4EI_{bc}/l_{bc})$ at joint $b$, a moment force of $2EI_{bc}/l_{bc}$ at joint $c$, and a lateral force of $-6EI_{ab}/l_{ab}^2$ (acting to the left) at $b$. The rest can similarly be

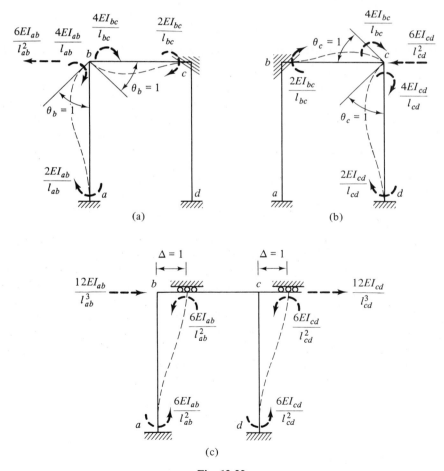

Fig. 12-22

explained. These are stiffness coefficients which construct the stiffness matrix of Eq. 12-59.

3. Steps 1 and 2 accomplish the geometric configuration (compatibility). It remains for us to say that the sum of joint forces thus obtained must be equal to the actual joint loads (equilibrium). This completes our formulation.

Using $2EI_{ab}/l_{ab} = 2EI_{cd}/l_{cd} = 1$, $2EI_{bc}/l_{bc} = 2$, and $l_{ab} = l_{cd} = 2l_{bc} = 20$, we reduce Eq. 12-59 to

$$\begin{Bmatrix} 25 \\ 0 \\ -5 \end{Bmatrix} + \begin{bmatrix} 6 & 2 & -\frac{3}{20} \\ 2 & 6 & -\frac{3}{20} \\ -\frac{3}{20} & -\frac{3}{20} & \frac{12}{400} \end{bmatrix} \begin{Bmatrix} \theta_b \\ \theta_c \\ \Delta \end{Bmatrix} = \begin{Bmatrix} 0 \\ 0 \\ 0 \end{Bmatrix} \qquad (12\text{-}61)$$

Replacing $\Delta$ with $R$ ($R = \Delta/20$) in Eq. 12-61 gives

$$\begin{Bmatrix} 25 \\ 0 \\ -5 \end{Bmatrix} + \begin{bmatrix} 6 & 2 & -3 \\ 2 & 6 & -3 \\ -\frac{3}{20} & -\frac{3}{20} & \frac{12}{20} \end{bmatrix} \begin{Bmatrix} \theta_b \\ \theta_c \\ R \end{Bmatrix} = \begin{Bmatrix} 0 \\ 0 \\ 0 \end{Bmatrix} \qquad (12\text{-}62)$$

which is identical to the set of simultaneous equations given in Eqs. 12-21 to 12-23.

## PROBLEMS

**12-1.** Analyze the beam shown in Fig. 12-23 by slope deflection. Draw the shear and moment diagrams.

*EI* constant

**Fig. 12-23**

**12-2.** Figure 12-24 shows a frame of uniform cross section. Find all the end moments by slope deflection, and sketch the deformed structure.

**Fig. 12-24**

**12-3.** Analyze the beam shown in Fig. 12-25 by slope deflection. Draw the shear and moment diagrams.

**Fig. 12-25**

**12-4.** Find all the end moments by slope deflection for the rigid frame shown in Fig. 12-26. Draw the moment diagram, and sketch the deformed structure.

**Fig. 12-26**

**12-5.** In Fig. 12-25 remove all the loads, and assume that the support $b$ settles vertically 0.5 in. Find all the end moments by slope deflection. $E = 30,000$ kips/in.$^2$.

**12-6.** In Fig. 12-26 remove the load, and assume that a rotational yield of 0.002 radian clockwise and a linear yield downward of 0.1 in. occur at support $a$. Find the moment diagram. $EI = 10,000$ kips-ft$^2$.

**12-7.** For the system and load shown in Fig. 12-27, use the method of slope deflection to find the spring force if $EI = 2 \times 10^8$ kN-cm$^2$ and the axial stiffness of the spring $k = 8$ kN/cm.

**Fig. 12-27**

**12-8.** For the load and frame shown in Fig. 12-28, use the method of slope deflection to find the reaction at supports *a* and *c*, and sketch the deformed structure.

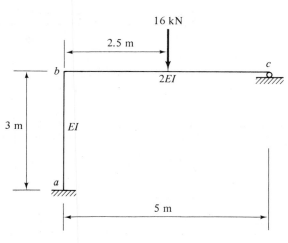

**Fig. 12-28**

**12-9.** Find all the end moments by slope deflection for the rigid frame shown in Fig. 12-29. Draw the moment diagram, and sketch the deformed structure. [*Hint:* There are two slopes at *c* (i.e., $\theta_{cb}$ and $\theta_{cd}$). Use the condition $M_{cb} = M_{cd} = 0$ to evaluate the two unknowns.]

**Fig. 12-29**

**12-10.** For the system shown in Fig. 12-30, use the method of slope deflection to find the reaction at support *d* if $E = 20,000$ kN/cm², $I = 40,000$ cm⁴, and the axial stiffness of the spring $k = 5$ kN/cm.

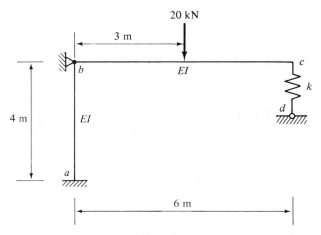

**Fig. 12-30**

**12-11.** Analyze each of the frames shown in Fig. 12-31 by slope deflection, and draw the moment diagram.

(a)

(b)

(c)

(d)

(e)

**Fig. 12-31**

265

**12-12.** Find the fixed-end moments for the beams shown in Fig. 12-32 by slope deflection. Assume that $w = 3$ kN/m, $l = 4$ m.

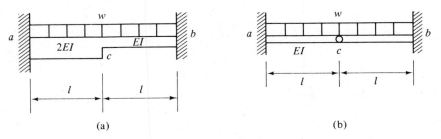

(a)                                              (b)

**Fig. 12-32**

**12-13.** Repeat Prob. 12-3 by the matrix stiffness procedure.

**12-14.** Repeat Prob. 12-11 shown in Fig. 12-31(b) by the matrix stiffness procedure.

**12-15.** Use the matrix stiffness method to analyze the frame shown in Fig. 12-33.

**Fig. 12-33**

# 13

## AN INTRODUCTION TO MOMENT

## DISTRIBUTION: MOMENT DISTRIBUTION

## WITHOUT JOINT TRANSLATION

### 13-1. GENERAL

The method of *moment distribution* was originated by Hardy Cross in 1930 in a paper entitled "Analysis of Continuous Frames by Distributing Fixed-End Moments." It is the method normally used to analyze all types of statically indeterminate beams and rigid frames in which the members are primarily subjected to bending. The process of moment distribution is initiated by the basic slope-deflection equation (see Sec. 12-2); the moment acting on the end of a member is the algebraic sum of four effects:

1. The moment due to the loads on the member if the member is considered as a fixed-end beam (the fixed-end moment)

2. The moment due to the rotation of the near end (this end) while the far end (the other end) is fixed

3. The moment due to the rotation of the far end, the near end being fixed

4. The moment due to the relative translation between the two ends of the member

This suggests that one line of attack might be to allow these effects to take place separately through a series of steps, first *locking* the joints and then *unlocking* them. For instance, in a rigid-joint structure without joint translation, once the joints are locked (held against rotation), each member is in the state of a fixed-end beam. By unlocking (releasing) a joint, we find that resisting moments will be developed or distributed at the near ends of the members meeting at the joint in proportion to their stiffnesses or according to their distribution factors. At the same time moments will be induced or carried over to the far ends of these members according to their carry-over factors. Joints may be successively released and reheld, one by one, as many times as necessary until each joint will have rotated into its actual, or nearly actual, position. Thus, the process is essentially one of successive approximations which can be carried to any degree of accuracy desired. We shall define the terms *stiffness*, *distribution factor*, and *carry-over factor* and explain them step by step in the following sections.

It must be noted that the method of moment distribution, although it depends on solving slope-deflection equations, is nevertheless a new approach in structural analysis. For instance, determining the end moments of a highly indeterminate rigid frame with joint translations prevented does not require solving any simultaneous equations in contrast to the previously discussed methods. Even in analyzing rigid frames having joint translations, the method of moment distribution usually does not involve as many simultaneous equations as are required by any of the methods already discussed. Furthermore, this method is adaptable to computer programming, since it is cyclic and the same operations are repeated several times over. As far as hand calculations are concerned, it is regarded as the most ingenious and convenient method contributed to structural engineering.

The method of moment distribution can be applied to structures composed of prismatic members or nonprismatic members—with or without joint translation. To simplify the discussion, the present chapter is confined to beams and frames of prismatic members without joint translation.

The sign convention adopted is the same as that previously suggested for the slope-deflection method: clockwise end moment and rotation of a member are considered positive.

### 13-2. FIXED-END MOMENT

The application of the method of moment distribution requires knowledge of the moments developed at the ends of loaded beams with both ends built in. These moments are called fixed-end moments, often denoted by the symbol $M^F$ or F.E.M. in tables and illustrations.

The determination of fixed-end moments was discussed in Sec. 9-2 (consistent deformations) and in Sec. 10-2 (least work). For a straight prismatic member the fixed-end moments due to common types of loading were given in Table 12-1.

### 13-3. STIFFNESS, DISTRIBUTION FACTOR, ### AND DISTRIBUTION OF EXTERNAL MOMENT ### APPLIED TO A JOINT

For a member of uniform section (constant $EI$), the *stiffness* (or more specifically the *rotational stiffness*) is defined as the end moment required to produce a unit rotation at one end of the member while the other end is fixed.

Consider member *ab* in Fig. 13-1 with a constant section. End *b* is fixed, and end *a* is allowed to rotate. The end moment required at end *a* to rotate

**Fig. 13-1**

$\theta_a = 1$, while $\theta_b = 0$ is given by

$$M_{ab} = 2E\frac{I}{l}(2\theta_a + \theta_b) = 2E\frac{I}{l}(2 + 0) = 4E\frac{I}{l}$$

$$= 4EK$$

This moment is defined as *stiffness* and will be denoted by $S$. Thus,

$$S = 4E\frac{I}{l}$$

$$= 4EK \qquad (13\text{-}1)$$

$I/l$ or $K$ being the *stiffness factor*.

Let us turn to Fig. 13-2(a), which shows a frame composed of four members, each with one end fixed and the other end rigidly connected at joint $j$ whose translation is prevented. If a clockwise moment $M$ is applied to the joint, it will cause the joint to rotate clockwise through an angular deformation $\theta$, as shown in Fig. 13-2(b). Since $j$ is a rigid joint, each tangent to the elastic curve of the connected end rotates the same angle $\theta$. The applied moment $M$ is resisted by the four members meeting at the joint. The resisting moments $M_{ja}$, $M_{jb}$, $M_{jc}$, and $M_{jd}$ will be induced at the ends of the four members to balance the effect of the external moment $M$, as shown in Fig. 13-2(c).

Equilibrium of the joint requires that

$$M_{ja} + M_{jb} + M_{jc} + M_{jd} = M \qquad (13\text{-}2)$$

and the slope-deflection equations for the four members give

$$M_{ja} = 4E\frac{I_{ja}}{l_{ja}}\theta = 4EK_{ja}\theta = S_{ja}\theta$$

$$M_{jb} = 4E\frac{I_{jb}}{l_{jb}}\theta = 4EK_{jb}\theta = S_{jb}\theta$$

$$M_{jc} = 4E\frac{I_{jc}}{l_{jc}}\theta = 4EK_{jc}\theta = S_{jc}\theta \qquad (13\text{-}3)$$

$$M_{jd} = 4E\frac{I_{jd}}{l_{jd}}\theta = 4EK_{jd}\theta = S_{jd}\theta$$

Equation 13-3 shows that, when an external moment is applied to a joint, the resisting moments developed at the near ends of the members meeting at the

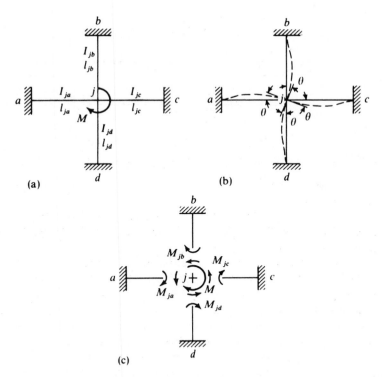

(a)        (b)

(c)

**Fig. 13-2**

joint, while the other ends are all fixed, are in direct proportion to the rotational stiffnesses.

Substituting Eq. 13-3 in Eq. 13-2, we obtain

$$(S_{ja} + S_{jb} + S_{jc} + S_{jd})\theta = M$$

or

$$4E(K_{ja} + K_{jb} + K_{jc} + K_{jd})\theta = M$$

Thus,

$$\theta = \frac{M}{4E \sum K} \tag{13-4}$$

where $\sum K = K_{ja} + K_{jb} + K_{jc} + K_{jd}$.

From Eqs. 13-3 and 13-4, we see that

$$M_{ja} = \frac{K_{ja}}{\sum K} M = D_{ja}M$$

$$M_{jb} = \frac{K_{jb}}{\sum K} M = D_{jb}M$$

$$M_{jc} = \frac{K_{jc}}{\sum K} M = D_{jc}M$$

$$M_{jd} = \frac{K_{jd}}{\sum K} M = D_{jd}M$$

$$\tag{13-5}$$

in which the ratio $K_{ji}/\sum K$ or $D_{ji}$ $(i = a, b, c, d)$ is defined as the *distribution factor*. Thus, a moment resisted by a joint will be distributed among the connecting members in proportion to their distribution factors. In determining the distribution factors, only the relative $K$ values for connected members are needed. Thus, in most cases we are concerned with the *relative stiffness* rather than the absolute stiffness (Eq. 13-1).

### 13-4. CARRY-OVER FACTOR AND CARRY-OVER MOMENT

Referring to Fig. 13-2, we evaluate the moments at the far ends (fixed ends) of the four members by the slope-deflection method as

$$M_{aj} = 2E\frac{I_{ja}}{l_{ja}}\theta = \left(\frac{1}{2}\right)M_{ja}$$

$$M_{bj} = 2E\frac{I_{jb}}{l_{jb}}\theta = \left(\frac{1}{2}\right)M_{jb}$$

$$M_{cj} = 2E\frac{I_{jc}}{l_{jc}}\theta = \left(\frac{1}{2}\right)M_{jc} \qquad (13\text{-}6)$$

$$M_{dj} = 2E\frac{I_{jd}}{l_{jd}}\theta = \left(\frac{1}{2}\right)M_{jd}$$

Equation 13-6 indicates that the moment induced at the far end (fixed) of a prismatic member equals *one-half* the distributed moment at the near end. The ratio $(\frac{1}{2})$ is called the *carry-over factor*, and the induced moments $M_{aj}$, $M_{bj}$, $M_{cj}$ and $M_{dj}$ are called the *carry-over moments*.

In general, the *carry-over factor* may be defined as the ratio of the induced moment at the far end, which is fixed, to the applied moment at the near end, which is prevented from translation but allowed to rotate. Consider Fig. 13-3. If an end moment $M_{ab}$ is applied at the near end $a$, then the moment induced at the far end $b$, called $M_{ba}$, is given by

$$M_{ba} = C_{ab}M_{ab} \qquad (13\text{-}7)$$

where $C_{ab}$ is the carry-over factor from $a$ to $b$. For a member of uniform $EI$,

$$M_{ab} = 2E\frac{I}{l}(2\theta_a)$$

$$M_{ba} = 2E\frac{I}{l}(\theta_a) = \frac{1}{2}M_{ab}$$

Therefore,                              $C_{ab} = \frac{1}{2}$

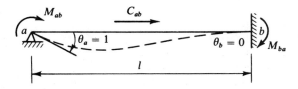

**Fig. 13-3**

If we consider end $a$ as the far end (fixed), and end $b$ as the near end (allowed to rotate), we can, in a like manner, prove that

$$C_{ba} = \tfrac{1}{2}$$

Thus for a member of uniform section,

$$C_{ab} = C_{ba} = \tfrac{1}{2} \tag{13-8}$$

We recapitulate some of the main points of Secs. 13-3 and 13-4 as follows. When an external moment is applied to a joint for which translation is prevented, the joint rotates; but the rotation is checked by members meeting at the joint. The resisting moment is then distributed to the near ends of the connected members according to their distribution factors, provided that all the far ends of these members are fixed. The distributed moment to the near end for each member based on the free body of the member (not the joint) equals the applied moment times the distribution factor bearing the same sign as that of the applied moment. Meanwhile, moment is carried over to the far end of each member, which equals one-half the distributed moment to the near end and bears the same sign.

In subsequent illustrations and tables we often use the symbols D.M. to denote the distributed moment; D.F., the distribution factor; C.O.M., the carry-over moment; and C.O.F., the carry-over factor.

**Example 13-1.** For the loaded frame shown in Fig. 13-4(a), find the end moments at $a$ and $c$.

We begin by putting the portion $abc$ of Fig. 13-4(a) into its equivalent [Fig. 13-4(b)] and obtaining the relative $K$ values for members $ab$ and $bc$ (circled). It is then readily seen that the end moments at $a$ and $c$ are the carry-over moments due to the external moment 60 kN·m applied to joint $b$. Thus,

$$M_{ab} = \frac{1}{2}M_{ba} = \left(\frac{1}{2}\right)(-60)\left(\frac{1}{3+1}\right) = -7.5 \text{ kN·m}$$

$$M_{cb} = \frac{1}{2}M_{bc} = \left(\frac{1}{2}\right)(-60)\left(\frac{3}{3+1}\right) = -22.5 \text{ kN·m}$$

The negative signs indicate counterclockwise moments.

(a)                              (b)

**Fig. 13-4**

## 13-5. THE PROCESS OF LOCKING AND UNLOCKING: ONE JOINT

The essence of moment distribution lies in locking and unlocking the joints based on the principle of superposition; that is, the effect of an artificial moment applied to a rigid joint of the frame and then eliminated is the same as no effect on the actual structure, since the two actions are neutralized. In this section our attention is confined to those frames that have all joints (including supported ends) fixed except one, which is allowed to rotate. Figure 13-5(a) shows the case which may then be considered as the superposition of effects of Fig. 13-5(b) and (c).

**Fig. 13-5**

In connection with the setup in Fig. 13-5, we note the following:

1. Suppose that the artificial moment $M$ imposed on joint $e$ [see Fig.. 13-5(b)] is so chosen as just to lock the joint against rotation (keeping $\theta_e = 0$) under the original loading (in the present case, the uniform load over span $ae$). Each member of the frame is then in the state of a fixed-end beam, and consequently, fixed-end moments will be developed at the ends of member $ae$:

$$M_{ea}^F = -M_{ae}^F = \frac{(1.2)(10)^2}{12} = 10 \text{ ft-kips}$$

all other member ends being subjected to no moment. The results are shown in row 1 of Table 13-1.

Referring to Fig. 13-5(b)', we find that the equilibrium condition

$$\sum M_{\text{joint } e} = 0$$

requires that the locking moment

$$M = 10 \text{ ft-kips}$$

act clockwise on joint $e$.

2. Next, let us release the joint $e$ from the artificial restraint, that is, apply to it an unlocking moment equal and opposite to the locking moment. Referring to Fig. 13-5(c), we see a counterclockwise $M$ equal to the value of 10 ft-kips:

273

TABLE 13-1

| | | \multicolumn{2}{c}{ae} | | \multicolumn{2}{c}{be} | | \multicolumn{2}{c}{ce} | | \multicolumn{2}{c}{de} |
|---|---|---|---|---|---|---|---|---|---|

<table>
<tr><td colspan="10" align="center">End Moment (ft-kips)</td></tr>
<tr><td></td><td></td><td colspan="2" align="center">ae</td><td colspan="2" align="center">be</td><td colspan="2" align="center">ce</td><td colspan="2" align="center">de</td></tr>
<tr><td>Row</td><td>Step</td><td>$M_{ae}$</td><td>$M_{ea}$</td><td>$M_{be}$</td><td>$M_{eb}$</td><td>$M_{ce}$</td><td>$M_{ec}$</td><td>$M_{de}$</td><td>$M_{ed}$</td></tr>
<tr><td>1</td><td>F.E.M.</td><td>−10</td><td>+10</td><td>0</td><td>0</td><td>0</td><td>0</td><td>0</td><td>0</td></tr>
<tr><td>2</td><td>D.M.</td><td>0</td><td>−1</td><td>0</td><td>−2</td><td>0</td><td>−3</td><td>0</td><td>−4</td></tr>
<tr><td>3</td><td>C.O.M.</td><td>−0.5</td><td>0</td><td>−1</td><td>0</td><td>−1.5</td><td>0</td><td>−2</td><td>0</td></tr>
<tr><td>4</td><td>Σ</td><td>−10.5</td><td>+9</td><td>−1</td><td>−2</td><td>−1.5</td><td>−3</td><td>−2</td><td>−4</td></tr>
</table>

$$M = -10 \text{ ft-kips}$$

is thus applied to joint $e$.

As a result of this unlocking moment, the resisting moment will be distributed at the near ends and carried over to the far ends, as described in preceding sections. Thus,

$$M_{ea} = (-10)\left(\frac{1}{10}\right) = -1.0 \text{ ft-kip} \qquad M_{ae} = (-1)\left(\frac{1}{2}\right) = -0.5 \text{ ft-kip}$$

$$M_{eb} = (-10)\left(\frac{2}{10}\right) = -2.0 \text{ ft-kips} \qquad M_{be} = (-2)\left(\frac{1}{2}\right) = -1.0 \text{ ft-kip}$$

$$M_{ec} = (-10)\left(\frac{3}{10}\right) = -3.0 \text{ ft-kips} \qquad M_{ce} = (-3)\left(\frac{1}{2}\right) = -1.5 \text{ ft-kips}$$

$$M_{ed} = (-10)\left(\frac{4}{10}\right) = -4.0 \text{ ft-kips} \qquad M_{de} = (-4)\left(\frac{1}{2}\right) = -2.0 \text{ ft-kips}$$

They are shown in rows 2 and 3 of Table 13-1, respectively.

3. The sum of the results from steps 1 and 2 gives the solution, as shown in row 4 of Table 13-1.

In analyzing problems like this, with all ends fixed except one rigid joint allowed to rotate, the method of moment distribution provides a very rapid tool, since it involves only one round of locking and unlocking. More examples to illustrate this process follow.

**Example 13-2.** The end moments at $a$, $b$, and $c$ for the beam and loading shown in Fig. 13-6(a) may be obtained by locking joint $b$ and then unlocking it, as indicated in Fig. 13-6(b) and (c). The complete analysis is shown in Fig. 13-6(d), which contains the following steps:

1. The values of stiffness for members $ab$ and $bc$ are found to be

$$K_{ab} = \frac{4EI}{60} \qquad K_{bc} = \frac{4EI}{40}$$

Multiplying each by $30/EI$ gives the relative $K$ values (circled).

**Fig. 13-6**

2. The distribution factors are computed according to $K/\sum K$. Thus,

$$D_{ba} = \frac{2}{2+3} = 0.4 \qquad D_{bc} = \frac{3}{2+3} = 0.6$$

We consider the immovable supports at $a$ and $c$ with infinite stiffness so that

$$D_{ab} = D_{cb} = 0$$

The distribution factors are indicated in the attached box at each of the joints.

3. The locking joint $b$ artificially puts members $ab$ and $bc$ in the state of fixed-end beams. We write the fixed-end moments as

$$M_{ab}^F = -M_{ba}^F = -\frac{(1)(60)^2}{12} = -300 \text{ ft-kips}$$

$$M_{bc}^F = -M_{cb}^F = -\frac{(20)(40)}{8} = -100 \text{ ft-kips}$$

Note that locking joint $b$ means applying an external clockwise moment equal to 200 ft-kips, required by the equilibrium of moments for that joint [see Fig. 13-6(b)'].

4. Unlocking joint $b$ (i.e., eliminating the artificial restraint acting on joint $b$) means applying an external counterclockwise moment equal to 200 ft-kips.

We write the distributed moment for each of near ends according to the distribution factor.

5. Write down the carry-over moment for each of the far ends equal to one-half the distributed moment of the near end.

6. The sum of the results from steps 3, 4, and 5 gives the solution.

**Example 13-3.** Find the end moments for the frame shown in Fig. 13-7(a) resulting from the rotational yield of support $a$ 0.0016 radian clockwise. $EI = 10,000$ kips-ft². Note that this problem was solved by slope deflection in Example 12-4.

**Fig. 13-7**

The moment required to produce a rotation of 0.0016 radian at $a$ is given by

$$M = \frac{4EI_{ab}\theta_a}{l_{ab}} = \frac{(4)(10,000)(0.0016)}{10} = 6.4 \text{ ft-kips}$$

if joint $b$ is temporarily fixed. Half the amount of this moment will be carried over to end $b$ of member $ab$. By releasing joint $b$, a process of distribution and carry-over takes place, as recorded in Fig. 13-7(b). This gives

$$M_{ab} = 5.6 \text{ ft-kips} \qquad M_{ba} = -M_{bc} = 1.6 \text{ ft-kips} \qquad M_{cb} = -0.8 \text{ ft-kip}$$

### 13-6. THE PROCESS OF LOCKING AND UNLOCKING: TWO OR MORE JOINTS

If a rigid frame or continuous beam involves more than one joint permitted to rotate, then the process of moment distribution consists of the repeated application of the principle of superposition, as stated briefly in the following steps:

1. The joints are first locked; all members, accordingly, are fixed-end. Write the fixed-end moments for all members.

2. The joints are then unlocked. Only one joint at a time is selected to be

unlocked. While one joint is unlocked, the rest of joints are assumed to be held against rotation.

Calculate the unlocking moment at this joint, and write distributed moments for the near ends of the members meeting at this joint.

3. Also write down the carry-over moments at the far ends of these members. Note that the carry-over moments constitute a new set of fixed-end moments for the far ends.

4. Relock the joint, and select the next joint to be unlocked. Repeat steps 2 and 3.

Note that after a joint is unlocked and the moments at a joint distributed, the joint is in balance, or in equilibrium, since the artificial restraint is removed. However, there are other joints still locked by external means; hence, the next step is to relock the joint and then proceed to unlock the next joint. The process of locking and unlocking each joint only once constitutes *one cycle* of moment distribution.

5. Joints are unlocked and relocked one by one; therefore, steps 2 and 3 are repeated several times. The process can be halted as soon as the carry-over moments are so small that we are willing to neglect them.

6. Sum up the moments to obtain the final result.

We see that the analysis starts from an alteration of the original structure by locking all joints against rotation. This means that artificial restraints are actually applied to the original structure. The altered structure, consisting of a number of fixed-end members, is then modified by unlocking and relocking joints one by one until all artificial restraints are removed or diminished to a sufficiently small amount. Thus, moment distribution is a method of successive approximations by which the exact results can be approached with the desired degree of precision.

The complete analysis of a loaded three-span continuous beam by moment distribution, shown in Fig. 13-8, will serve to illustrate the foregoing procedure.

The presentation of moment distribution for the preceding illustration may be rearranged as shown in Fig. 13-9. At first glance it seems as if joints $b$ and $c$ were locked and then unlocked simultaneously. However, the performance can still be considered under the restriction of unlocking one joint at a time. For the loads given, the fixed-end moments are recorded in step 1 (see Fig. 13-9). Next, we may consider joint $c$ as being held against rotation and joint $b$ as being unlocked first. The unlocking moment $+10$ at $b$ is then equally distributed to the near ends of members $ba$ and $bc$ as indicated in step 2, and one-half of the amount is carried over to the far ends of these members as indicated in step 3. Next, we consider joint $b$ as locked and joint $c$ as released, but only partially, by applying an unlocking moment of $-(40 - 30)$ or $-10$ to it, since the complete releasing of joint $c$ would require an unlocking moment of $-(40 - 30 + 2.5)$ or $-12.5$ for the time being. This unlocking moment $-10$ is equally distributed to the near ends of members $cb$ and $cd$, and one-

**Fig. 13-8**

half of the amount is carried over to the far ends of these members as indicated in steps 2 and 3, leaving the just received carry-over moment $+2.5$ to be handled later.

Referring to step 3 of Fig. 13-9, we find the carry-over moments form a new set of fixed-end moments for the beam, and a process of unlocking joints can be carried out in a similar manner. The process will thus be repeated in a cyclic fashion until the carry-over moments are neglected.

In fact, if the carry-over moments are neglected, the joints, after being unlocked, are in balance (i.e., no external constraint exists). This gives the approximate solution for the analysis. For instance, the first approximation may be obtained from the sum of steps 1 and 2, the result of the first cycle; the second approximation may be obtained from the sum of steps 1 to 4, the result up to the second cycle; and so on. It may be interesting to point out that, in this particular problem, even the first cycle yields a good approximation of the exact solution. After two or three cycles, the carry-over values become negligible.

Note that the beam and the loading shown in Fig. 13-9 are symmetrical, the data presented on each side of the line of symmetry are equal in magnitude

278

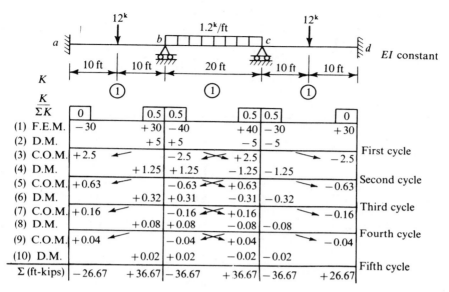

| K | 0 | | 0.5 | 0.5 | | 0.5 | 0.5 | | 0 | |
|---|---|---|---|---|---|---|---|---|---|---|
| $\dfrac{K}{\Sigma K}$ | | | | | | | | | | |
| (1) F.E.M. | −30 | | +30 | −40 | | +40 | −30 | | +30 | |
| (2) D.M. | | | +5 | +5 | | −5 | −5 | | | |
| (3) C.O.M. | +2.5 | ← | | −2.5 | ⤫ | +2.5 | | ↘ | −2.5 | First cycle |
| (4) D.M. | | | +1.25 | +1.25 | | −1.25 | −1.25 | | | |
| (5) C.O.M. | +0.63 | ← | | −0.63 | ⤫ | +0.63 | | ↘ | −0.63 | Second cycle |
| (6) D.M. | | | +0.32 | +0.31 | | −0.31 | −0.32 | | | |
| (7) C.O.M. | +0.16 | ← | | −0.16 | ⤫ | +0.16 | | ↘ | −0.16 | Third cycle |
| (8) D.M. | | | +0.08 | +0.08 | | −0.08 | −0.08 | | | |
| (9) C.O.M. | +0.04 | ← | | −0.04 | ⤫ | +0.04 | | ↘ | −0.04 | Fourth cycle |
| (10) D.M. | | | +0.02 | +0.02 | | −0.02 | −0.02 | | | |
| Σ (ft-kips) | −26.67 | | +36.67 | −36.67 | | +36.67 | −36.67 | | +26.67 | Fifth cycle |

**Fig. 13-9**

but opposite in sign. This special display suggests that some modification could be made in order to facilitate the process of moment distribution by working with only half the structure. See Sec. 13-7 for *modified stiffness*. This problem will be re-solved in Example 13-6 by using modified stiffness.

More examples are given to illustrate the cyclic process.

**Example 13-4.** Analyze the two-span continuous beam shown in Fig. 13-10 by moment distribution.

The convergence in the solution shown in Fig. 13-10 is rather slow because the hinged end $c$ continuously sends back sizable carry-over moments to joint $b$. However, since the final result $M_c = 0$ is known beforehand, the convergence may be improved by using modified stiffness as suggested in Sec. 13-7. This problem will be re-solved in Example 13-8.

**Example 13-5.** Analyze the frame in Fig. 13-11 by moment distribution. The relative stiffnesses for the frame members are computed first as

$$K_{ab}_{(cd)} = \frac{2I}{30} \quad \text{say} \quad 4$$

$$K_{bc} = \frac{2I}{40} \quad \text{say} \quad 3$$

$$K_{be}_{(cf)} = \frac{I}{20} \quad \text{say} \quad 3$$

|  | 0 | 0.4 | 0.6 | 1 |
|---|---|---|---|---|
| F.E.M. | −300 | +300 | −100 | +100 |
| D.M. |  | −80 | −120 | −100 |
| C.O.M. | −40 |  | −50 | −60 |
| D.M. |  | +20 | +30 | +60 |
| C.O.M. | +10 |  | +30 | +15 |
| D.M. |  | −12 | −18 | −15 |
| C.O.M. | −6 |  | −7.5 | −9 |
| D.M. |  | +3 | +4.5 | +9 |
| C.O.M. | +1.5 |  | +4.5 | +2.25 |
| D.M. |  | −1.8 | −2.7 | −2.25 |
| C.O.M. | −0.9 |  | −1.13 | −1.35 |
| D.M. |  | +0.45 | +0.68 | +1.35 |
| C.O.M. | +0.23 |  | +0.67 | +0.34 |
| D.M. |  | −0.27 | −0.40 | −0.34 |
| C.O.M. | −0.13 |  | −0.17 | −0.20 |
| D.M. |  | +0.07 | +0.10 | +0.20 |
| Σ (ft-kips) | −335.3 | +229.5 | −229.5 | 0 |

**Fig. 13-10**

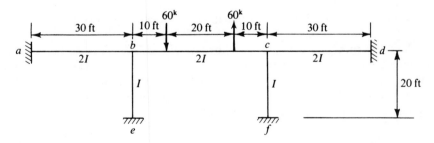

**Fig. 13-11**

Next, the fixed-end moments for the loaded member $bc$ are found to be

$$M_{bc}^F = -\frac{(60)(10)(30)^2}{(40)^2} + \frac{(60)(30)(10)^2}{(40)^2} = -225 \text{ ft-kips}$$

$$M_{cb}^F = +\frac{(60)(30)(10)^2}{(40)^2} - \frac{(60)(10)(30)^2}{(40)^2} = -225 \text{ ft-kips}$$

The complete analysis is shown in Fig. 13-12. Note that the intermediate

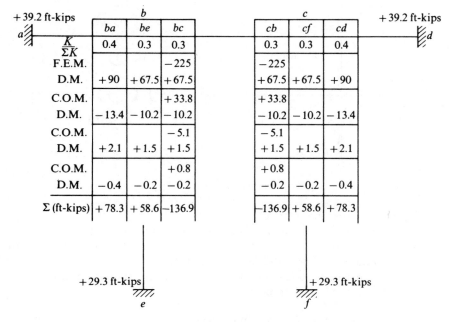

+39.2 ft-kips ... b ... c ... +39.2 ft-kips

| | ba | be | bc | | cb | cf | cd |
|---|---|---|---|---|---|---|---|
| $\dfrac{K}{\Sigma K}$ | 0.4 | 0.3 | 0.3 | | 0.3 | 0.3 | 0.4 |
| F.E.M. | | | −225 | | −225 | | |
| D.M. | +90 | +67.5 | +67.5 | | +67.5 | +67.5 | +90 |
| C.O.M. | | | +33.8 | | +33.8 | | |
| D.M. | −13.4 | −10.2 | −10.2 | | −10.2 | −10.2 | −13.4 |
| C.O.M. | | | −5.1 | | −5.1 | | |
| D.M. | +2.1 | +1.5 | +1.5 | | +1.5 | +1.5 | +2.1 |
| C.O.M. | | | +0.8 | | +0.8 | | |
| D.M. | −0.4 | −0.2 | −0.2 | | −0.2 | −0.2 | −0.4 |
| Σ (ft-kips) | +78.3 | +58.6 | −136.9 | | −136.9 | +58.6 | +78.3 |

+29.3 ft-kips ... e ... +29.3 ft-kips ... f

**Fig. 13-12**

values of the moments at ends $a$, $d$, $e$, and $f$ are not shown. Since members $ab$, $cd$, $be$, and $cf$ are not loaded, it is evident that

$$M_{ab} = \tfrac{1}{2}M_{ba} = 39.2 \text{ ft-kips} \qquad M_{dc} = \tfrac{1}{2}M_{cd} = 39.2 \text{ ft-kips}$$
$$M_{eb} = \tfrac{1}{2}M_{be} = 29.3 \text{ ft-kips} \qquad M_{fc} = \tfrac{1}{2}M_{cf} = 29.3 \text{ ft-kips}$$

From Fig. 13-12 the values of the moments obtained on the left side of the center line of the structure are exactly the same as those on the right. Such a special display is referred to as *antisymmetry*, which yields

$$\theta_b = \theta_c$$

An adjustment can be made to the stiffness of the center beam $bc$ that will permit the analyst to work with only half the structure. See Sec. 13-7 for modified stiffness. This problem will be re-solved in Example 13-9.

### 13-7. MODIFIED STIFFNESSES

The examples given in Sec. 13-6 have illustrated three special cases that suggest that some modifications for simplifying the moment-distribution process might be found by recognizing certain known conditions.

Consider the frame subjected to a clockwise external moment $M$ applied to the connecting joint as shown in Fig. 13-13(a) for which we note the following:

**Fig. 13-13**

1. $\theta_a = 0$; that is, the member is fixed at end $a$.
2. $M_b = 0$; that is, the member is simply supported at end $b$.
3. $\theta_c = -\theta$; that is, the member rotates through an equal but opposite angle at the other end $c$ as in the case of symmetry.
4. $\theta_d = \theta$; that is, the member rotates through the same angle at the other end $d$ as in the case of antisymmetry.

Referring to Fig. 13-13(b), we find the equilibrium of joint $j$ requires that

$$M_{ja} + M_{jb} + M_{jc} + M_{jd} = M \qquad (13\text{-}9)$$

From the slope-deflection equations we obtain

$$M_{ja} = 4EK_{ja}\theta \qquad (13\text{-}10)$$

and

$$M_{jb} = 2EK_{jb}(2\theta + \theta_b)$$

but

$$M_{bj} = 2EK_{jb}(2\theta_b + \theta) = 0$$

or

$$\theta_b = -\frac{\theta}{2}$$

Hence,

$$M_{jb} = 2EK_{jb}\left(2\theta - \frac{\theta}{2}\right) = 4E\left(\frac{3}{4}K_{jb}\right)\theta$$

$$= 4EK'_{jb}\theta \qquad (13\text{-}11)$$

where we let

$$K'_{jb} = \tfrac{3}{4}K_{jb} \qquad (13\text{-}12)$$

$K'_{jb}$ being called the *modified stiffness factor* for member $jb$. Similarly,

$$M_{jc} = 2EK_{jc}(2\theta - \theta) = 4E(\tfrac{1}{2}K_{jc})\theta$$

$$= 4EK'_{jc}\theta \qquad (13\text{-}13)$$

where

$$K'_{jc} = \tfrac{1}{2}K_{jc} \qquad (13\text{-}14)$$

$$M_{jd} = 2EK_{jd}(2\theta + \theta) = 4E(\tfrac{3}{2}K_{jd})\theta$$

$$= 4EK'_{jd}\theta \qquad (13\text{-}15)$$

where
$$K'_{jd} = \tfrac{3}{2}K_{jd} \tag{13-16}$$

Substituting Eqs. 13-10, 13-11, 13-13, and 13-15 in Eq. 13-9 yields
$$4E(K_{ja} + K'_{jb} + K'_{jc} + K'_{jd})\,\theta = M$$

or
$$\theta = \frac{M}{4E\sum K'} \tag{13-17}$$

where
$$\sum K' = K_{ja} + K'_{jb} + K'_{jc} + K'_{jd}$$

Substituting Eq. 13-17 in Eqs. 13-10, 13-11, 13-13, and 13-15 yields

$$M_{ja} = \frac{K_{ja}}{\sum K'}\,M$$

$$M_{jb} = \frac{K'_{jb}}{\sum K'}\,M$$

$$M_{jc} = \frac{K'_{jc}}{\sum K'}\,M \tag{13-18}$$

$$M_{jd} = \frac{K'_{jd}}{\sum K'}\,M$$

Thus, it is seen that, when an external moment is applied to joint $j$, the distributed moments to the near ends of the members meeting at the joint are in direct proportion to their modified stiffness factors if the conditions of the far ends are known. By using the modified stiffness factor for one end, we actually eliminate the carry-over to the other end except writing down the final result for that part.

Let us recapitulate the modified stiffness factors for various end conditions:

1. If the one end is simply supported, the modified stiffness factor for the other end is given by
$$K' = \tfrac{3}{4}K \tag{13-19}$$

2. If the one end is symmetrical to the other end, then
$$K' = \tfrac{1}{2}K \tag{13-20}$$

3. If the one end is antisymmetrical to the other end, then
$$K' = \tfrac{3}{2}K \tag{13-21}$$

$K'$ denotes the modified stiffness factor. In parallel with Eq. 13-1, we may have
$$S' = 4EK' \tag{13-22}$$

$S'$ being called the *modified stiffness*. A general definition for *modified stiffness* is the end moment required to produce a unit rotation at this end (simple end), while the other end remains in the actual condition. The modified stiffnesses, hence the modified stiffness factors, for various end conditions may be secured by direct application of this definition, as discussed in Sec. 17-8.

The following examples illustrate the process of moment distribution by using the modified $K$ values.

**Example 13-6.** Analyze the symmetrical beam in Fig. 13-14 by using modified $K$ value in the center span.

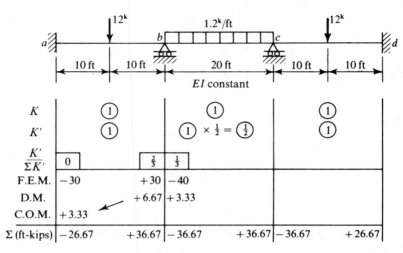

**Fig. 13-14**

**Example 13-7.** Analyze the symmetrical frame in Fig. 13-15, which was solved by the methods of consistent deformations, least work, and slope deflection. Once again the method of moment distribution demonstrates its superiority over any other method previously discussed.

**Fig. 13-15**

**Example 13-8.** Analyze the beam in Fig. 13-16 by using modified $K$ for end $b$ of member $bc$.

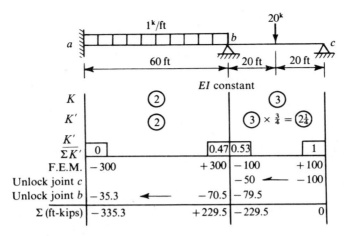

Fig. 13-16

**Example 13-9.** Analyze the frame in Fig. 13-11 by using the modified $K$ value in the center span for the antisymmetrical system. The solution is given in Fig. 13-17.

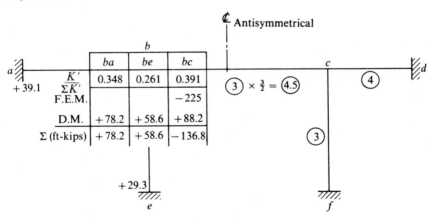

Fig. 13-17

## 13-8. MATRIX FORMULATION OF MOMENT-DISTRIBUTION PROCEDURE

The moment distribution is a cyclic process and tends to be convergent to an exact solution. This can best be illustrated by the matrix formulation for the moment-distribution process listed in Fig. 13-9.

We begin with a column matrix $A$ representing the fixed-end moments. In this case,

$$A = \begin{Bmatrix} M_{ab}^F \\ M_{ba}^F \\ M_{bc}^F \\ M_{cb}^F \\ M_{cd}^F \\ M_{dc}^F \end{Bmatrix} = \begin{Bmatrix} -30 \\ 30 \\ -40 \\ 40 \\ -30 \\ 30 \end{Bmatrix} \qquad (13\text{-}23)$$

See the first row of Fig. 13-9.

Next, we form a distribution matrix $B$ with distribution factors as its constituents in such a way that the matrix product $BA$ (a column matrix) will give the respective distributed moments. Thus,

$$BA = \begin{bmatrix} 0 & 0 & 0 & 0 & 0 & 0 \\ 0 & -\frac{1}{2} & -\frac{1}{2} & 0 & 0 & 0 \\ 0 & -\frac{1}{2} & -\frac{1}{2} & 0 & 0 & 0 \\ 0 & 0 & 0 & -\frac{1}{2} & -\frac{1}{2} & 0 \\ 0 & 0 & 0 & -\frac{1}{2} & -\frac{1}{2} & 0 \\ 0 & 0 & 0 & 0 & 0 & 0 \end{bmatrix} \begin{Bmatrix} -30 \\ 30 \\ -40 \\ 40 \\ -30 \\ 30 \end{Bmatrix} = \begin{Bmatrix} 0 \\ 5 \\ 5 \\ -5 \\ -5 \\ 0 \end{Bmatrix} \qquad (13\text{-}24)$$

See the second row of Fig. 13-9. Apparently, the resulting moment of the first cycle is

$$M_1 = A + BA = (I + B)A \qquad (13\text{-}25)$$

in which $I$ is a unit matrix.

Next, let the matrix $C$ perform as a carry-over operator so that the matrix product $CBA$, a column matrix, will represent the respective carry-over moments. Using the result of Eq. 13-24, we have in this case:

$$CBA = \begin{bmatrix} 0 & \frac{1}{2} & 0 & 0 & 0 & 0 \\ \frac{1}{2} & 0 & 0 & 0 & 0 & 0 \\ 0 & 0 & 0 & \frac{1}{2} & 0 & 0 \\ 0 & 0 & \frac{1}{2} & 0 & 0 & 0 \\ 0 & 0 & 0 & 0 & 0 & \frac{1}{2} \\ 0 & 0 & 0 & 0 & \frac{1}{2} & 0 \end{bmatrix} \begin{Bmatrix} 0 \\ 5 \\ 5 \\ -5 \\ -5 \\ 0 \end{Bmatrix} = \begin{Bmatrix} 2.5 \\ 0 \\ -2.5 \\ 2.5 \\ 0 \\ -2.5 \end{Bmatrix} \qquad (13\text{-}26)$$

See the third row of Fig. 13-9. The carry-over moments constitute a new set of fixed-end moments. We repeat the same process by replacing $A$ with $CBA$ in Eq. 13-25 to obtain the moments of the second cycle. Thus,

$$M_2 = (I + B)CBA \qquad (13\text{-}27)$$

For simplicity, we introduce $D = CB$, and Eq. 13-27 becomes

$$M_2 = (I + B)DA \tag{13-28}$$

Similarly, we can have

$$M_3 = (I + B)(D)^2 A$$
$$M_4 = (I + B)(D)^3 A \tag{13-29}$$
$$\vdots$$

in which $(D)^2 = (D)(D)$, $(D)^3 = (D)(D)(D)$, and so on. (Note that at the end of first cycle, the total moment is $M_1$; at the end of second cycle, the total moment is $M_1 + M_2$; and so on.) Summing up $M_1, M_2 \ldots$, we obtain the final resulting moment as

$$M = (I + B)[A + DA + (D)^2 A + (D)^3 A + \cdots]$$
$$= (I + B)\left(\frac{1}{I - D}\right)(A)$$
$$= (I + B)(I - D)^{-1}(A) \tag{13-30}$$

In our case

$$I + B = \begin{bmatrix} 1 & 0 & 0 & 0 & 0 & 0 \\ 0 & \frac{1}{2} & -\frac{1}{2} & 0 & 0 & 0 \\ 0 & -\frac{1}{2} & \frac{1}{2} & 0 & 0 & 0 \\ 0 & 0 & 0 & \frac{1}{2} & -\frac{1}{2} & 0 \\ 0 & 0 & 0 & -\frac{1}{2} & \frac{1}{2} & 0 \\ 0 & 0 & 0 & 0 & 0 & 1 \end{bmatrix} \tag{13-31}$$

$$D = CB = \begin{bmatrix} 0 & \frac{1}{2} & 0 & 0 & 0 & 0 \\ \frac{1}{2} & 0 & 0 & 0 & 0 & 0 \\ 0 & 0 & 0 & \frac{1}{2} & 0 & 0 \\ 0 & 0 & \frac{1}{2} & 0 & 0 & 0 \\ 0 & 0 & 0 & 0 & 0 & \frac{1}{2} \\ 0 & 0 & 0 & 0 & \frac{1}{2} & 0 \end{bmatrix} \begin{bmatrix} 0 & 0 & 0 & 0 & 0 & 0 \\ 0 & -\frac{1}{2} & -\frac{1}{2} & 0 & 0 & 0 \\ 0 & -\frac{1}{2} & -\frac{1}{2} & 0 & 0 & 0 \\ 0 & 0 & 0 & -\frac{1}{2} & -\frac{1}{2} & 0 \\ 0 & 0 & 0 & -\frac{1}{2} & -\frac{1}{2} & 0 \\ 0 & 0 & 0 & 0 & 0 & 0 \end{bmatrix}$$

$$= \begin{bmatrix} 0 & -\frac{1}{4} & -\frac{1}{4} & 0 & 0 & 0 \\ 0 & 0 & 0 & 0 & 0 & 0 \\ 0 & 0 & 0 & -\frac{1}{4} & -\frac{1}{4} & 0 \\ 0 & -\frac{1}{4} & -\frac{1}{4} & 0 & 0 & 0 \\ 0 & 0 & 0 & 0 & 0 & 0 \\ 0 & 0 & 0 & -\frac{1}{4} & -\frac{1}{4} & 0 \end{bmatrix}$$

$$(I - D)^{-1} = \begin{bmatrix} 1 & \frac{1}{4} & \frac{1}{4} & 0 & 0 & 0 \\ 0 & 1 & 0 & 0 & 0 & 0 \\ 0 & 0 & 1 & \frac{1}{4} & \frac{1}{4} & 0 \\ 0 & \frac{1}{4} & \frac{1}{4} & 1 & 0 & 0 \\ 0 & 0 & 0 & 0 & 1 & 0 \\ 0 & 0 & 0 & \frac{1}{4} & \frac{1}{4} & 1 \end{bmatrix}^{-1}$$

$$= \begin{bmatrix} 1 & -\frac{4}{15} & -\frac{4}{15} & \frac{1}{15} & \frac{1}{15} & 0 \\ 0 & 1 & 0 & 0 & 0 & 0 \\ 0 & \frac{1}{15} & \frac{16}{15} & -\frac{4}{15} & -\frac{4}{15} & 0 \\ 0 & -\frac{4}{15} & -\frac{4}{15} & \frac{16}{15} & \frac{1}{15} & 0 \\ 0 & 0 & 0 & 0 & 1 & 0 \\ 0 & \frac{1}{15} & \frac{1}{15} & -\frac{4}{15} & -\frac{4}{15} & 1 \end{bmatrix} \qquad (13\text{-}32)$$

Substituting Eqs. 13-23, 13-31, and 13-32 in Eq. 13-30 gives

$$M = \begin{Bmatrix} M_{ab} \\ M_{ba} \\ M_{bc} \\ M_{cb} \\ M_{cd} \\ M_{dc} \end{Bmatrix} = \begin{Bmatrix} -26.67 \\ 36.67 \\ -36.67 \\ 36.67 \\ -36.67 \\ 26.67 \end{Bmatrix} \qquad (13\text{-}33)$$

which can be checked with the result shown in the last row of Fig. 13-9.

## PROBLEMS

**13-1.** Solve the end moments for Prob. 12-1 by moment distribution.

**13-2.** Solve the end moments for Prob. 12-2 by moment distribution.

**13-3.** Solve the end moments for Prob. 12-3 by moment distribution.

**13-4.** Solve the end moments for Prob. 12-4 by moment distribution: (a) not using modified stiffness; (b) using a modified $K$ value in the center span.

**13-5.** Analyze the box shown in Fig. 13-18 by moment distribution. Assume constant $EI$.

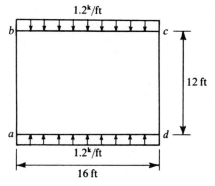

Fig. 13-18

**13-6.** Analyze the continuous beam in Fig. 13-19 by moment distribution. Assume constant *EI*.

Fig. 13-19

**13-7.** Analyze the continuous beam in Fig. 13-20 by moment distribution. Take advantage of modified stiffnesses, replacing the unsymmetrical loading system with a symmetrical and an antisymmetrical system. Assume constant *EI*.

Fig. 13-20

**13-8.** Analyze the frame in Fig. 13-21 by moment distribution and find the reaction at support *c*.

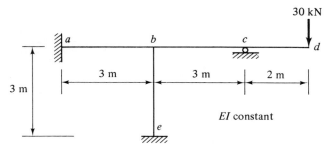

Fig. 13-21

289

**13-9.** With reference to Fig. 13-22, find the stiffness for end $a$ of member $ab$ if a hinge connection is inserted in the member at $c$ as shown.

**Fig. 13-22**

**13-10.** Use the matrix formulation of moment distribution to solve Probs. 13-1 to 13-3.

# 14

## MOMENT DISTRIBUTION WITH JOINT

## TRANSLATIONS

### 14-1. GENERAL

The procedure of moment distribution discussed in Chapter 13 is based on the restriction that the joints of the structure do not move. However, many frames encountered in practice undergo joint translations, and therefore, the process of moment distribution, previously described, cannot be directly applied without certain adjustments. There are, in general, two methods available for frames with joint translations in an analysis by moment distribution. One is the direct method, which includes the balancing of end shears; the other is the indirect method, which requires the set up and elimination of restraints to joint translation and, thus, involves the solution of simultaneous equations for frames with several degrees of joint translation. However, it is the latter method which seems, in most cases, more acceptable to a beginner and, therefore, will be discussed in this chapter.

Let us consider two types of loaded frames mounted on nonyielding supports in which joint translations are involved. The first is a frame under the action of a lateral force applied at a joint, such as the one shown in Fig. 14-1(a); the second is a frame that, together with loads acting on its members, forms an unsymmetrical system, such as the one shown in Fig. 14-1(b). To handle the latter type of frame, we may resort to the principle of superposition. Artificial holding force to prevent joint translation is imposed on the structure and subsequently eliminated. Thus the frame in Fig. 14-1(b) can be considered as the superposed effect of the two separate systems indicated in Fig. 14-1(c) and (d). In the first place [Fig. 14-1(c)], the translation of joints is prevented by providing an artificial support at the top of the column so that moment distribution can be carried out in the usual manner. The required holding force $R$ is then obtained by statics. The next step [Fig. 14-1(d)] is to eliminate the artificial restraint by applying to the top of column a lateral force equal to $R$ but opposite in direction. The resulting configuration is the same as that shown in Fig. 14-1(a). Therefore, the problem now reduces to dealing with frame under lateral forces applied at the joints.

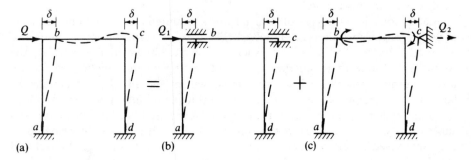

**Fig. 14-1**

To handle this type of frame, we consider the frame in Fig. 14-2(a) in which joints $a$ and $d$ are fixed, whereas joints $b$ and $c$ undergo both translation and rotation because of the lateral force $Q$ applied at the top of column. It is obvious that if the joint translation of $b$ is specified as $\delta$, then the joint translation of $c$ is determined and, in this case, is also $\delta$. Now the distortion imposed on joints $b$ and $c$ may be regarded as the superposed effect of the two following separate steps:

1. Translation without rotation [see Fig. 14-2(b)]
2. Rotation without translation [see Fig. 14-2(c)]

In step 1 joints $b$ and $c$ are locked against rotation ($\theta_b = \theta_c = 0$), and the joint translation $\delta$ is produced by applying a lateral force $Q_1$. It is clear that some external restraints (i.e., *locking moments*) are required at joints $b$ and $c$ in order to hold both joints against rotation. Also, end moments will be induced in the columns from which the lateral force $Q_1$ can be figured.

**Fig. 14-2**

In step 2 further joint translations are checked by providing an artificial support at the top of the column. Joints $b$ and $c$ are then unlocked so that they finally rotate to their actual position. Recall that to unlock a joint means to apply a moment equal but opposite to the locking moment at the joint. From the resulting end moments, the holding force $Q_2$ at the artificial support can be figured.

From the point of view of moment distribution step 1, which consists of locking all joints against rotation and performing joint translations by the force applied to joint, gives the fixed-end moment; and step 2, unlocking all joints and allowing no further joint translation, gives distributed moments and carry-over moments. These two steps parallel the procedures for analyzing frames without joint translation discussed in Chapter 13, which involve first locking all joints and applying loads to the members—fixed-end moment—and then unlocking all joints—distributed moments and carry-over moments.

Refer to Fig. 14-2. Since the analysis of the frame in Fig. 14-2(a) can be carried out by superposing the effects given in Fig. 14-2(b) and (c), the values of $Q_1$ and $Q_2$, determined from the resulting end moments in each of the separate cases, must satisfy $Q_1 + Q_2 = Q$. However, we usually do not know the value of $\delta$ from which the fixed-end moment is derived at the outset. After moment distribution, the structure response (end moments, shear forces, etc.) remains a function of the unknown $\delta$, which is yet to be determined from structure equilibrium with $Q$. (In this case, we may use $\sum F_x = 0$ to evaluate $\delta$.) However to make the calculation easier, we actually start by applying an arbitrary lateral joint translation, say $\Delta$, to the same structure. From this a set of fixed-end moments can be determined [Fig. 14-3(a)]. After the joints are unlocked, the distributed moments and carry-over moments are obtained [Fig. 14-3(b)]. In this manner, we reach an independent set of balanced force systems corresponding to the assumed joint translation $\Delta$ or the laterally applied force $P$ [Fig. 14-3(c)] which is figured from the end moments obtained in Fig. 14-3(a) and (b). With this done, and since the structure is linear, the solution for the frame in Fig. 14-2(a) can be found by the direct proportion $Q/P$. In the sections to follow, we discuss the fixed-end moment due to joint translation and illustrate the procedures discussed above.

**Fig. 14-3**

## 14-2. FIXED-END MOMENT DUE TO JOINT TRANSLATION

Figure 14-4(a) shows a frame with the length and moment of inertia for each member indicated. We begin by locking all joints against rotation so that

$$\theta_a = \theta_b = \theta_c = \theta_d = 0$$

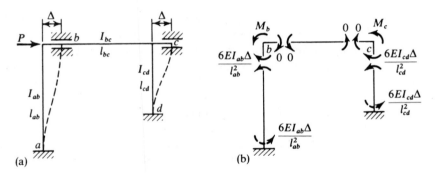

**Fig. 14-4**

Next, we apply a lateral force $P$ to the top of column under which force the frame will deflect sidewise, and the column tops will move a horizontal distance $\Delta$. End moments are then set up in the columns $ab$ and $cd$ because of the relative end deflection $\Delta$. These end moments, often referred to as *fixed-end moments due to joint translation* (sidesway), can be obtained easily by applying the slope-deflection equation:

$$M_{ab} = M_{ba} = 2E\frac{I_{ab}}{l_{ab}}\left(-\frac{3\Delta}{l_{ab}}\right) = -\frac{6EI_{ab}\,\Delta}{l_{ab}^2} \tag{14-1}$$

$$M_{cd} = M_{dc} = 2E\frac{I_{cd}}{l_{cd}}\left(-\frac{3\Delta}{l_{cd}}\right) = -\frac{6EI_{cd}\,\Delta}{l_{cd}^2} \tag{14-2}$$

Replacing $I/l$ with $K$ and $\Delta/l$ with $R$, we obtain

$$M_{ab} = M_{ba} = -6EK_{ab}R_{ab} \tag{14-1a}$$

$$M_{cd} = M_{dc} = -6EK_{cd}R_{cd} \tag{14-2a}$$

It is apparent that
$$M_{bc} = M_{cb} = 0$$

since $\theta_b = \theta_c = 0$ and there is no member rotation for $bc$.

These end moments are shown in Fig. 14-4(b), for which we note that the equilibrium of the moments at joint $b$ requires a counterclockwise external moment at $b$, called $M_b$, of the value $6EI_{ab}\,\Delta/l_{ab}^2$ in order to lock joint $b$ against rotation. Similarly, a counterclockwise external moment $M_c$ equal to $6EI_{cd}\,\Delta/l_{cd}^2$ is needed at $c$ to keep $\theta_c = 0$ during sidesway of the bent.

In general, the fixed-end moments developed in a member as the result of a pure relative end displacement $\Delta$ are given by

$$M^F = -\frac{6EI\,\Delta}{l^2} \tag{14-3}$$

or
$$M^F = -6EKR \tag{14-3a}$$

if we assume that the member rotates in a positive direction (clockwise).

It should be pointed out that, when applying Eq. 14-3, in most cases the exact value of $\Delta$ is not necessary, and the relative value of $I/l^2$ or $K/l$ is usually used in frame analysis.

## 14-3. ANALYSIS OF STATICALLY INDETERMINATE RIGID FRAMES WITH ONE DEGREE OF FREEDOM OF JOINT TRANSLATION BY MOMENT DISTRIBUTION

Figure 14-5(a) shows a frame composed of members of uniform cross section and subjected to a transverse load of 30 kips acting to the right. The analysis of it by the method of moment distribution is contained in the following steps:

1. Suppose that the structure of Fig. 14-5(a) is held against rotation at all joints. An arbitrary lateral force $P_1$ is then introduced so that the frame deflects to the right with a horizontal displacement $\Delta$ at the column tops as indicated in Fig. 14-5(b). The fixed-end moments resulting from sidesway are then induced in columns $ab$ and $cd$ according to Eq. 14-3. Hence,

$$M_{ab}^F = M_{ba}^F = -\frac{6EI\,\Delta}{400}$$

$$M_{cd}^F = M_{dc}^F = -\frac{6EI\,\Delta}{100}$$

Note that $\Delta$ is an arbitrary horizontal displacement corresponding to some lateral force $P_1$; the value of $\Delta$ is not known at the outset. Let us obtain a set (any set) of balanced force systems. Then we are free to assign any convenient value to $\Delta$ or in this case any convenient value to $6EI\,\Delta$. In other words, we assume any set of fixed-end moments inversely proportional to the square length of the member. Thus, if we choose

$$M_{ab}^F = M_{ba}^F = -100 \text{ ft-kips}$$

then
$$M_{cd}^F = M_{dc}^F = -400 \text{ ft-kips}$$

Having recorded these as in Fig. 14-5(c), we can calculate the end shears at $a$ and $d$ as indicated and figure

$$P_1 = 90 \text{ kips}$$

from the equilibrium condition $\sum F_x = 0$ for the whole frame. Meanwhile, from the equilibrium of joints $b$ and $c$, we observe that we must apply an artificial locking moment of $-100$ ft-kips at joint $b$ and $-400$ ft-kips at joint $c$ to keep $\theta_b = \theta_c = 0$. In this manner, we reach a system of balanced forces under the condition of joint translation without rotation, as shown in Fig. 14-5(c).

2. Next, let us introduce an imaginary lateral support at $c$ so as to prevent the frame from further joint translation. We unlock joints $b$ and $c$ by applying a couple of $+100$ ft-kips at joint $b$ and a couple of $+400$ ft-kips at joint $c$, as shown in Fig. 14-5(d). We then distribute moments at the near ends and carry over moments to the far ends—the result of unlocking moments following the usual moment-distribution procedures, as recorded in Fig. 14-5(e). The resulting moments are found to be consistent with a lateral force $P_2$ equal to $-30$ kips (acting to the left at $c$) in order to keep the frame from joint translation during

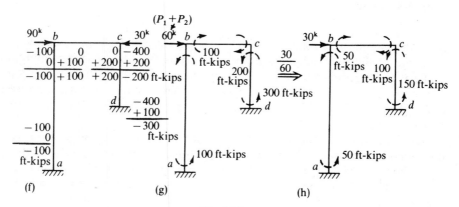

**Fig. 14-5**

the process of unlocking. In this manner, we obtain a set of balanced forces under the condition of joint rotation without translation.

3. The sum of the results of steps 1 and 2 is given in Fig. 14-5(f) or (g).

4. Multiplying this set of balance forces [Fig. 14-5(g)] with a ratio of 30/60 will give the solution for the frame of Fig. 14-5(a), as shown in Fig. 14-5(h).

This procedure for solving the frame in Fig. 14-5(a) is rather lengthy; however, the data can be presented in a more compact form as shown in Fig. 14-6, which involves the following:

1. Assume fixed-end moments for those members having relative end deflections in proportion to their values of $-6EI\Delta/l^2$. In this problem, we use

$$M^F_{ab} = M^F_{ba} = -100 \qquad M^F_{cd} = M^F_{dc} = -400$$

2. Distribute, carry over, and add together to obtain all the end moments.

3. Find the lateral force consistent with the end moments obtained.

4. Proportioning the result with a ratio of *given lateral force* to *obtained lateral force* will give the solution.

**Fig. 14-6**

If one of the ends, say end $d$, of the frame is hinged [see Fig. 14-7(a)], then $M_d = 0$. The procedure of moment distribution can be simplified by using modified stiffness for member $cd$:

$$K'_{cd} = \tfrac{3}{4}K_{cd}$$

as indicated.

Meanwhile, by releasing end $d$ and rendering it to its actual condition $M_d = 0$, one-half the amount of the distributed moment will be carried over to end $c$.

Fig. 14-7

The modified fixed-end moment at $c$ of member $cd$ due to joint translation $\Delta$ is therefore equal to $-3EI\,\Delta/l_{cd}^2$ [see Fig. 14-7(b)].

Generally, for a member pin-connected at one end, the modified fixed-end moment at the other end due to lateral joint translation $\Delta$ is given by

$$M^F = -\frac{3EI\,\Delta}{l^2} = -3EKR \qquad (14\text{-}4)$$

which is one-half the value previously obtained by assuming both ends fixed.

The analysis of the frame in Fig. 14-7(a) is shown in Fig. 14-7(c). We start by assuming $M_{ab}^F = M_{ba}^F = -100$ and $M_{cd}^F = -200$ consistent with the values given in Fig. 14-7(b). From the resulting end moments, we find that the end shear at $a$ is 8.8 kips and that at $d$ is 12.1 kips, both acting to the left. These correspond to a lateral force of 20.9 kips on the top of column acting to the right. Multiply the obtained set of balanced force system by the ratio 30/20.9 to get the solution. More illustrative examples follow.

**Example 14-1.** In Fig. 14-8(a) is shown a loaded one-story bent with an inclined leg, the relative $K$ value for each member is circled. The end moments were solved by slope deflection in Example 12-6. Let us now re-solve them by moment distribution.

**Fig. 14-8**

We begin by finding the relative end displacement for each member as shown in Fig. 14-8(b). Next, we assume fixed-end moments, due to consistent joint translations, in proportion to the value of $-6EK\Delta/l$:

$$M_{ab}^F = M_{ba}^F = -\frac{6E(1)(\Delta)}{15} \qquad \text{say} \quad -100$$

$$M_{bc}^F = M_{cb}^F = +\frac{6E(2)[(3/4)\,\Delta]}{15} \qquad \text{say} \quad +150$$

$$M_{cd}^F = M_{dc}^F = -\frac{6E(2)[(5/4)\,\Delta]}{25} \qquad \text{say} \quad -150$$

The process of moment distribution for this particular set up is shown in Fig. 14-8(c). Referring to Fig. 14-8(d), by applying $\sum M_o = 0$ for the entire frame, we find the resulting moments are consistent with a horizontal force $P = 42.5$ kips acting at $b$:

$$(15.15)(35) + (11.5)(50) - 109.1 - 145.5 - 20P = 0$$

$$P = 42.5 \text{ kips}$$

The final result, which is obtained by multiplying all the moments in Fig. 14-8(d) by the ratio 100/42.5, is shown in Fig. 14-8(e).

**Example 14-2.** Determine all the end moments of the loaded frame in Fig. 14-9(a).

The complete analysis is contained in the following steps:

1. Hold the loaded frame at the top of the column, say at joint $c$, against sidesway, and obtain the end moments by the usual moment-distribution procedures [Fig. 14-9(b)].

2. From $\sum F_x = 0$ for the entire frame, calculate the holding force needed to prevent sidesway. In this case, it is

$$12 - (4.71 + 2.06) = 5.23 \text{ kips}$$

acting to the left, as indicated in Fig. 14-9(c).

3. Remove the artificial holding force by the application of an equal and opposite force at the top of the column, and find the resulting end moments [see Fig. 14-9(d)].

4. Add the end moments from steps 1 and 3 to obtain the final solution [see Fig. 14-9(e)].

Note that the analysis for step 3 [Fig. 14-9(d)] can easily be accomplished by taking advantage of the antisymmetry of the system (i.e., by using modified stiffness for member $bc$). The complete analysis, which begins by assuming an arbitrary fixed-end moment equal to $-100$ ft-kips for the column because of joint translation, is shown in Fig. 14-10.

The technique employed in the preceding examples can be used in analyzing any frame having one degree of freedom of joint translation.

**Fig. 14-9**

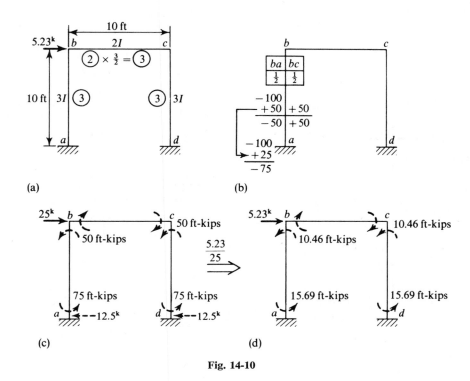

(a)

(b)

(c)

(d)

**Fig. 14-10**

## 14-4. ANALYSIS OF STATICALLY INDETERMINATE
RIGID FRAMES WITH TWO DEGREES
OF FREEDOM OF JOINT TRANSLATION
BY MOMENT DISTRIBUTION

A rigid frame having two degrees of freedom of joint translation can be analyzed by breaking it down to two independent cases in each of which only one degree of freedom of joint translation is allowed to occur.

Consider the two-story frame in Fig. 14-11(a). To handle it, let us refer to two separate cases, as shown in Fig. 14-11(b) and (c), each involving only one degree of freedom of joint translation. Each of these cases can be analyzed by the method of moment distribution previously described. For example, in the case in Fig. 14-11(b), joints $c$ and $d$ are held from translation by providing a lateral support at $d$, whereas joints $b$ and $e$ are displaced horizontally because of a force $X_1$ applied laterally at $b$. A moment-distribution solution can then be obtained, and all the end moments, shearing forces, and reactions are in terms of $X_1$. Let $mX_1$ denote the corresponding reaction of the lateral support at $d$, $m$ being a constant of proportionality. Note that the deflected shape (dashed line) indicates the initial position of the frame where the joint displace-

302

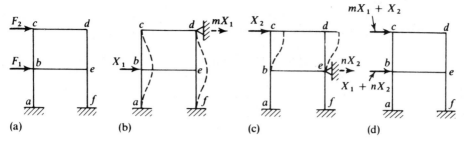

**Fig. 14-11**

ment of $b$ and $e$ has been introduced with all joints held against rotation. The final elastic curve after the joints have been released is not shown.

A similar solution can be carried out for the case in Fig. 14-11(c), where the pushing force is $X_2$ and the reaction of the lateral support at $e$ is $nX_2$, $n$ being a constant of proportionality.

The superposition of effects in the cases shown in Fig. 14-11(b) and (c) results in the case shown in Fig. 14-11(d), in which all the internal forces and reactions can be found as the linear combination of $X_1$ and $X_2$. Comparing Fig. 14-11(d) with Fig. 14-11(a), we see that (d) will be the solution of (a) by solving $X_1$ and $X_2$ from

$$X_1 + nX_2 = F_1 \tag{14-5}$$

$$mX_1 + X_2 = F_2 \tag{14-6}$$

A similar procedure can be used to analyze the two-stage bent shown in Fig. 14-12(a) or the gable bent shown in Fig. 14-12(b).

**Fig. 14-12**

303

**Example 14-3.** Find all the end moments for the frame in Fig. 14-13. Assume the same $EI$ for all members. This is the same problem as Example 12-8, which was solved by the method of slope deflection.

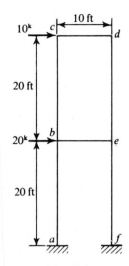

**Fig. 14-13**

To determine all the end moments by the method of moment distribution, we begin with the analysis of the frame in Fig. 14-14(a). The antisymmetry of the system permits us to use modified stiffnesses for members *be* and *cd* and to work with only half the structure while balancing moments, as shown in Fig. 14-14(b). In this case joints *c* and *d* are held from translation by the lateral support at *d*, and joints *b* and *e* are displayed horizontally an equal distance because of the lateral action applied at *b*. Consistent with the joint translation and the property of structure, we may assume the fixed-end moments for the columns as follows:

$$M^F_{ab} = M^F_{ba} = M^F_{ef} = M^F_{fe} = -100$$
$$M^F_{bc} = M^F_{cb} = M^F_{de} = M^F_{ed} = +100$$

After releasing the joints and balancing the moments, we obtain a set of end moments [see Fig. 14-14(b)] consistent with a pushing force of 36.13 applied at *b* and a lateral reaction of 16.5 acting to the left at *d* [see Fig. 14-14(c)]. Note that the reaction at *d* is first obtained from the balance of shear of the upper story (i.e., from $\sum F_x = 0$ for the portion of frame just above the level *be*). The pushing force at *b* is then determined by applying $\sum F_x = 0$ for the entire frame. Multiplying the result of Fig. 14-14(c) by the ratio $X_1/36.13$ gives the solution of Fig. 14-14(a), as shown in Fig. 14-14(d).

Next, let us follow the same procedure and analyze the frame in Fig. 14-15(a), in which joints *b* and *e* are held from translation owing to the lateral support

Fig. 14-14

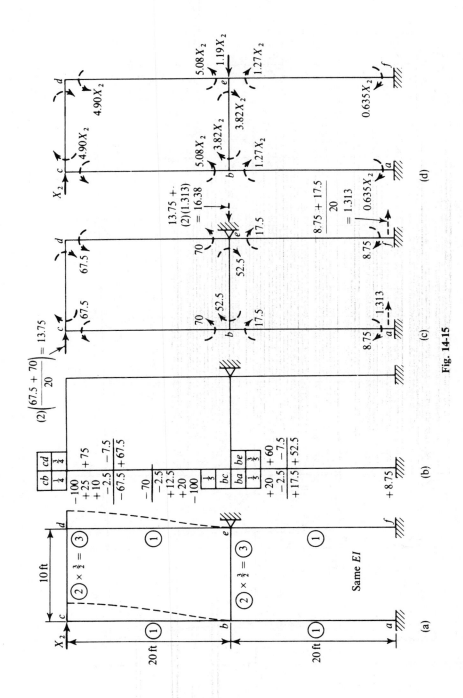

Fig. 14-15

at $e$, whereas joints $c$ and $d$ are displaced horizontally an equal distance due to a pushing force applied at $c$. Consistent with the sidesway and the property of structure, the fixed-end moments for the columns $bc$ and $de$ may be assumed to be

$$M_{bc}^F = M_{cb}^F = M_{de}^F = M_{ed}^F = -100$$

Taking advantage of antisymmetry by using modified stiffnesses in members $be$ and $cd$ and working with only half the structure, we complete the process of moment distribution as shown in Fig. 14-15(b). The resulting end moments are found to be consistent with a pushing force of 13.75 applied at $c$ and a lateral reaction of 16.38 acting to the left at $e$ [see Fig. 14-15(c)]. In this case the force at $c$ is first obtained from the balance of shear of the upper story, and the reaction at $e$ is then determined from $\sum F_x = 0$ for the entire frame.

Multiplying the result of Fig. 14-15(c) by the ratio $X_2/13.75$ gives the solution for the frame in Fig. 14-15(a), as shown in Fig. 14-15(d).

Let us now sum up the findings in the two steps above [Figs. 14-14(d) and 14-15(d)] as shown in Fig. 14-16(a). Moments and forces are taken positive following the indicated directions.

For solving the frame in Fig. 14-13, we set

$$X_1 - 1.19X_2 = 20 \qquad (14\text{-}7)$$

$$X_2 - 0.455X_1 = 10 \qquad (14\text{-}8)$$

from which

$$X_1 = 69.55$$

$$X_2 = 41.64$$

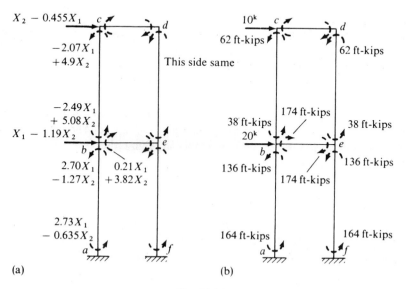

(a)                                 (b)

**Fig. 14-16**

Substituting these in Fig. 14-16(a) yields the desired solution, as shown in Fig. 14-16(b).

The small discrepancy between the foregoing result and the more exact solution obtained by the slope-deflection method (see Example 12-8) is due to the fact that we complete the moment distribution in each of the foregoing cases with only two cycles. However, two cycles have provided sufficient accuracy for practical purposes.

**Example 14-4.** Find all the end moments for the frame in Fig. 14-17(a). Assume the same *EI* for all members.

The analysis is contained in the following steps:

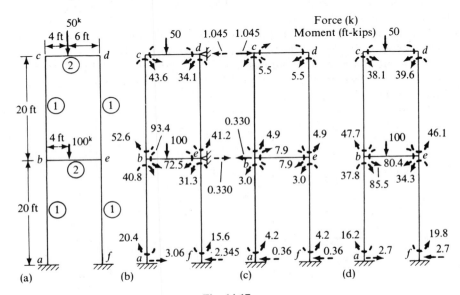

**Fig. 14-17**

1. The frame is completely prevented from sidesway by introducing lateral supports at joints *d* and *e*. Following the usual procedures for moment distribution, we determine all the end moments and find that they are consistent with a lateral holding force of 1.045 kips acting to the left at *d* and a lateral holding force of 0.330 kip acting to the right at *e*. The results are recorded, as in Fig. 14-17(b), by the dashed symbols. (See also Fig. 14-18 for the process of moment distribution in this step.)

2. Eliminate these artificial restraints by applying a set of equal and opposite forces to the corresponding joints, as shown in Fig. 14-17(c). To analyze it, we take advantage of the results of the preceding example [see Fig. 14-16(a)] and set

$$\frac{52.6 + 43.6 - 34.1 - 41.2}{20} = 1.045$$

|  | c |  |  | d |  |
|---|---|---|---|---|---|
| cb | cd |  | dc | de |  |
| $\frac{1}{3}$ | $\frac{2}{3}$ |  | $\frac{2}{3}$ | $\frac{1}{3}$ |  |
| 0 | −72 |  | +48 | 0 |  |
| +24 | +48 |  | −32 | −16 |  |
| +18 | −16 |  | +24 | −12 |  |
| −0.7 | −1.3 |  | −8 | −4 |  |
| +1.5 | −4 |  | −0.7 | −3.5 |  |
| +0.8 | +1.7 |  | +2.8 | +1.4 |  |
| +43.6 | −43.6 |  | +34.1 | −34.1 |  |
| +52.6 |  |  |  | −41.2 |  |
| +1.9 |  |  |  | −0.2 |  |
| −0.3 |  |  |  | −2 |  |
| +3 |  |  |  | −7 |  |
| +12 |  |  |  | −8 |  |
| +36 |  |  |  | −24 |  |
| 0 |  |  |  | 0 |  |

$$1.045 + 2.345 - 3.06 = 0.330$$

| $\frac{1}{4}$ |  |  | $\frac{1}{4}$ |  |
|---|---|---|---|---|
| bc | b |  | e | ed |
| ba | be |  | eb | ef |
| $\frac{1}{4}$ | $\frac{2}{4}$ |  | $\frac{2}{4}$ | $\frac{1}{4}$ |
| 0 | −144 |  | +96 | 0 |
| +36 | +72 |  | −48 | −24 |
| 0 | −24 |  | +36 | 0 |
| +3 | +6 |  | −14 | −7 |
| 0 | −7 |  | +3 | 0 |
| +1.8 | +3.6 |  | −0.5 | −0.3 |
| +40.8 | −93.4 |  | +72.5 | −31.3 |

+20.4  a

−15.6  f

$$\frac{40.8 + 20.4}{20} = 3.06 \qquad \frac{31.3 + 15.6}{20} = 2.345$$

**Fig. 14-18**

$$X_1 - 1.190X_2 = -0.330 \qquad (14\text{-}9)$$

$$-0.455X_1 + X_2 = 1.045 \qquad (14\text{-}10)$$

and solve for $\qquad X_1 = 1.99 \qquad X_2 = 1.95$

Substituting these in Fig. 14-16(a) yields the rseult recorded in Fig. 14-17(c) by the dashed symbols.

3. The superposition of steps 1 and 2 will give the final solution of the moments, as in Fig. 14-17(d).

## 14-5. ANALYSIS OF STATICALLY INDETERMINATE RIGID FRAMES WITH SEVERAL DEGREES OF FREEDOM OF JOINT TRANSLATION BY MOMENT DISTRIBUTION

The procedures of moment distribution used in the analysis of frames having two degrees of joint translation can be extended to the analysis having multiple degrees of freedom with respect to sidesway. For a frame with $n$ degrees of freedom of joint translation, the moment-distribution solution may be broken down into $(n + 1)$ separate cases including:

  1. A case in which the frame is held from joint translation by introducing $n$ artificial supports
  2. $n$ independent cases in each of which only one degree of freedom of joint translation is allowed to occur by maintaining $(n - 1)$ holding forces at other points where sidesway would take place

The superposition of these $(n + 1)$ cases will give the final solution provided that the artificial holding forces in step 1 are all eliminated by applying the results obtained in step 2. This necessitates the solution of $n$ simultaneous equations.

To illustrate this procedure, consider the four-story, two-bay building bent, which has four degrees of freedom of joint translation, shown in Fig. 14-19(a). The first step is to introduce artificial lateral supports at all the floors, as in Fig. 14-19(b), to prevent the frame from sidesway. All the loads are then applied, and a regular moment distribution without joint translation is performed to obtain all the end moments from which the reactions (the holding forces) at these supports can be found. We let $F_1, F_2, F_3$ and $F_4$ denote these forces.

The next step is to eliminate all of these artificial forces by applying a set of equal and opposite forces at the corresponding points, as shown in Fig. 14-19(c). The sum of the two steps of Fig. 14-19(b) and (c) will give the final solution.

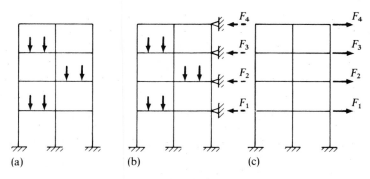

**Fig. 14-19**

However, to obtain the solution for the frame in Fig. 14-19(c) would require the complete analysis of four independent cases, as shown in Fig. 14-20. In each of the four cases, only one degree of freedom of sidesway is involved. We have, therefore, no difficulty in determining all the end moments and the reactions of the frame in terms of the lateral force applied to the floor level where the sidesway is not inhibited. For instance, in Fig. 14-20(a) a lateral force $X_1$ is applied to the first floor level which deflects sidewise. Note that the deflected shape (dashed line) indicates only the initial position of the frame before the balancing moments. Following the procedure described in Sec. 14-4, we find the reactions of the lateral supports at the second, third, and fourth floor levels to be $r_{21}X_1, r_{31}X_1$, and $r_{41}X_1$, respectively; $r_{21}, r_{31}, r_{41}$ being constants of proportionality. The other cases shown in Fig. 14-20 are similarly conducted. It now remains to find $X_1, X_2, X_3$, and $X_4$ by solving the four simultaneous equations,

$$X_1 + r_{12}X_2 + r_{13}X_3 + r_{14}X_4 = F_1 \tag{14-11}$$

$$r_{21}X_1 + X_2 + r_{23}X_3 + r_{24}X_4 = F_2 \tag{14-12}$$

$$r_{31}X_1 + r_{32}X_2 + X_3 + r_{34}X_4 = F_3 \tag{14-13}$$

$$r_{41}X_1 + r_{42}X_2 + r_{43}X_3 + X_4 = F_4 \tag{14-14}$$

With $X_1, X_2, X_3$, and $X_4$ determined and substituted in the cases of Fig. 14-20(a), (b), (c), and (d), respectively, we superpose the results to reach the solution for the frame in Fig. 14-19(c).

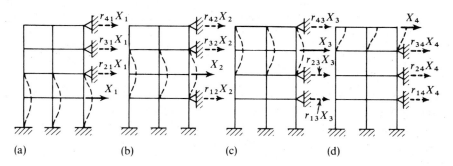

(a)   (b)   (c)   (d)

**Fig. 14-20**

## 14-6. INFLUENCE LINES BY MOMENT DISTRIBUTION

The Müller–Breslau procedures for constructing quantitative influence lines of indeterminate structures given in Secs. 11-5 and 11-6 are often laborious without the aid of a calculating machine. The method of moment distribution, which provides an easy way of determining the end moments of indeterminate beams or frames, can, therefore, be used to establish the influence lines for the end moments in all members and from them deduce all the other influence lines (reaction, shear, bending moment, etc.) by statics.

Consider the three-span uniform beam shown in Fig. 14-21(a). Find the ordinates, at intervals of 3 ft, of the influence line for $M_{bc}$ (moment over support $b$).

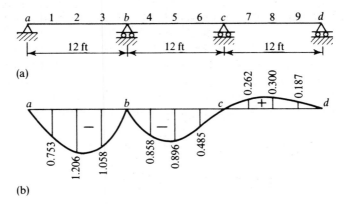

(a)

(b)

**Fig. 14-21**

The first step is to determine the fixed-end moment coefficients, the end moments in all members caused by applying a unit fixed-end moment at the end of each of the members. For example, when the fixed-end moment of unity occurs to end $a$, the resulting end moments in all members can be obtained by moment-distribution procedures, as shown in Fig. 14-22(a). Likewise, when a unit fixed-end moment is applied at end $b$ of member $ab$, all the end moments are computed as in Fig. 14-22(b).

Once these are done, the effects of a fixed-end moment of unity applied at each of the other ends may be obtained by inspection for the given beam. The complete fixed-end moment coefficients are shown in Fig. 14-23.

With the fixed-end moment coefficients determined, we may rather easily find any of the end moments of the beam due to any loading acting on the span. First compute the fixed-end moments caused by the loading; then multiply each of them by the corresponding fixed-end moment coefficient for the desired end moment; and finally sum up to obtain the result.

For instance, to find the influence ordinate for $M_{bc}$ at point 1, that is, 3 feet from the left end [Fig. 14-21(a)], we place a unit load there and compute the fixed-end moments for span $ab$ as

$$M_{ab}^F = \frac{(1)(3)(9)^2}{(12)^2} = 1.69 \qquad \text{(counterclockwise)}$$

$$M_{ba}^F = \frac{(1)(9)(3)^2}{(12)^2} = 0.56 \qquad \text{(clockwise)}$$

(a)

(b)

Fig. 14-22

Fig. 14-23

Multiplying each of these by the corresponding fixed-end moment coefficient for $M_{bc}$ and summing up gives

$$M_{bc} = (1.69)(-0.268) + (0.56)(-0.537) = -0.753$$

which is the influence ordinate for $M_{bc}$ at point 1.

Likewise, to obtain the influence ordinate for $M_{bc}$ at point 8, that is, 6 feet from the right end, we place a unit load at the midpoint of span $cd$ and compute

$$M_{cd}^F = \frac{(1)(12)}{8} = 1.5 \qquad \text{(counterclockwise)}$$

$$M_{dc}^F = 1.5 \qquad \text{(clockwise)}$$

Thus, $\qquad M_{bc} = (1.5)(+0.133) + (1.5)(+0.066) = +0.300$

In this manner, we complete the influence line for $M_{bc}$ as shown in Fig. 14-21(b).

## 14-7. SECONDARY STRESSES IN TRUSSES SOLVED BY MOMENT DISTRIBUTION

Recall that the elementary stress analysis of a truss (Chapter 4) is based on the following assumptions:

1. The ends of all members of a truss are jointed by smooth pins.
2. The external loads and reactions lie in the same plane as the truss and act only at the pins.
3. The centroidal axis of each member is straight and coincides with the line connecting the joint centers.
4. The weight of each member is neglected.

From these assumptions, it follows that each member in a truss is under purely axial load—tension or compression. The stresses so obtained are called *primary stresses*. In computing primary stresses, we use the original dimensions of the truss since the changes in the bar length are small and negligible. Stresses caused by conditions not taken in the primary stress analysis are called *secondary stresses*. Of these the most significant is due to the fact that in the frameworks used in present-day construction, the joints are seldom pinned but bolted to gusset plates or welded. The angles between the members cannot change freely; hence, the members are actually bent. As a result of this, end moments are set up and thus give rise to the secondary stresses.

In most light frameworks, the effect of the rigidity of the joints is not sufficient to cause any appreciable modification in the axial forces, and it is common practice to analyze them as if they were pin-connected. However, for the heavy trusses of bridges or industrial buildings, the stiffness of the joints often introduces secondary stresses too great to be neglected in the design. One way of obtaining approximate values of the secondary stresses is by the moment-distribution procedures, which may be outlined as follows:

1. Calculate the primary stresses by assuming the framework to be pin-connected.

2. Find the relative end deflection $\Delta$ or rotation $R$ for each member (see Example 8-8 for the rotation of a member).

3. Assume all joints are locked against rotation, and compute the fixed-end moment set up in each member based on $-6EI\Delta/l^2$ or $-6EKR$.

4. As in the moment-distribution method, the joints are then released until joint balance is obtained.

5. The sum of the results obtained for steps 3 and 4 gives the first approximation of the secondary end moments from which the secondary stresses can be determined.

6. The new stresses so obtained will cause additional joint deflections which cause additional end moments, stresses, and so on. However, in most practical cases, only the first approximation of secondary stresses is significant and is used in the design.

Other secondary effects, such as the weight of members and the eccentricity of bar forces, can be handled similarly by moment distribution, that is, by first locking all joints (obtaining fixed-end moments) then releasing them (balancing moments).

# PROBLEMS

**14-1.** Use the method of moment distribution to find the fixed-end moment and the spring force for the beam shown in Fig. 14-24.

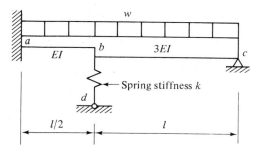

**Fig. 14-24**

**14-2.** Analyze each of the frames in Fig. 14-25 by moment distribution.

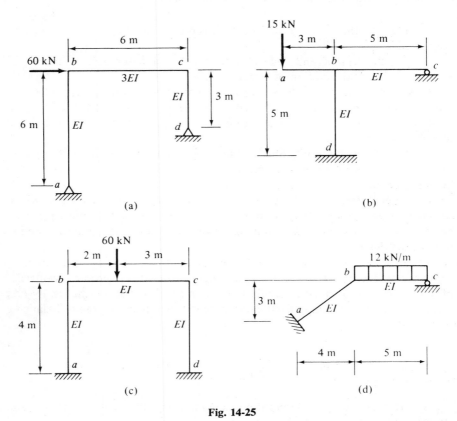

(a)

(b)

(c)

(d)

**Fig. 14-25**

**14-3.** Use the method of moment distribution to find the fixed-end moments for each of the beams of Prob. 12-12.

**14-4.** Solve the frame in Fig. 14-26 by moment distribution.

**Fig. 14-26**

**14-5.** Solve each frame of Prob. 12-11 by moment distribution.

**14-6.** Analyze the frame in Fig. 14-27 by moment distribution.

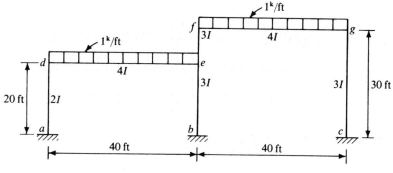

**Fig. 14-27**

**14-7.** Use the moment-distribution procedures to compute the ordinates, at 4-ft intervals, of the influence line for the moment over the center support of the four-span uniform beam. The length of each span equals 16 ft.

**14-8.** Use the method of moment distribution to construct the influence lines of Prob. 11-3.

# 15

## MATRIX ANALYSIS OF STRUCTURES

## BY THE FINITE ELEMENT METHOD

## PART I: THE FORCE METHOD

### 15-1. GENERAL

During the past decades, the rapid development of computers and the growing demand for better methods of analysis for complex and lightweight structures led to the development of methods for matrix analysis of structures. The use of matrix notation in expressing structural theory is in itself simple and elegant, but the practical value of matrix analysis would not have become clear without the invention of the high-speed digital computer.

It is true that the classical methods of structural analysis, such as the method of consistent deformations, Castigliano's theorems, or the slope-deflection method, which had only limited use in the past because of operational difficulties, have now regained their strength because of the invention of the digital computer. Indeed, solving a set of 100 simultaneous equations with a modern computer would hardly take a minute, and the solution of simultaneous equations is equivalent to inverting a matrix. However, the matrix method, to be discussed in this and the subsequent chapter, has its unique theoretical basis and particular procedures. According to its basic principle, it is sometimes better known as the *finite element* method, because a structure, instead of being a continuation of differential elements, is idealized as a composition of a number of finite pieces. This idea enables the step-by-step buildup of the force-displacement relationship of a structure from those basic elements of which the structure is composed.

Many complicated problems in various fields are thus solved by this new computational technique, including:

1. Trusses, beams, and rigid frames
2. Plates and shells of arbitrary shape and loading
3. Composite structures
4. Pressure vessels
5. Torsion in members with irregular section

6. Dynamic analysis of framework

and so on.

Since this book is an elementary treatment, we will confine the discussion to plane structures of the type listed as item 1.

The principal matrix procedures based on a finite element representation commonly fall into two categories: the force method and the displacement method. The force method treats the member forces as the basic unknowns; whereas the displacement method regards the nodal displacements as the basic unknowns. As will be seen, a duality exists between the two approaches. We discuss the force method in this chapter and the displacement method in Chapter 16.

## 15-2. BASIC CONCEPTS OF STRUCTURES

Structures, such as trusses, beams, and rigid frames, are defined as an assemblage of structural elements jointed together at a finite number of discrete points, called *nodes* or *nodal points*, and loaded only at these points. The term "node" is used instead of "joint" because often a concentratedly loaded point, not the conventional joint, is taken as a node. Distributed loads or any other type of member loads acting between the nodes can be replaced by equivalent loads at the nodes (see Sec. 15-8).

Any complicated structure can be cut into simpler components. For instance, a truss may be considered as composed of many two-force members pin-connected at their ends. A rigid frame may be taken as a composition of a number of three-force members or other convenient units. The behavior of the subdivided elements, as well as the whole structure, must satisfy the following basic conditions:

1. Equilibrium of forces
2. Compatibility of displacements
3. Force–displacement relationship specified by the geometric and elastic properties of the elements

These conditions are generally required by a linear structure no matter what method is used.

In the systematic analysis of structures, it is essential in the first place to introduce the notation used for force and displacement in this and Chapter 16. Consider the frame in Fig. 15-1(a), which is composed of three elements (members $a$, $b$, and $c$) subjected to external loads (nodal forces) denoted by $R_1$, $R_2$, and $R_3$ with the corresponding nodal displacements $r_1$, $r_2$, and $r_3$, respectively.

A typical member element is shown in Fig. 15-1(b) and is generally subjected to the internal forces of the end moments, denoted by $Q_i$ and $Q_j$, and the axial forces, denoted by $Q_k$. The member is also subjected to end shears;

(a)　　　　　　　　　　　　(b)

Fig. 15-1

however, the end shears can be expressed in terms of the end moments and therefore are not considered independent forces. Associated with the end moments and axial forces are the end rotations, $q_i$ and $q_j$, and the axial elongation $q_k$. The signs shown in Fig. 15-1 are considered as positive. A superscript is used for these quantities to identify their belonging to a particular member. For instance, $Q_i^a$, $Q_j^a$, and $Q_k^a$ indicate the internal forces for member $a$, and $q_i^a$, $q_j^a$, and $q_k^a$ their corresponding deformations.

In matrix representation, we use a column matrix $R$

$$R = \begin{Bmatrix} R_1 \\ R_2 \\ R_3 \\ \cdot \\ \cdot \\ \cdot \end{Bmatrix}$$

to denote all the nodal forces, and a column matrix $r$

$$r = \begin{Bmatrix} r_1 \\ r_2 \\ r_3 \\ \cdot \\ \cdot \\ \cdot \end{Bmatrix}$$

to denote the corresponding nodal displacements.
Also, we use a column matrix $Q$,

$$Q = \begin{Bmatrix} Q_i^a \\ Q_j^a \\ \cdot \\ \cdot \\ \cdot \\ Q_i^b \\ Q_j^b \\ \cdot \\ \cdot \\ \cdot \end{Bmatrix}$$

to denote all the internal end forces for members $a, b, \ldots$, and a column matrix $q$,

$$
q = \begin{Bmatrix} q_i^a \\ q_j^a \\ \cdot \\ \cdot \\ \cdot \\ q_i^b \\ q_j^b \\ \cdot \\ \cdot \end{Bmatrix}
$$

to denote the corresponding internal displacements.

Frequently involved in the subsequent discussion is the principle of virtual work, which serves in many cases as an effective substitute for the equations of equilibrium or compatibility. The principle simply states that for an elastic structure in equilibrium, the external virtual work is equal to the internal virtual work (virtual strain energy). The virtual work may be the result of either a virtual displacement or a virtual force. Let us use an overbar to denote each of the virtual elements. Thus, if virtual displacements are used, we have

$$
\bar{r}^T R = \bar{q}^T Q \tag{15-1}
$$

On the other hand, if virtual forces are used,

$$
\bar{R}^T r = \bar{Q}^T q \tag{15-2}
$$

## 15-3. EQUILIBRIUM, FORCE TRANSFORMATION MATRIX

For a statically determinate structure, each of the member forces may be expressed in terms of the external nodal loads by using the equilibrium conditions of the system alone. Thus,

$$
\begin{aligned}
Q_1 &= b_{11}R_1 + b_{12}R_2 + \cdots + b_{1n}R_n \\
Q_2 &= b_{21}R_1 + b_{22}R_2 + \cdots + b_{2n}R_n \\
&\quad \cdot \\
&\quad \cdot \\
&\quad \cdot \\
Q_m &= b_{m1}R_1 + b_{m2}R_2 + \cdots + b_{mn}R_n
\end{aligned} \tag{15-3}
$$

in which $Q_1 = Q_i^a$, $Q_2 = Q_j^a$, $\ldots$. Observe that $R_1, R_2, \ldots, R_n$ represent the total set of applied loads and $Q_1, Q_2, \ldots, Q_m$ the total set of member forces. No connection between the subscripts on $R$ and $Q$ is implied.

The matrix form for Eq. 15-3 is

$$
Q = bR \tag{15-4}
$$

where

$$
b = \begin{bmatrix} b_{11} & b_{12} & \cdots & b_{1n} \\ b_{21} & b_{22} & \cdots & b_{2n} \\ \cdot & & & \\ \cdot & & & \\ \cdot & & & \\ b_{m1} & b_{m2} & \cdots & b_{mn} \end{bmatrix}
\tag{15-5}
$$

called the *force transformation matrix*, which relates the internal forces to the external forces. Matrix $b$ is usually a rectangular matrix in which the typical element $b_{ij}$ is the value of the internal force component $Q_i$ caused by a unit value of external load $R_j$. Note that $b$ is merely an expression of equilibrium for the system.

As for a statically indeterminate structure, the internal member forces cannot be expressed in terms of the external loads by equilibrium alone. However, as previously stated (see Sec. 9-1), a statically indeterminate structure can be made determinate by removing the redundant elements. The statically determinate and stable structure that remains after the removal of the extra restraints is called a *primary structure*. We then consider the original structure as equivalent to the primary structure subjected to the combined influences of the applied loads and the unknown redundant forces, thereby treating the redundants as a part of the external loads of unknown magnitude. In this way, we can express member forces in terms of the original applied loads $R$ and the redundant forces $X$ as

$$
Q = b_R R + b_X X
\tag{15-6}
$$

or

$$
Q = [b_R \,|b_X] \left\{ \frac{R}{X} \right\}
\tag{15-7}
$$

where $b_R$ and $b_X$ are force transformation matrices representing the separate influences of the known applied loads $R$ and the unknown redundants $X$ on the member forces. They are generally rectangular matrices.

## 15-4. COMPATIBILITY

Compatibility is a continuity condition on the displacements of the structure after the external loads are applied to the structure. Compatibility must be brought into the analysis of statically indeterminate structures since the equilibrium equations alone do not suffice to solve the problem.

The compatibility conditions used in the force method for solving static structures mounted on rigid supports are that the displacements at all the cuts, or releases of redundant points caused by the original applied loads, and the redundant forces must be made to vanish in order that the continuity of the structures can be maintained. That is, the gap in the displacements resulting from the applied loads is precisely removed by the redundant forces.

If we use $r_x$ to denote the displacement matrix (a column matrix) at the releases of the redundant points of an indeterminate structure, we have

$$r_x = 0 \tag{15-8}$$

## 15-5. FORCE–DISPLACEMENT RELATIONSHIP, FLEXIBILITY COEFFICIENT, FLEXIBILITY MATRIX

A *flexibility coefficient* $f_{ij}$ is the displacement at point $i$ due to a unit action at point $j$, all other points being unloaded. Apparently, the flexibility coefficient constitutes a relationship between deformation and force. Applying the principle of superposition, we may express the deformation at any point of a system caused by a set of forces in terms of the flexibility coefficients.

Our intention is, first of all, to establish the relationship between the member displacements and the member forces of a structure. Consider a typical member $a$ taken from a plane structure as shown in Fig. 15-2. As before, the member forces are represented by a column matrix $Q^a$,

$$Q^a = \begin{Bmatrix} Q_i^a \\ Q_j^a \\ Q_k^a \end{Bmatrix}$$

and the corresponding member deformations are represented by a column matrix $q^a$,

$$q^a = \begin{Bmatrix} q_i^a \\ q_j^a \\ q_k^a \end{Bmatrix}$$

Note that the clockwise end moments and rotations and the tensile axial forces and elongation are considered as positive.

**Fig. 15-2**

Using the flexibility coefficient $f_{ij}^a$, we may express each of the member deformations in terms of the separate influences of the whole set of member forces

$$\begin{aligned}
q_i^a &= f_{ii}^a Q_i^a + f_{ij}^a Q_j^a + f_{ik}^a Q_k^a \\
q_j^a &= f_{ji}^a Q_i^a + f_{jj}^a Q_j^a + f_{jk}^a Q_k^a \\
q_k^a &= f_{ki}^a Q_i^a + f_{kj}^a Q_j^a + f_{kk}^a Q_k^a
\end{aligned} \tag{15-9}$$

or in matrix form

$$q^a = f^a Q^a \tag{15-10}$$

in which

$$f^a = \begin{bmatrix} f^a_{ii} & f^a_{ij} & f^a_{ik} \\ f^a_{ji} & f^a_{jj} & f^a_{jk} \\ f^a_{ki} & f^a_{kj} & f^a_{kk} \end{bmatrix} \tag{15-11}$$

defined as the *element flexibility matrix*. Clearly, the coefficient, for instance $f^a_{ii}$, is given by

$$f^a_{ii} = q^a_i \quad \text{as} \quad Q^a_i = 1 \qquad Q^a_j = Q^a_k = 0$$

The rest can similarly be defined.

The descriptions above refer to an individual element. For a structure consisting of $a, b, \ldots$ elements, we have

$$q^a = f^a Q^a$$
$$q^b = f^b Q^b$$
$$\cdot$$
$$\cdot$$
$$\cdot$$

Let

$$q = \begin{Bmatrix} q^a \\ q^b \\ \cdot \\ \cdot \end{Bmatrix} \quad \text{and} \quad Q = \begin{Bmatrix} Q^a \\ Q^b \\ \cdot \\ \cdot \end{Bmatrix}$$

These equations can be put in matrix form as

$$q = fQ \tag{15-12}$$

where

$$f = \begin{bmatrix} f^a & & \\ & f^b & \\ & & \cdot \\ & & & \cdot \end{bmatrix} \tag{15-13}$$

which is a diagonal matrix with element flexibility matrices as its constituents.

Since the flexibility coefficients of Eq. 15-11 serve to relate the member deformations to the member forces, they are certainly governed by the geometric and material properties of the member. Suppose that the member is prismatic with length $L$, cross-sectional area $A$, moment of inertia $I$, and modulus of elasticity $E$ and regarded as simply supported. The elements in the first column of $f^a$ are, by definition, the member deformation resulting from $Q^a_i = 1$. These are found to be

$$f^a_{ii} = \text{rotation of the left end} = \frac{L}{3EI}$$

$$f^a_{ji} = \text{rotation at the right end} = -\frac{L}{6EI}$$

$$f^a_{ki} = \text{elongation of the member} = 0$$

Note that $f_{ii}^a$ and $f_{ji}^a$ can easily be determined by the conjugate-beam method and that $f_{ki}^a = 0$ is apparent. All the other elements can be obtained similarly. Thus, the member flexibility matrix is given by

$$f^a = \begin{bmatrix} \dfrac{L}{3EI} & -\dfrac{L}{6EI} & 0 \\[2ex] -\dfrac{L}{6EI} & \dfrac{L}{3EI} & 0 \\[2ex] 0 & 0 & \dfrac{L}{AE} \end{bmatrix} \tag{15-14}$$

Note that the member flexibility matrix is symmetric because of reciprocity.

If the effect of axial forces in the member is neglected, as is usually done in rigid-frame analysis, then

$$f^a = \frac{L}{6EI}\begin{bmatrix} 2 & -1 \\ -1 & 2 \end{bmatrix} \tag{15-15}$$

For a truss member subjected to axial forces only

$$f^a = \begin{bmatrix} \dfrac{L}{AE} \end{bmatrix} \tag{15-16}$$

We have already established the relationship between the member deformations and the member forces (the internal relationship). Next we must establish the relationship between the nodal displacements $r$ and the nodal forces $R$ (the external relationship). This may be accomplished by using the technique of virtual work. We start by Eq. 15-2:

$$\bar{R}^T r = \bar{Q}^T q \tag{15-17}$$

Based on Eqs. 15-4 and 15-12, namely

$$Q = bR \qquad \text{and} \qquad q = fQ$$

we have
$$\bar{Q} = b\bar{R}$$

or
$$\bar{Q}^T = \bar{R}^T b^T \tag{15-18}$$

and
$$q = fbR \tag{15-19}$$

Substituting Eqs. 15-18 and 15-19 in Eq. 15-17 yields

$$\bar{R}^T r = \bar{R}^T b^T f b R$$

form which
$$r = b^T f b R \tag{15-20}$$

If we let

$$F = b^T f b \tag{15-21}$$

$F$ being called the *total flexibility matrix* or *flexibility matrix of the structure*, then

$$r = FR \tag{15-22}$$

For a statically determinate structure Eq. 15-22 gives a direct solution of all the nodal displacements in terms of the external nodal forces.

If the structure is statically indeterminate, then the external work must include the work caused by the redundant forces; also, Eq. 15-6 must be used instead of Eq. 15-4. To relate $r$ to $R$, we begin the derivation by

$$\bar{R}^T r + \bar{X}^T r_X = \bar{Q}^T q \tag{15-23}$$

Using equilibrium Eq. 15-6 and virtual force, we have

$$\bar{Q} = b_R \bar{R} + b_X \bar{X}$$

so that

$$\bar{Q}^T = \bar{R}^T b_R^T + \bar{X}^T b_X^T \tag{15-24}$$

Also, because of Eqs. 15-12 and 15-6,

$$q = fQ = fb_R R + fb_X X \tag{15-25}$$

Substituting Eqs. 15-24 and 15-25 in Eq. 15-23 yields

$$\bar{R}^T r + \bar{X}^T r_X = (\bar{R}^T b_R^T + \bar{X}^T b_X^T)(fb_R R + fb_X X)$$

or     $$\bar{R}^T r + \bar{X}^T r_X = \bar{R}^T(b_R^T fb_R R + b_R^T fb_X X) + \bar{X}^T(b_X^T fb_R R + b_X^T fb_X X)$$

Comparing the virtual forces on the left and the right sides of the equation above we have

$$r = b_R^T fb_R R + b_R^T fb_X X$$

$$r_X = b_X^T fb_R R + b_X^T fb_X X$$

These may be arranged as

$$r = F_{RR}R + F_{RX}X \tag{15-26}$$

$$r_X = F_{XR}R + F_{XX}X \tag{15-27}$$

if we let

$$F_{RR} = b_R^T fb_R \qquad F_{RX} = b_R^T fb_X$$
$$F_{XR} = b_X^T fb_R \qquad F_{XX} = b_X^T fb_X \tag{15-28}$$

Equations 15-26 and 15-27 may be put in matrix form as

$$\left\{ \frac{r}{r_X} \right\} = \left[ \begin{array}{c|c} F_{RR} & F_{RX} \\ \hline F_{XR} & F_{XX} \end{array} \right] \left\{ \frac{R}{X} \right\} \tag{15-29}$$

For structures on rigid supports Eq. 15-29 becomes

$$\left\{ \frac{r}{r_X = 0} \right\} = \left[ \begin{array}{c|c} F_{RR} & F_{RX} \\ \hline F_{XR} & F_{XX} \end{array} \right] \left\{ \frac{R}{X} \right\} \tag{15-30}$$

The compatibility condition is, therefore,

$$F_{XR}R + F_{XX}X = 0 \tag{15-31}$$

from which

$$X = -F_{XX}^{-1}F_{XR}R \tag{15-32}$$

Equation 15-32 expresses the solution for the redundants.

Substituting Eq. 15-32 in Eq. 15-26, we finally relate the nodal displacements to the nodal forces,

$$r = (F_{RR} - F_{RX}F_{XX}^{-1}F_{XR})R$$

or simply

$$r = F'R \tag{15-33}$$

if we let

$$F' = F_{RR} - F_{RX}F_{XX}^{-1}F_{XR} \tag{15-34}$$

$F'$ is called the *flexibility matrix of the indeterminate structure,* which relates directly the nodal displacements to the nodal loads covering the effects of the redundants.

With the redundants $X$ found to be $-F_{XX}^{-1}F_{XR}R$, the member forces are then solved by equilibrium:

$$Q = b_R R + b_X X = (b_R - b_X F_{XX}^{-1} F_{XR})R$$

That is,

$$Q = b'R \tag{15-35}$$

if we let

$$b' = b_R - b_X F_{XX}^{-1} F_{XR} \tag{15-36}$$

$b'$ being the *force transformation matrix of the indeterminate structure,* relating directly the member forces to the applied nodal loads covering the effects of the redundants.

An alternative form of $F'$ may be obtained in terms of $b'$:

$$F' = b_R^T f b' \tag{15-37}$$

since

$$F' = F_{RR} - F_{RX} F_{XX}^{-1} F_{XR}$$

$$= b_R^T f b_R - b_R^T f b_X F_{XX}^{-1} F_{XR} = b_R^T f (b_R - b_X F_{XX}^{-1} F_{XR})$$

which yields Eq. 15-37. Note that Eq. 15-34 for finding $F'$ is quite general but that the alternative form given by Eq. 15-37 is more convenient if $b'$ is first determined.

The following identity is useful for checking results:

$$b_X^T f b' = 0 \tag{15-38}$$

This can easily be proved as follows:

$$b_X^T f b' = b_X^T f (b_R - b_X F_{XX}^{-1} F_{XR}) = F_{XR} - F_{XX} F_{XX}^{-1} F_{XR}$$

$$= F_{XR} - F_{XR} = 0$$

## 15-6. ANALYSIS OF STATICALLY DETERMINATE STRUCTURES BY THE FORCE METHOD

As developed in Sec. 15-3, for a statically determinate structure, the internal forces $Q$ can be solved by equilibrium alone:

$$Q = bR$$

See Eq. 15-4.

Also the nodal displacements $r$ can be solved by

$$r = b^T f b R$$

See Eq. 15-20.

Assume that the purely statical task of evaluating the force transformation matrix $b$ is not difficult, although this phase of analysis may cost a considerable amount of labor in complicated problems.

The procedures for analyzing a statically determinate structure by the force method are contained in the following steps:

1. Define the external nodal loads $R$.
2. Define the internal member forces $Q$.
3. Determine the force transformation matrix $b$.

Consider the elements of the first column of $b$. If we let

$$R_1 = 1 \qquad R_2 = R_3 = \cdots = R_n = 0$$

it is readily seen from Eq. 15-3 that $Q_1, Q_2, \ldots, Q_m$ are the elements of the first column. The rest can be obtained similarly.

4. The internal member forces $Q$ are then solved by

$$Q = bR$$

5. Determine individual element flexibility matrices $f^a, f^b, \ldots$ according to Eq. 15-15 or 15-16, and assemble them as a diagonal matrix

$$f = \begin{bmatrix} f^a & & & \\ & f^b & & \\ & & \cdot & \\ & & & \cdot \end{bmatrix}$$

6. Compute the flexibility matrix of the structure

$$F = b^T f b$$

7. Find the nodal displacements $r$,
$$r = FR$$

**Example 15-1.** Find the bar forces of the truss shown in Fig. 15-3. Find also the deflections corresponding to the applied loads $R_1$ and $R_2$. Assume that $L/A = 1$ for all members.

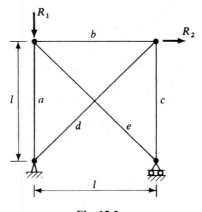

**Fig. 15-3**

The load matrix is

$$R = \begin{Bmatrix} R_1 \\ R_2 \end{Bmatrix}$$

The truss has five bars designated by $a, b, c, d$, and $e$. The member-force matrix is

$$Q = \begin{Bmatrix} Q^a \\ Q^b \\ Q^c \\ Q^d \\ Q^e \end{Bmatrix}$$

The force transformation matrix $b$ is given by

$$b = \begin{bmatrix} -1 & 0 \\ 0 & 0 \\ 0 & -1 \\ 0 & \sqrt{2} \\ 0 & 0 \end{bmatrix}$$

in which the first column contains the bar forces of the truss in Fig. 15-3 in the order $a, b, c, d, e$, resulting from $R_1 = 1, R_2 = 0$. The second column contains the corresponding bar forces resulting from $R_2 = 1, R_1 = 0$. From equilibrium

$$\begin{Bmatrix} Q^a \\ Q^b \\ Q^c \\ Q^d \\ Q^e \end{Bmatrix} = \begin{bmatrix} -1 & 0 \\ 0 & 0 \\ 0 & -1 \\ 0 & \sqrt{2} \\ 0 & 0 \end{bmatrix} \begin{Bmatrix} R_1 \\ R_2 \end{Bmatrix}$$

For individual members the flexibility matrices are found to be

$$f^a = f^b = f^c = f^d = f^e = \frac{1}{E}$$

since $L/A = 1$ for all members. Thus, the diagonal matrix

$$f = \begin{bmatrix} f^a & & & & \\ & f^b & & & \\ & & f^c & & \\ & & & f^d & \\ & & & & f^e \end{bmatrix} = \frac{1}{E} \begin{bmatrix} 1 & & & & \\ & 1 & & & \\ & & 1 & & \\ & & & 1 & \\ & & & & 1 \end{bmatrix}$$

The total flexibility matrix is then determined:

$$F = b^T f b$$

$$= \begin{bmatrix} -1 & 0 & 0 & 0 & 0 \\ 0 & 0 & -1 & \sqrt{2} & 0 \end{bmatrix} \frac{1}{E} \begin{bmatrix} 1 & & & & \\ & 1 & & & \\ & & 1 & & \\ & & & 1 & \\ & & & & 1 \end{bmatrix} \begin{bmatrix} -1 & 0 \\ 0 & 0 \\ 0 & -1 \\ 0 & \sqrt{2} \\ 0 & 0 \end{bmatrix}$$

$$= \frac{1}{E} \begin{bmatrix} 1 & 0 \\ 0 & 3 \end{bmatrix}$$

The nodal displacements $r$ are solved by

$$r = FR$$

$$\begin{Bmatrix} r_1 \\ r_2 \end{Bmatrix} = \frac{1}{E} \begin{bmatrix} 1 & 0 \\ 0 & 3 \end{bmatrix} \begin{Bmatrix} R_1 \\ R_2 \end{Bmatrix}$$

or $$\qquad r_1 = \frac{R_1}{E} \qquad r_2 = \frac{3R_2}{E}$$

**Example 15-2.** Find the deflections corresponding to the applied loads for the cantilever beam shown in Fig. 15-4(a). Assume constant $EI$.

(a)                                      (b)

**Fig. 15-4**

Since the loaded point of $R_1$ must be considered as a nodal point, it divides the beam into two segments, designated as member $a$ and member $b$ in Fig. 15-4(b). The internal member forces are shown by dashed lines. From equilibrium

$$R_1 = 1 \qquad R_2 = 1 \qquad R_3 = 1$$

$$Q = bR \qquad \begin{Bmatrix} Q_i^a \\ Q_j^a \\ Q_i^b \\ Q_j^b \end{Bmatrix} = \begin{bmatrix} -L_1 & -(L_1 + L_2) & -1 \\ 0 & L_2 & 1 \\ 0 & -L_2 & -1 \\ 0 & 0 & 1 \end{bmatrix} \begin{Bmatrix} R_1 \\ R_2 \\ R_3 \end{Bmatrix}$$

Note that the elements of the first column of matrix $b$ are the member forces caused by $R_1 = 1, R_2 = R_3 = 0$ for the beam shown in Fig. 15-4(b). This gives

$$Q_i^a = -L_1 \qquad Q_j^a = Q_i^b = Q_j^b = 0$$

The second column of matrix $b$ contains the member forces resulting from

$R_2 = 1$, $R_1 = R_3 = 0$. Thus,

$$Q_i^a = -(L_1 + L_2) \qquad Q_j^a = L_2 \qquad Q_i^b = -L_2 \qquad Q_j^b = 0$$

And the third column of matrix $b$ contains the member forces due to a unit couple applied only at the free end of the beam (i.e., $R_3 = 1$, $R_1 = R_2 = 0$). This gives

$$Q_i^a = -1 \qquad Q_j^a = 1 \qquad Q_i^b = -1 \qquad Q_j^b = 1$$

The individual member flexibility matrices are

$$f^a = \frac{1}{6EI}\begin{bmatrix} 2L_1 & -L_1 \\ -L_1 & 2L_1 \end{bmatrix} \qquad f^b = \frac{1}{6EI}\begin{bmatrix} 2L_2 & -L_2 \\ -L_2 & 2L_2 \end{bmatrix}$$

from which

$$f = \frac{1}{6EI}\begin{bmatrix} 2L_1 & -L_1 & 0 & 0 \\ -L_1 & 2L_1 & 0 & 0 \\ 0 & 0 & 2L_2 & -L_2 \\ 0 & 0 & -L_2 & 2L_2 \end{bmatrix}$$

The total flexibility matrix $F$ is obtained from

$$F = b^T f b = \begin{bmatrix} -L_1 & 0 & 0 & 0 \\ -(L_1 + L_2) & L_2 & -L_2 & 0 \\ -1 & 1 & -1 & 1 \end{bmatrix}\left(\frac{1}{6EI}\right)$$

$$\cdot \begin{bmatrix} 2L_1 & -L_1 & 0 & 0 \\ -L_1 & 2L_1 & 0 & 0 \\ 0 & 0 & 2L_2 & -L_2 \\ 0 & 0 & -L_2 & 2L_2 \end{bmatrix}\begin{bmatrix} -L_1 & -(L_1 + L_2) & -1 \\ 0 & L_2 & 1 \\ 0 & -L_2 & -1 \\ 0 & 0 & 1 \end{bmatrix}$$

$$= \begin{bmatrix} \dfrac{L_1^3}{3EI} & \dfrac{2L_1^3 + 3L_1^2 L_2}{6EI} & \dfrac{L_1^2}{2EI} \\ \dfrac{2L_1^3 + 3L_1^2 L_2}{6EI} & \dfrac{(L_1 + L_2)^3}{3EI} & \dfrac{(L_1 + L_2)^2}{2EI} \\ \dfrac{L_1^2}{2EI} & \dfrac{(L_1 + L_2)^2}{2EI} & \dfrac{L_1 + L_2}{EI} \end{bmatrix}$$

Thus,

$$\begin{Bmatrix} r_1 \\ r_2 \\ r_3 \end{Bmatrix} = \begin{bmatrix} \dfrac{L_1^3}{3EI} & \dfrac{2L_1^3 + 3L_1^2 L_2}{6EI} & \dfrac{L_1^2}{2EI} \\ \dfrac{2L_1^3 + 3L_1^2 L_2}{6EI} & \dfrac{(L_1 + L_2)^3}{3EI} & \dfrac{(L_1 + L_2)^2}{2EI} \\ \dfrac{L_1^2}{2EI} & \dfrac{(L_1 + L_2)^2}{2EI} & \dfrac{L_1 + L_2}{EI} \end{bmatrix}\begin{Bmatrix} R_1 \\ R_2 \\ R_3 \end{Bmatrix}$$

or
$$r_1 = \frac{R_1 L_1^3}{3EI} + \frac{R_2(2L_1^3 + 3L_1^2 L_2)}{6EI} + \frac{R_3 L_1^2}{2EI} \qquad (15\text{-}39)$$

$$r_2 = \frac{R_1(2L_1^3 + 3L_1^2 L_2)}{6EI} + \frac{R_2(L_1 + L_2)^3}{3EI} + \frac{R_3(L_1 + L_2)^2}{2EI} \qquad (15\text{-}40)$$

$$r_3 = \frac{R_1 L_1^2}{2EI} + \frac{R_2(L_1 + L_2)^2}{2EI} + \frac{R_3(L_1 + L_2)}{EI} \qquad (15\text{-}41)$$

As a particular problem, find the vertical deflection and the rotation at the free end of the loaded cantilever beam shown in Fig. 15-5.

$$EI \quad \text{constant}$$

**Fig. 15-5**

To do this, we set $R_1 = P$, $R_2 = R_3 = 0$ in Eq. 15-40 to obtain

$$r_2 = \frac{P(2L_1^3 + 3L_1^2 L_2)}{6EI}$$

which is the resulting vertical deflection of the end of the beam, and we set $R_1 = P$, $R_2 = R_3 = 0$ in Eq. 15-41 to obtain

$$r_3 = \frac{PL_1^2}{2EI}$$

which is the resulting rotation of the end of the beam.

## 15-7. ANALYSIS OF STATICALLY INDETERMINATE STRUCTURES BY THE FORCE METHOD

As developed in Secs. 15-3 to 15-5, the procedures for analyzing a statically indeterminate structure by the force method are given as follows:

1. Define the external loads $R$.
2. Define the internal member forces $Q$, and specify the redundants $X$.
3. Calculate the force transformation matrices $b_R$ and $b_X$ from equilibrium:

$$Q = [b_R \mid b_X] \left\{ \begin{matrix} R \\ X \end{matrix} \right\}$$

4. Determine the individual element flexibility matrices $f^a, f^b, \ldots,$ and assemble them to obtain $f$:

$$f = \begin{bmatrix} f^a & & & \\ & f^b & & \\ & & \cdot & \\ & & & \cdot \end{bmatrix}$$

5. Calculate $F_{XR}$:

$$F_{XR} = b_X^T f b_R$$

6. Calculate $F_{XX}$:

$$F_{XX} = b_X^T f b_X$$

7. Find the inverse of $F_{XX}$.
8. Solve the redundants $X$ by

$$X = -F_{XX}^{-1} F_{XR} R$$

and substitute $X$ in the equilibrium equation to obtain the member forces $Q$.
9. Alternatively, we may find $b'$ by

$$b' = b_R - b_X F_{XX}^{-1} F_{XR}$$

and obtain the member forces $Q$ by

$$Q = b'R$$

10. If the nodal displacements are desired, calculate $F'$ by

$$F' = b_R^T f b'$$

and find $r$ by                                    $$r = F'R$$

As seen in the latter part of Example 15-2, if the points where the displacements are desired are not actually loaded, then we must apply fictitious loads of zero value at these points in order to carry out the procedures listed above.

**Example 15-3.** Find the bar forces of the truss in Fig. 15-6(a) by the force method. Also find the nodal displacement corresponding to the applied load. Assume that $E = 30,000$ kips/in.² and $L(\text{ft})/A(\text{in.}^2) = 1$ for all members.

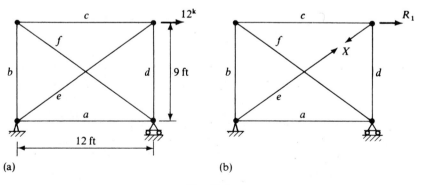

(a)                                    (b)

**Fig. 15-6**

The truss shown in Fig. 15-6(a) is statically indeterminate to the first degree. Let us select bar $e$ as the redundant and denote the external load of 12 kips by $R_1$, as shown in Fig. 15-6(b). The bar forces are denoted by $Q^a, Q^b, \ldots, Q^f$. From equilibrium based on the primary structure of Fig. 15-6(b),

$$R_1 = 1 \quad X = 1$$

$$\begin{Bmatrix} Q^a \\ Q^b \\ Q^c \\ Q^d \\ Q^e \\ Q_f \end{Bmatrix} = \begin{bmatrix} 1 & -\frac{4}{5} \\ \frac{3}{4} & -\frac{3}{5} \\ 1 & -\frac{4}{5} \\ 0 & -\frac{3}{5} \\ 0 & 1 \\ -\frac{5}{4} & 1 \end{bmatrix} \begin{Bmatrix} R_1 \\ X \end{Bmatrix}$$

$$\qquad\qquad b_R \qquad b_X$$

since $L/A = 1$ for all members

$$f = \frac{1}{E} \begin{bmatrix} 1 & & & & \\ & 1 & & & \\ & & 1 & & \\ & & & 1 & \\ & & & & 1 \end{bmatrix}$$

Thus,

$$F_{XR} = b_X^T f b_R$$

$$= [-\tfrac{4}{5} \quad -\tfrac{3}{5} \quad -\tfrac{4}{5} \quad -\tfrac{3}{5} \quad 1 \quad 1]\left(\frac{1}{E}\right)\begin{Bmatrix} 1 \\ \frac{3}{4} \\ 1 \\ 0 \\ 0 \\ -\frac{5}{4} \end{Bmatrix} = -\frac{3.3}{E}$$

$$F_{XX} = b_X^T f b_X$$

$$= [-\tfrac{4}{5} \quad -\tfrac{3}{5} \quad -\tfrac{4}{5} \quad -\tfrac{3}{5} \quad 1 \quad 1]\left(\frac{1}{E}\right)\begin{Bmatrix} -\frac{4}{5} \\ -\frac{3}{5} \\ -\frac{4}{5} \\ -\frac{3}{5} \\ 1 \\ 1 \end{Bmatrix} = \frac{4}{E}$$

$$F_{XX}^{-1} = \frac{E}{4}$$

The redundant force $X$ is then solved by

$$X = -F_{XX}^{-1} F_{XR} R$$

$$= -\left(\frac{E}{4}\right)\left(-\frac{3.3}{E}\right)(12) = 9.9 \text{ kips}$$

Substituting in the equilibrium equation, we obtain

$$\begin{Bmatrix} Q^a \\ Q^b \\ Q^c \\ Q^d \\ Q^e \\ Q^f \end{Bmatrix} = \begin{bmatrix} 1 & -\tfrac{4}{5} \\ \tfrac{3}{4} & -\tfrac{3}{5} \\ 1 & -\tfrac{4}{5} \\ 0 & -\tfrac{3}{5} \\ 0 & 1 \\ -\tfrac{5}{4} & 1 \end{bmatrix} \begin{Bmatrix} 12 \\ 9.9 \end{Bmatrix} = \begin{Bmatrix} 4.08 \\ 3.06 \\ 4.08 \\ -5.94 \\ 9.90 \\ -5.10 \end{Bmatrix} \text{ kips}$$

Alternatively, we find

$$b' = b_R - b_X F_{XX}^{-1} F_{XR}$$

$$= \begin{Bmatrix} 1 \\ \tfrac{3}{4} \\ 1 \\ 0 \\ 0 \\ -\tfrac{5}{4} \end{Bmatrix} - \begin{Bmatrix} -\tfrac{4}{5} \\ -\tfrac{3}{5} \\ -\tfrac{4}{5} \\ -\tfrac{3}{5} \\ 1 \\ 1 \end{Bmatrix} \left(\frac{E}{4}\right)\left(-\frac{3.3}{E}\right) = \begin{Bmatrix} 0.340 \\ 0.255 \\ 0.340 \\ -0.495 \\ 0.825 \\ -0.425 \end{Bmatrix}$$

and obtain $Q$ by $Q = b'R$:

$$\begin{Bmatrix} Q^a \\ Q^b \\ Q^c \\ Q^d \\ Q^e \\ Q^f \end{Bmatrix} = \begin{Bmatrix} 0.340 \\ 0.255 \\ 0.340 \\ -0.495 \\ 0.825 \\ -0.425 \end{Bmatrix} (12) = \begin{Bmatrix} 4.08 \\ 3.06 \\ 4.08 \\ -5.94 \\ 9.90 \\ -5.10 \end{Bmatrix} \text{ kips}$$

To find $r_1$, we first calculate the flexibility matrix of structure $F'$:

$$F' = b_R^T f b'$$

$$= \begin{bmatrix} 1 & \tfrac{3}{4} & 1 & 0 & 0 & -\tfrac{5}{4} \end{bmatrix} \left(\frac{1}{E}\right) \begin{Bmatrix} 0.340 \\ 0.255 \\ 0.340 \\ -0.495 \\ 0.825 \\ -0.425 \end{Bmatrix} = \frac{1.4}{E}$$

The displacement $r_1$ is then solved as

$$r_1 = F'R_1$$

$$= \left(\frac{1.4}{E}\right)(12) = \frac{(1.4)(12)}{30,000} = 0.00056 \text{ ft}$$

in the direction of the applied load.

**Example 15-4.** Find the member forces (end moments) of the rigid frame in Fig. 15-7(a) by the force method. $E$ is constant.

The frame shown in Fig. 15-7(a) is statically indeterminate to the second degree. It may be made determinate by inserting two pins as in Fig. 15-7(b). Then the structure is subjected to the original applied loads denoted by $R_1$ and $R_2$ together with the redundant couples $X_1$ and $X_2$. The member forces (end moments) in Fig. 15-7(b), $Q_i^a$, $Q_j^a$, ..., are shown by dashed lines.

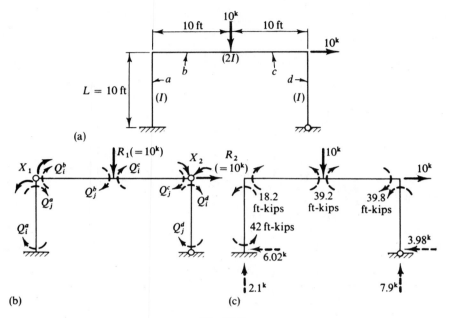

Fig. 15-7

The force transformation matrix is obtained by considering the influences of $R_1 = 1$, $R_2 = 1$, $X_1 = 1$, and $X_2 = 1$ successively and separately, as shown in Fig. 15-8.

$$
\begin{Bmatrix} Q_i^a \\ Q_j^a \\ Q_i^b \\ Q_j^b \\ Q_i^c \\ Q_j^c \\ Q_i^d \\ Q_j^d \end{Bmatrix}
=
\left[
\begin{array}{cc|cc}
0 & -L & 1 & 1 \\
0 & 0 & -1 & 0 \\
0 & 0 & 1 & 0 \\
-\dfrac{L}{2} & 0 & -\dfrac{1}{2} & \dfrac{1}{2} \\
\dfrac{L}{2} & 0 & \dfrac{1}{2} & -\dfrac{1}{2} \\
0 & 0 & 0 & 1 \\
0 & 0 & 0 & -1 \\
0 & 0 & 0 & 0
\end{array}
\right]
\begin{Bmatrix} R_1 \\ R_2 \\ X_1 \\ X_2 \end{Bmatrix}
$$

$$\underbrace{\phantom{\begin{matrix} 0 & -L \end{matrix}}}_{b_R} \quad \underbrace{\phantom{\begin{matrix} 1 & 1 \end{matrix}}}_{b_X}$$

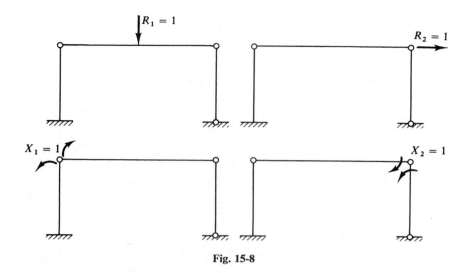

**Fig. 15-8**

From individual member flexibility matrices, we form

$$f = \frac{L}{6EI} \begin{bmatrix} 2 & -1 & & & & & & \\ -1 & 2 & & & & & & \\ & & 1 & -\frac{1}{2} & & & & \\ & & -\frac{1}{2} & 1 & & & & \\ & & & & 1 & -\frac{1}{2} & & \\ & & & & -\frac{1}{2} & 1 & & \\ & & & & & & 2 & -1 \\ & & & & & & -1 & 2 \end{bmatrix}$$

Using $b_R$, $b_X$, and $f$ found previously, we obtain

$$F_{XR} = b_X^T f b_R = \frac{L^2}{6EI} \begin{bmatrix} \frac{3}{4} & -3 \\ -\frac{3}{4} & -2 \end{bmatrix}$$

$$F_{XX} = b_X^T f b_X = \frac{L}{6EI} \begin{bmatrix} 8 & 2 \\ 2 & 6 \end{bmatrix}$$

and

$$F_{XX}^{-1} = \frac{6EI}{L} \frac{\begin{bmatrix} 6 & -2 \\ -2 & 8 \end{bmatrix}^T}{\begin{vmatrix} 8 & 2 \\ 2 & 6 \end{vmatrix}} = \left(\frac{6EI}{L}\right)\left(\frac{1}{44}\right) \begin{bmatrix} 6 & -2 \\ -2 & 8 \end{bmatrix}$$

The force transformation matrix of the indeterminate structure is

$$b' = b_R - b_X F_{XX}^{-1} F_{XR}$$

$$= \begin{bmatrix} 0 & -L \\ 0 & 0 \\ 0 & 0 \\ -\dfrac{L}{2} & 0 \\ \dfrac{L}{2} & 0 \\ 0 & 0 \\ 0 & 0 \\ 0 & 0 \end{bmatrix} - \begin{bmatrix} 1 & 1 \\ -1 & 0 \\ 1 & 0 \\ -\dfrac{1}{2} & \dfrac{1}{2} \\ \dfrac{1}{2} & -\dfrac{1}{2} \\ 0 & 1 \\ 0 & -1 \\ 0 & 0 \end{bmatrix}$$

$$\cdot \left(\dfrac{6EI}{L}\right)\left(\dfrac{1}{44}\right) \begin{bmatrix} 6 & -2 \\ -2 & 8 \end{bmatrix} \left(\dfrac{L^2}{6EI}\right) \begin{bmatrix} \frac{3}{4} & -3 \\ -\frac{3}{4} & -2 \end{bmatrix}$$

$$= \dfrac{L}{88} \begin{bmatrix} 0 & -88 \\ 0 & 0 \\ 0 & 0 \\ -44 & 0 \\ 44 & 0 \\ 0 & 0 \\ 0 & 0 \\ 0 & 0 \end{bmatrix} - \dfrac{L}{88} \begin{bmatrix} -3 & -48 \\ -12 & 28 \\ 12 & -28 \\ -13.5 & 4 \\ 13.5 & -4 \\ -15 & -20 \\ 15 & 20 \\ 0 & 0 \end{bmatrix} = \dfrac{L}{88} \begin{bmatrix} 3 & -40 \\ 12 & -28 \\ -12 & 28 \\ -30.5 & -4 \\ 30.5 & 4 \\ 15 & 20 \\ -15 & -20 \\ 0 & 0 \end{bmatrix}$$

The end moments are then solved by $Q = b'R$:

$$\begin{Bmatrix} Q_i^a \\ Q_j^a \\ Q_i^b \\ Q_j^b \\ Q_i^c \\ Q_j^c \\ Q_i^d \\ Q_j^d \end{Bmatrix} = \dfrac{L}{88} \begin{bmatrix} 3 & -40 \\ 12 & -28 \\ -12 & 28 \\ -30.5 & -4 \\ 30.5 & 4 \\ 15 & 20 \\ -15 & -20 \\ 0 & 0 \end{bmatrix} \begin{Bmatrix} R_1 \\ R_2 \end{Bmatrix}$$

Using $L = 10$ ft and $R_1 = R_2 = 10$ kips, we obtain

$$Q_i^a = -42 \text{ ft-kips}$$
$$Q_j^a = -Q_i^b = -18.2 \text{ ft-kips}$$
$$Q_j^b = -Q_i^c = -39.2 \text{ ft-kips}$$
$$Q_j^c = -Q_i^d = 39.8 \text{ ft-kips}$$
$$Q_j^d = 0$$

To check, we find that the identity

$$b_x f b' = 0$$

is satisfied by substituting in the values of $b_x, f,$ and $b'$ previously found. The answer diagram for the end moments together with the reactions at the supports found by statics is shown by the dashed line in Fig. 15-7(c).

## 15-8. THE TREATMENT OF DISTRIBUTED LOAD

As stated in Sec. 15-2, in matrix analysis only *nodal loads* (concentrated loads or moments) are considered (i.e., external loads are only applied to nodes). Because of this restriction, distributed loads cannot be handled directly. When distributed loads are involved, one way to handle them is to divide the distributed loads into a series of closely spaced concentrated loads and to consider each concentrated load point as a node. The disadvantage of this procedure is obvious. It increases the number of elements of the structure and, therefore, the size of the matrices that must be used in the analysis. The alternative is to fix the loaded beam and to apply to its two ends (nodes) the reverse of the fixed-end moments and shears, as illustrated in Fig. 15-9. The final forces and displacements in the loaded member must be obtained by superposing the effects of the fixed-end beam and those resulting from the nodal-force analysis of the original structure.

**Fig. 15-9**

**Example 15-5.** Find the end moments for the rigid frame shown in Fig. 15-10(a) by the force method. Assume constant $EI$.

The equivalent form of the given loaded frame is shown in Fig. 15-10(b). Because of symmetry, the vertical reaction at each support of the frame is known to be 6 kips acting upward, as indicated. If only flexural deformation is considered, then the nodal axial forces, shown in the frame in Fig. 15-10(b), only increase the compression in the two columns but cause no effect on the end moments of the frame and can, therefore, be neglected in the nodal-force analysis for obtaining end moments. The primary structure may be chosen as the one shown in Fig. 15-11, subjected to nodal moments $R_1$ and $R_2$ and redundant reaction components of the left support, denoted by $X_1$ and $X_2$. Those shown by dashed lines are member end moments $Q_i^a, Q_j^a, \ldots$. They can be expressed in terms of $R$ and $X$ as

Fig. 15-10

Fig. 15-11

$$\begin{Bmatrix} Q_i^a \\ Q_j^a \\ Q_i^b \\ Q_j^b \\ Q_i^c \\ Q_j^c \end{Bmatrix} = \begin{matrix} R_1 = 1 & R_2 = 1 & X_1 = 1 & X_2 = 1 \\ \left[ \begin{matrix} 0 & 0 & 0 & 1 \\ 0 & 0 & L & -1 \\ 1 & 0 & -L & 1 \\ -1 & 0 & L & -1 \\ 1 & -1 & -L & 1 \\ -1 & 1 & 0 & -1 \end{matrix} \right] \end{matrix} \begin{Bmatrix} R_1 \\ R_2 \\ \overline{X_1} \\ X_2 \end{Bmatrix}$$

$$\underbrace{\phantom{xxxx}}_{b_R} \qquad \underbrace{\phantom{xxxx}}_{b_X}$$

340

From the member flexibility matrices, we form

$$f = \frac{L}{6EI} \begin{bmatrix} 2 & -1 & & & & \\ -1 & 2 & & & & \\ & & 2 & -1 & & \\ & & -1 & 2 & & \\ & & & & 2 & -1 \\ & & & & -1 & 2 \end{bmatrix}$$

Thus,

$$F_{XR} = b_X^T f b_R$$

$$= \begin{bmatrix} 0 & L & -L & L & -L & 0 \\ 1 & -1 & 1 & -1 & 1 & -1 \end{bmatrix} \left(\frac{L}{6EI}\right)$$

$$\cdot \begin{bmatrix} 2 & -1 & & & & \\ -1 & 2 & & & & \\ & & 2 & -1 & & \\ & & -1 & 2 & & \\ & & & & 2 & -1 \\ & & & & -1 & 2 \end{bmatrix} \begin{bmatrix} 0 & 0 \\ 0 & 0 \\ 1 & 0 \\ -1 & 0 \\ 1 & -1 \\ -1 & 1 \end{bmatrix}$$

$$= \frac{L}{6EI} \begin{bmatrix} -9L & 3L \\ 12 & -6 \end{bmatrix}$$

Similarly,

$$F_{XX} = b_X^T f b_X$$

$$= \frac{L}{6EI} \begin{bmatrix} 10L^2 & -12L \\ -12L & 18 \end{bmatrix}$$

from which

$$F_{XX}^{-1} = \frac{6EI}{L} \frac{\begin{bmatrix} 18 & 12L \\ 12L & 10L^2 \end{bmatrix}^T}{(180L^2 - 144L^2)} = \frac{6EI}{L} \frac{\begin{bmatrix} 18 & 12L \\ 12L & 10L^2 \end{bmatrix}}{36L^2}$$

The force transformation matrix of the indeterminate structure is

$$b' = b_R - b_X F_{XX}^{-1} F_{XR}$$

$$= \begin{bmatrix} 0 & 0 \\ 0 & 0 \\ 1 & 0 \\ -1 & 0 \\ 1 & -1 \\ -1 & 1 \end{bmatrix} - \begin{bmatrix} 0 & 1 \\ L & -1 \\ -L & 1 \\ L & -1 \\ -L & 1 \\ 0 & -1 \end{bmatrix} \left(\frac{1}{36L^2}\right) \begin{bmatrix} 18 & 12L \\ 12L & 10L^2 \end{bmatrix} \begin{bmatrix} -9L & 3L \\ 12 & -6 \end{bmatrix}$$

$$
= \begin{bmatrix} 0 & 0 \\ 0 & 0 \\ 1 & 0 \\ -1 & 0 \\ 0 & -1 \\ -1 & 1 \end{bmatrix} - \begin{bmatrix} \frac{1}{3} & -\frac{2}{3} \\ -\frac{5}{6} & \frac{1}{6} \\ \frac{5}{6} & -\frac{1}{6} \\ -\frac{5}{6} & \frac{1}{6} \\ \frac{5}{6} & -\frac{1}{6} \\ -\frac{1}{3} & \frac{2}{3} \end{bmatrix} = \begin{bmatrix} -\frac{1}{3} & \frac{2}{3} \\ \frac{5}{6} & -\frac{1}{6} \\ \frac{1}{6} & \frac{1}{6} \\ -\frac{1}{6} & -\frac{1}{6} \\ \frac{1}{6} & -\frac{5}{6} \\ -\frac{2}{3} & \frac{1}{3} \end{bmatrix}
$$

The end moments based on the nodal-force analysis are then solved by $Q = b'R$:

$$
\begin{Bmatrix} Q_i^a \\ Q_j^a \\ Q_i^b \\ Q_j^b \\ Q_i^c \\ Q_j^c \end{Bmatrix} = \begin{bmatrix} -\frac{1}{3} & \frac{2}{3} \\ \frac{5}{6} & -\frac{1}{6} \\ \frac{1}{6} & \frac{1}{6} \\ -\frac{1}{6} & -\frac{1}{6} \\ \frac{1}{6} & -\frac{5}{6} \\ -\frac{2}{3} & \frac{1}{3} \end{bmatrix} \begin{Bmatrix} 10 \\ 10 \end{Bmatrix} = \begin{Bmatrix} \frac{10}{3} \\ \frac{20}{3} \\ \frac{10}{3} \\ -\frac{10}{3} \\ -\frac{20}{3} \\ -\frac{10}{3} \end{Bmatrix} \text{ ft-kips}
$$

The final result is obtained by adding the fixed-end moments [see upper part of Fig. 15-10(b)] to the end moments of member $b$. Thus,

$$
\begin{Bmatrix} Q_i^a \\ Q_j^a \\ Q_i^b \\ Q_j^b \\ Q_i^c \\ Q_j^c \end{Bmatrix} = \begin{Bmatrix} \frac{10}{3} \\ \frac{20}{3} \\ \frac{10}{3} - 10 \\ -\frac{10}{3} + 10 \\ -\frac{20}{3} \\ -\frac{10}{3} \end{Bmatrix} = \begin{Bmatrix} \frac{10}{3} \\ \frac{20}{3} \\ -\frac{20}{3} \\ \frac{20}{3} \\ -\frac{20}{3} \\ -\frac{10}{3} \end{Bmatrix} \text{ ft-kips}
$$

**Example 15-6.** Use the matrix force method to find the reaction at support $C$, and the deflection and slope at $B$, for the beam shown in Fig. 15-12(a). Assume that the spring flexibility is $f_s$.

We consider the spring as a member and therefore the system is composed of the beam portion, denoted as member $a$, and the spring, denoted as member $b$. The whole can be separated into two parts: the fixed-end beam under a uniform load, and the system subjected to nodal forces $R_1$, $R_2$ at $B$ and redundant reaction $X$ at $C$ as shown in Fig. 15-12(b). To obtain the nodal displacements at $B$ and reaction at $C$, it is only necessary to analyze the nodal-load system of Fig. 15-12(b).

We first relate the end moments $Q_i^a$, $Q_j^a$ of member $a$, and the spring force $Q^b$ to the external loads $R_1$, $R_2$ and $X$ as

$$
\begin{Bmatrix} Q_i^a \\ Q_j^a \\ Q^b \end{Bmatrix} = \begin{bmatrix} -L & 1 & \vdots & L \\ 0 & -1 & \vdots & 0 \\ 0 & 0 & \vdots & 1 \end{bmatrix} \begin{Bmatrix} R_1 \\ R_2 \\ X \end{Bmatrix}
$$

$$
\qquad\qquad\qquad b_R \qquad\quad b_X
$$

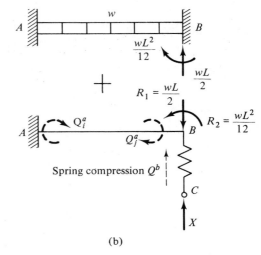

(a)                                           (b)

**Fig. 15-12**

The flexibility matrix of member $a$ is

$$f^a = \begin{bmatrix} \dfrac{L}{3EI} & -\dfrac{L}{6EI} \\[2mm] -\dfrac{L}{6EI} & \dfrac{L}{3EI} \end{bmatrix}$$

The flexibility of member $b$ (spring) is $f_s$. Therefore, the assembled flexibility is

$$f = \begin{bmatrix} \dfrac{L}{3EI} & -\dfrac{L}{6EI} & \\[2mm] -\dfrac{L}{6EI} & \dfrac{L}{3EI} & \\[2mm] & & f_s \end{bmatrix}$$

With $b_R$, $b_X$, and $f$ obtained, we have

$$F_{XR} = b_X^T f b_R = \begin{bmatrix} -\dfrac{L^3}{3EI} & \dfrac{L^2}{2EI} \end{bmatrix}$$

$$F_{XX} = b_X^T f b_X = \dfrac{L^3}{3EI} + f_s$$

$$F_{XX}^{-1} = \dfrac{1}{L^3/3EI + f_s}$$

The redundant force $X$, which is equal to the spring force $Q^b$, is determined by

$$X = -F_{XX}^{-1} F_{XR} R$$

$$= -\left( \dfrac{1}{L^3/3EI + f_s} \right) \begin{bmatrix} -\dfrac{L^3}{3EI} & \dfrac{L^2}{2EI} \end{bmatrix} \begin{Bmatrix} \dfrac{wL}{2} \\[2mm] \dfrac{wL^2}{12} \end{Bmatrix}$$

$$= \dfrac{3}{8} wL \left( \dfrac{1}{1 + 3EI f_s / L^3} \right)$$

343

Apparently, if $f_s = 0$,

$$X = \tfrac{3}{8} wL$$

To find the deflection and slope at $B$, we use

$$r = F_{RR}R + F_{RX}X$$

Now since

$$F_{RR} = b_R^T f b_R = \begin{bmatrix} \dfrac{L^3}{3EI} & -\dfrac{L^2}{2EI} \\[2ex] -\dfrac{L^2}{2EI} & \dfrac{L}{EI} \end{bmatrix}$$

$$F_{RX} = b_R^T f b_X = \begin{bmatrix} -\dfrac{L^3}{3EI} \\[2ex] \dfrac{L^2}{2EI} \end{bmatrix}$$

we reach

$$\left\{ \begin{matrix} r_1 \\ r_2 \end{matrix} \right\} = \begin{bmatrix} \dfrac{L^3}{3EI} & -\dfrac{L^2}{2EI} \\[2ex] -\dfrac{L^2}{2EI} & \dfrac{L}{EI} \end{bmatrix} \left\{ \begin{matrix} R_1 \\ R_2 \end{matrix} \right\} + \begin{bmatrix} -\dfrac{L^3}{3EI} \\[2ex] \dfrac{L^2}{2EI} \end{bmatrix}(X)$$

Using

$$R_1 = \frac{wL}{2} \qquad R_2 = \frac{wL^2}{12} \qquad X = \frac{3}{8} wL \left( \frac{1}{1 + 3EIf_s/L^3} \right)$$

we obtain

$$r_1 \text{ (deflection)} = \frac{wL^4}{8EI} - \frac{L^3}{3EI} \left( \frac{\tfrac{3}{8} wL}{1 + 3EIf_s/L^3} \right)$$

$$= \frac{3}{8} wL \left( \frac{1}{3EI/L^3 + 1/f_s} \right)$$

$$r_2 \text{ (slope)} = -\frac{wL^3}{6EI} + \frac{L^2}{2EI} \left( \frac{\tfrac{3}{8} wL}{1 + 3EIf_s/L^3} \right)$$

$$= \left( -\frac{w}{2} + \frac{wL^3}{48EIf_s} \right) \left( \frac{1}{3EI/L^3 + 1/f_s} \right)$$

As a check, if $f_s = 0$ (rigid support), we have

$$r_1 = 0 \qquad r_2 = \frac{wL^3}{48EI} \quad \text{\reflectbox{$\curvearrowright$}}$$

If $f_s = \infty$ (free end),

$$r_1 = \frac{wL^4}{8EI} \quad \downarrow \qquad r_2 = -\frac{wL^3}{6EI} \quad \curvearrowright$$

The foregoing procedure for fixing a loaded beam is not limited to the case of distributed loads. The procedure can also be applied to members subjected to a set of concentrated loads, if reducing the number of nodes is desirable.

## 15-9. ON THE NOTION OF PRIMARY STRUCTURE

The procedures for the analysis of statically indeterminate structures by the force method already discussed are based on the concept of *primary structure* previously developed in the method of consistent deformations. The notion of primary structure serves a convenient means of setting up equilibrium equation. However, if we, without considering the notion of primary structure, examine the basic equation

$$Q = b_R R + b_X X$$

we observe that it merely states: $Q$ is linearly related to a set of applied forces $R$ and a set of unknown forces $X$. The equation itself does not necessarily suggest a primary structure. As a result, we may separate these two sets of influences from the two independent force systems imposed on the original structure. So doing does not violate the truth of the preceding equation but certainly broadens our view of handling the problem.

Now $b_R$ represents an array of member forces in equilibrium with unit applied loads based on the original structure. More specifically, each column of $b_R$ represents member forces in equilibrium with a certain unit load applied to the original structure. Since the original structure is statically indeterminate, many equilibrating systems may be chosen from to establish each column of $b_R$.

Likewise, each column of $b_X$ can be thought of as independent self-equilibrating internal force system for the original structure. For a structure indeterminate to the $n$th degree, $b_X$ will represent any group of $n$ independent self-equilibrating member force systems, one for each redundant.

If it is convenient, these member forces may be determined by introducing a primary structure. However, in a larger sense, the traditional notion of primary structure is not essential to the analysis of a statically indeterminate structure; rather, it introduces unnecessary restrictions to the analysis.

**Example 15-7.** Solve the bar forces of the truss in Fig. 15-6(a) (Example 15-3) by the preceding generalized procedures.

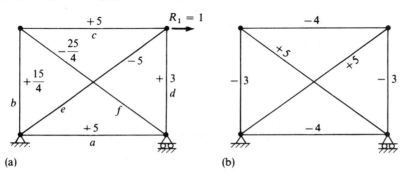

**Fig. 15-13**

345

*Solution 1* Disregarding the notion of a primary structure, we may choose a set of member forces in equilibrium with external load $R_1 = 1$, as shown in Fig. 15-13(a), and a set of self-equilibrating internal forces, as shown in Fig. 15-13(b). Thus,

$$b_R = \begin{Bmatrix} 5 \\ \frac{15}{4} \\ 5 \\ 3 \\ -5 \\ -\frac{25}{4} \end{Bmatrix} \qquad b_X = \begin{Bmatrix} -4 \\ -3 \\ -4 \\ -3 \\ 5 \\ 5 \end{Bmatrix}$$

There are, of course, many other choices that might be made.

Using $b_R$ and $b_X$ and $f$, obtained in Example 15-3, we have

$$F_{XR} = b_X^T f b_R$$

$$= [-4 \quad -3 \quad -4 \quad -3 \quad 5 \quad 5]\left(\frac{1}{E}\right)\begin{Bmatrix} 5 \\ \frac{15}{4} \\ 5 \\ 3 \\ -5 \\ -\frac{25}{4} \end{Bmatrix} = -\frac{116.5}{E}$$

$$F_{XX} = b_X^T f b_X$$

$$= [-4 \quad -3 \quad -4 \quad -3 \quad 5 \quad 5]\left(\frac{1}{E}\right)\begin{Bmatrix} -4 \\ -3 \\ -4 \\ -3 \\ 5 \\ 5 \end{Bmatrix} = \frac{100}{E}$$

$$F_{XX}^{-1} = \frac{E}{100}$$

The force transformation matrix is then determined:

$$b' = b_R - b_X F_{XX}^{-1} F_{XR}$$

$$= \begin{Bmatrix} 5 \\ \frac{15}{4} \\ 5 \\ 3 \\ -5 \\ -\frac{25}{4} \end{Bmatrix} - \begin{Bmatrix} -4 \\ -3 \\ -4 \\ -3 \\ 5 \\ 5 \end{Bmatrix}\left(\frac{E}{100}\right)\left(\frac{-116.5}{E}\right) = \begin{Bmatrix} 5 \\ 3.75 \\ 5 \\ 3 \\ -5 \\ -6.25 \end{Bmatrix} - \begin{Bmatrix} 4.660 \\ 3.495 \\ 4.660 \\ 3.495 \\ -5.825 \\ -5.825 \end{Bmatrix} = \begin{Bmatrix} 0.340 \\ 0.255 \\ 0.340 \\ -0.495 \\ 0.825 \\ -0.425 \end{Bmatrix}$$

This is the same $b'$ obtained in Example 15-3 and will lead to the same final results for the bar forces.

*Solution 2* It may be interesting to point out that when primary structure is used in analyzing an indeterminate structure, the same final results will be obtained if different primary structures are chosen in developing $b_R$ and $b_X$.

To illustrate, let us first take member $e$ as the redundant. The bar forces associated with the given primary structure due to external load $R_1 = 1$ are elements of $b_R$, as indicated in Fig. 15-14(a). Next, let member $a$ be chosen as the redundant. Setting the redundant force equal to unity, we obtain a set of internal forces in equilibrium [Fig. 15-14(b)] which forms $b_X$.

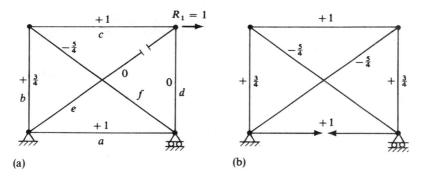

Fig. 15-14

$$b_R = \begin{Bmatrix} 1 \\ \frac{3}{4} \\ 1 \\ 0 \\ 0 \\ -\frac{5}{4} \end{Bmatrix} \qquad b_X = \begin{Bmatrix} 1 \\ \frac{3}{4} \\ 1 \\ \frac{3}{4} \\ -\frac{5}{4} \\ -\frac{5}{4} \end{Bmatrix}$$

$$F_{XR} = b_X^T f b_R$$

$$= \begin{bmatrix} 1 & \frac{3}{4} & 1 & \frac{3}{4} & -\frac{5}{4} & -\frac{5}{4} \end{bmatrix} \left(\frac{1}{E}\right) \begin{Bmatrix} 1 \\ \frac{3}{4} \\ 1 \\ 0 \\ 0 \\ -\frac{5}{4} \end{Bmatrix} = \frac{66}{16E}$$

$$F_{XX} = b_X^T f b_X$$

$$= [1 \quad \tfrac{3}{4} \quad 1 \quad \tfrac{3}{4} \quad -\tfrac{5}{4} \quad -\tfrac{5}{4}]\left(\frac{1}{E}\right)\begin{Bmatrix} 1 \\ \tfrac{3}{4} \\ 1 \\ \tfrac{3}{4} \\ -\tfrac{5}{4} \\ -\tfrac{5}{4} \end{Bmatrix} = \frac{100}{16E}$$

$$F_{XX}^{-1} = \frac{16E}{100}$$

The force transformation matrix $b'$ is found to be

$$b' = b_R - b_X F_{XX}^{-1} F_{XR} = \begin{Bmatrix} 1 \\ \tfrac{3}{4} \\ 1 \\ 0 \\ 0 \\ -\tfrac{5}{4} \end{Bmatrix} - \begin{Bmatrix} 1 \\ \tfrac{3}{4} \\ 1 \\ \tfrac{3}{4} \\ -\tfrac{5}{4} \\ -\tfrac{5}{4} \end{Bmatrix}\left(\frac{16E}{100}\right)\left(\frac{66}{16E}\right) = \begin{Bmatrix} 0.340 \\ 0.255 \\ 0.340 \\ -0.495 \\ 0.825 \\ -0.425 \end{Bmatrix}$$

the same as previously obtained.

## PROBLEMS

**15-1.** Use the force method to find the vertical deflection at each of the loaded points of the beam shown in Fig. 15-15. Assume constant $EI$.

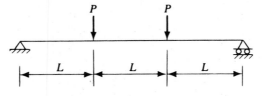

**Fig. 15-15**

**15-2.** Use the force method to find the slope and deflection at the loaded point of the beam shown in Fig. 15-16.

**Fig. 15-16**

**15-3.** Find, by the force method, all the bar forces and the vertical deflection at each of the loaded joints of the truss shown in Fig. 15-17. Assume that $A = 10$ in.$^2$, $E = 30,000$ kips/in.$^2$ for all members.

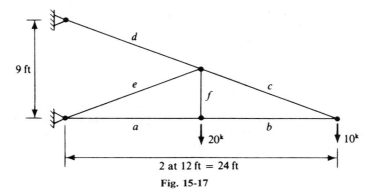

**Fig. 15-17**

**15-4.** Find, by the force method, all the member forces (end moments) and the nodal displacements corresponding to the applied loads for the frame in Fig. 15-18. Assume constant $EI$

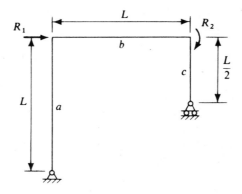

**Fig. 15-18**

**15-5.** Find, by the force method, the slope and deflection at the loaded end of the beam shown in Fig. 15-19. Assume constant $EI$.

**Fig. 15-19**

**15-6.** Find, by the force method, the bar forces and the deflection components at the loaded point of the truss in Fig. 15-20. Assume that $A = 10$ in.$^2$ and $E = 30,000$ kips/in.$^2$ for all members.

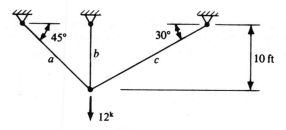

**Fig. 15-20**

**15-7.** The truss shown in Fig. 15-21 is statically indeterminate to the first degree. Choose the axial force in bar *c* as redundant. Use the force method to find the bar forces and the deflection components corresponding to the applied loads. Assume that $A = 50$ cm$^2$ and $E = 20,000$ kN/cm$^2$ for all bars.

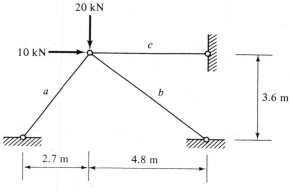

**Fig. 15-21**

**15-8.** Use the force method to obtain the member end moments for the frame shown in Fig. 15-22. Assume constant *EI*.

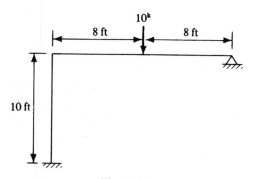

**Fig. 15-22**

**15-9.** Use the force method to find the member end moments for the frame shown in Fig. 15-23. Assume constant *EI*.

350

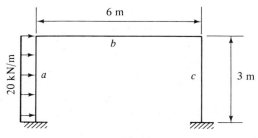

Fig. 15-23

**15-10.** Refer to Fig. 15-24. Find, by the force method, the slope and deflection of the beam at the point *B*.

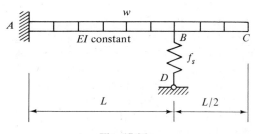

Fig. 15-24

**15-11.** Use the force method to find the member end moments for the gable bent in Fig. 15-25. Assume constant *EI*.

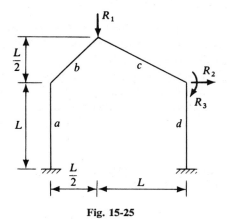

Fig. 15-25

**15-12.** Solve Prob. 15-6 disregarding the notion of primary structure.

**15-13.** Solve Prob. 15-7 disregarding the notion of primary structure.

**15-14.** Solve Prob. 15-8 by using different primary structures for developing the force transformation matrices $b_R$ and $b_X$.

# 16

## MATRIX ANALYSIS OF STRUCTURES

## BY THE FINITE ELEMENT METHOD

## PART II: THE DISPLACEMENT METHOD

### 16-1. GENERAL

As previously pointed out, the force method and the displacement method represent two different approaches to analyzing structures. The basic concepts of the structure remain the same (see Sec. 15-2). The fundamental difference between these two methods is that the force method chooses the member forces as the basic unknowns, whereas the displacement method chooses the nodal displacements as the basic unknowns. Like the force method, the basic equations of the displacement method are derived from

1. The equilibrium of forces
2. The compatibility of displacements
3. The force-displacement relationship

The compatibility condition is first satisfied by correlating the external nodal displacements to the end deformations of the members. The force-displacement relationship is then established between the member end forces and deformations and between the nodal forces and nodal displacements. Finally, using nodal equilibrium equations, we solve for the unknown nodal displacements and, therefore, for the member forces and deformations of the structure.

### 16-2. COMPATIBILITY, DISPLACEMENT
### TRANSFORMATION MATRIX

The compatibility used in the displacement method is that the geometry of deformation must be such that the elements of structure fit together at the nodal points; that is, the member deformations $q$ should be consistently related to the nodal displacements $r$. Let $a_{ij}$ represent the value of member deformation $q_i$ caused by a unit nodal displacement $r_j$. The total value of each member deformation caused by all the nodal displacements may be written as

$$q_1 = a_{11}r_1 + a_{12}r_2 + \cdots + a_{1n}r_n$$
$$q_2 = a_{21}r_1 + a_{22}r_2 + \cdots + a_{2n}r_n$$
$$\cdot$$
$$\cdot$$
$$\cdot \tag{16-1}$$
$$q_m = a_{m1}r_1 + a_{m2}r_2 + \cdots + a_{mn}r_n$$

in which $q_1 = q_i^a$, $q_2 = q_j^a$, ... represent the total set of member deformations and $r_1, r_2, \ldots, r_n$ the total set of nodal displacements. Note that no connection between the subscripts on $q$ and $r$ is implied. In matrix form,

$$\begin{Bmatrix} q_1 \\ q_2 \\ \cdot \\ \cdot \\ \cdot \\ q_m \end{Bmatrix} = \begin{bmatrix} a_{11} & a_{12} & \cdots & a_{1n} \\ a_{21} & a_{22} & \cdots & a_{2n} \\ \cdot \\ \cdot \\ \cdot \\ a_{m1} & a_{m2} & \cdots & a_{mn} \end{bmatrix} \begin{Bmatrix} r_1 \\ r_2 \\ \cdot \\ \cdot \\ \cdot \\ r_n \end{Bmatrix}$$

That is,

$$q = ar \tag{16-2}$$

where

$$a = \begin{bmatrix} a_{11} & a_{12} & \cdots & a_{1n} \\ a_{21} & a_{22} & \cdots & a_{2n} \\ \cdot \\ \cdot \\ \cdot \\ a_{m1} & a_{m2} & \cdots & a_{mn} \end{bmatrix} \tag{16-2a}$$

called the *displacement transformation matrix*, which relates the internal member deformations to the external nodal displacements. Matrix $a$ is usually a rectangular matrix. It is simply a geometric transformation of coordinates representing the compatibility of the displacements of a system.

For example, let us consider the truss in Fig. 16-1. The possible nodal displacements are one linear displacement at the roller support, denoted by $r_1$, and two linear displacement components at the top joint, denoted by $r_2$ and $r_3$. Also indicated in Fig. 16-1 are the member deformations of bar $c$ as the result of the separate influences of $r_1$, $r_2$, and $r_3$, which gives the total $q^c$ as

$$q^c = 0.6r_1 - 0.6r_2 + 0.8r_3$$

$q^a$ and $q^b$ can be similarly obtained. The result, put in matrix form, is

$$\begin{Bmatrix} q^a \\ q^b \\ q^c \end{Bmatrix} = \underbrace{\begin{bmatrix} 1 & 0 & 0 \\ 0 & 0.6 & 0.8 \\ 0.6 & -0.6 & 0.8 \end{bmatrix}}_{a} \begin{Bmatrix} r_1 \\ r_2 \\ r_3 \end{Bmatrix}$$

**Fig. 16-1**

Note that the displacement transformation matrix in this case happens to be a square matrix.

If the same forces and displacements are involved the relationship between the displacement transformation matrix and the force transformation matrix can be obtained by virtual work as follows:

$$R^T r = Q^T q$$

Because of Eq. 15-4,

$$Q = bR$$

or

$$Q^T = R^T b^T$$

and Eq. 16-2,

$$q = ar$$

we have

$$R^T r = R^T b^T ar$$

from which

$$b^T a = I \tag{16-3}$$

Similarly, we can prove

$$a^T b = I \tag{16-4}$$

## 16-3. FORCE-DISPLACEMENT RELATIONSHIP, STIFFNESS COEFFICIENT, STIFFNESS MATRIX

A *stiffness coefficient* $k_{ij}$ is defined as the force developed at point $i$ due to a unit displacement at point $j$, all other points (nodes) being fixed. Like the flexibility coefficient, the stiffness coefficient constitutes a relationship between force and displacement. Applying the principle of superposition, we may express the force component at any point of a system in terms of a set of prescribed displacements.

The first step in this analysis is to express the end forces in terms of the end deformations of an individual member. Using the stiffness coefficients and the

354

notation defined in Fig. 15-2, we have

$$Q_i^a = k_{ii}^a q_i^a + k_{ij}^a q_j^a + k_{ik}^a q_k^a$$
$$Q_j^a = k_{ji}^a q_i^a + k_{jj}^a q_j^a + k_{jk}^a q_k^a \qquad (16\text{-}5)$$
$$Q_k^a = k_{ki}^a q_i^a + k_{kj}^a q_j^a + k_{kk}^a q_k^a$$

in which $q_i^a$, $q_j^a$, and $q_k^a$ are the deformations of a particular member $a$ and $Q_i^a$, $Q_j^a$, and $Q_k^a$ the corresponding member forces. It is clear that the stiffness coefficient, say $k_{ii}^a$, is defined as

$$k_{ii}^a = Q_i^a \quad \text{as} \quad q_i^a = 1 \qquad q_j^a = q_k^a = 0$$

The rest can be defined similarly.

Equation 16-5 in matrix form is

$$Q^a = k^a q^a \qquad (16\text{-}6)$$

in which

$$k^a = \begin{bmatrix} k_{ii}^a & k_{ij}^a & k_{ik}^a \\ k_{ji}^a & k_{jj}^a & k_{jk}^a \\ k_{ki}^a & k_{kj}^a & k_{kk}^a \end{bmatrix} \qquad (16\text{-}6a)$$

defined as the *element stiffness matrix*.

If Eq. 16-6 is premultiplied by $(k^a)^{-1}$,

$$(k^a)^{-1} Q^a = (k^a)^{-1} k^a q^a$$

or

$$q^a = (k^a)^{-1} Q^a \qquad (16\text{-}7)$$

Comparing Eq. 16-7 with Eq. 15-10,

$$q^a = f^a Q^a$$

we see that

$$f^a = (k^a)^{-1} \qquad (16\text{-}8)$$

Thus, *the element flexibility matrix is the inverse of the element stiffness matrix, and vice versa.*

The descriptions above refer to an individual element. For the entire assemblage composed of $a, b, \ldots$ elements, we have

$$Q^a = k^a q^a$$
$$Q^b = k^b q^b$$
$$\vdots$$

Since

$$Q = \begin{Bmatrix} Q^a \\ Q^b \\ \vdots \end{Bmatrix} \qquad q = \begin{Bmatrix} q^a \\ q^b \\ \vdots \end{Bmatrix}$$

the equations above can be assembled as

$$Q = kq \tag{16-9}$$

where

$$k = \begin{bmatrix} k^a & & & \\ & k^b & & \\ & & \cdot & \\ & & & \cdot \end{bmatrix} \tag{16-9a}$$

which is a diagonal matrix with individual element stiffness matrices as its constituents.

Refer to Eqs. 16-5 and 16-6a, and consider a prismatic member with length $L$, cross-sectional area $A$, moment of inertia $I$, and modulus of elasticity $E$. The elements in the first column of $k^a$ are, by definition, the member forces resulting from $q_i^a = 1$, that is, a unit rotation at the left end of the member (see Fig. 15-2). Thus,

$$k_{ii}^a = \text{moment at the left end} = \frac{4EI}{L}$$

$$k_{ji}^a = \text{moment at the right end} = \frac{2EI}{L}$$

$$k_{ki}^a = \text{axial force of the member} = 0$$

Note that $k_{ii}^a$ and $k_{ji}^a$ can easily be obtained by the slope-deflection method; $k_{ki}^a = 0$ is apparent. All the other elements of $k^a$ are similarly determined. Thus,

$$k^a = \begin{bmatrix} \dfrac{4EI}{L} & \dfrac{2EI}{L} & 0 \\[2mm] \dfrac{2EI}{L} & \dfrac{4EI}{L} & 0 \\[2mm] 0 & 0 & \dfrac{AE}{L} \end{bmatrix} \tag{16-10}$$

which is a symmetric matrix. As a check, we note that

$$f^a k^a = \begin{bmatrix} \dfrac{L}{3EI} & -\dfrac{L}{6EI} & 0 \\[2mm] -\dfrac{L}{6EI} & \dfrac{L}{3EI} & 0 \\[2mm] 0 & 0 & \dfrac{L}{AE} \end{bmatrix} \begin{bmatrix} \dfrac{4EI}{L} & \dfrac{2EI}{L} & 0 \\[2mm] \dfrac{2EI}{L} & \dfrac{4EI}{L} & 0 \\[2mm] 0 & 0 & \dfrac{AE}{L} \end{bmatrix} = \begin{bmatrix} 1 & 0 & 0 \\ 0 & 1 & 0 \\ 0 & 0 & 1 \end{bmatrix}$$

When the effect of axial forces is disregarded, as is usually done in rigid frame analysis,

$$k^a = \frac{EI}{L}\begin{bmatrix} 4 & 2 \\ 2 & 4 \end{bmatrix} \tag{16-11}$$

For a pin-connected truss,

$$k^a = \left[\frac{AE}{L}\right] \tag{16-12}$$

The second step in the analysis is to express the nodal forces in terms of the corresponding nodal displacements. This can be accomplished easily by the method of vitual work. Let us start by Eq. 15-1:

$$\bar{r}^T R = \bar{q}^T Q \tag{16-13}$$

From Eqs. 16-2 and 16-9,

$$q = ar \quad \text{and} \quad Q = kq$$

we have                                   $\bar{q} = a\bar{r}$

or                                        $\bar{q}^T = \bar{r}^T a^T$ $\qquad$ (16-14)

and                                       $Q = kar$ $\qquad\qquad$ (16-15)

Substituting Eqs. 16-14 and 16-15 in Eq. 16-13 gives

$$\bar{r}^T R = \bar{r}^T a^T kar$$

from which                                $R = a^T kar$ $\qquad$ (16-16)

or                                        $R = Kr$ $\qquad\qquad$ (16-17)

if we make                                $K = a^T ka$ $\qquad\quad$ (16-18)

$K$ being called the *total stiffness matrix*, or *the stiffness matrix of structure*, which directly relates the nodal forces to the nodal displacements of a structure.

If we premultiply Eq. 16-17 with $K^{-1}$ on both sides, we have

$$K^{-1}R = K^{-1}Kr = r$$

or                                        $r = K^{-1}R$ $\qquad\quad$ (16-19)

Comparing Eq. 16-19 with Eq. 15-22,

$$r = FR$$

we have                                   $F = K^{-1}$ $\qquad\qquad$ (16-20)

It is thus seen that *the total flexibility matrix is the inverse of total stiffness matrix, and vice versa if the same forces and displacements are involved.*

## 16-4. EQUILIBRIUM

Refer to Eq. 16-17:

$$R = Kr$$

If $r$ denotes the elements of all possible unknown nodal displacements (not including the known support or boundary conditions), then $R$ must denote all the corresponding nodal forces. The equilibrium of each node requires that the possible nodal forces, expressed in terms of unknown nodal displacements, must be equal to the applied loads. Thus, if these nodal loads are given, we can solve for the unknown nodal displacements by Eq. 16-19,

$$r = K^{-1}R$$

and for the member forces by Eqs. 16-2 and 16-9,

$$Q = kq = kar$$

## 16-5. ANALYSIS OF STRUCTURES BY THE DISPLACEMENT METHOD

It is interesting that in the discussion of the displacement method the question of redundancy did not arise. The displacement method can apply with equal ease to statically determinate structures and statically indeterminate structures. The procedures of analysis by the displacement method are contained in the following steps:

1. Define all the possible unknown nodal displacements $r$.

Generally, a pin-connected node has two linear displacement components, with the rotation of the pin considered free of the connected members. A rigidly connected node has three displacement components, two linear and one rotational, but the linear displacements may be excepted if the axial deformations of the connected members are neglected and sidesway prevented. No displacements are assigned to the nodes that cannot move. Thus, in a pin-connected truss the hinged support is considered completely restrained; the roller support has one linear movement. In a rigid frame the built-in support undergoes no displacement; a hinged support can have only an angular displacement, whereas a roller support has one angular displacement and one linear displacement.

2. Determine the displacement transformation matrix $a$ from geometric configuration.

3. From the individual stiffness matrices $k^a$, $k^b$, ... obtain

$$k = \begin{bmatrix} k^a & & & \\ & k^b & & \\ & & \cdot & \\ & & & \cdot \\ & & & & \cdot \end{bmatrix}$$

4. Compute the stiffness matrix of structure $K$:

$$K = a^T k a$$

5. Obtain the inverse of $K$.
6. Compute nodal displacements $r$ by

$$r = K^{-1}R$$

Note that $R$ is in one-to-one correspondence with $r$. Some of the $R$ are the actual loads; others are zero if no load is applied there. All of the $R$ are known.

7. Compute the member forces $Q$ by

$$Q = kar$$

8. Distributed loads are handled indirectly by the procedure outlined in Sec. 15-8.

**Example 16-1.** Compute the nodal displacements and bar forces for the truss shown in Fig. 16-2(a). Assume that $E = 30,000$ kips/in.$^2$ and $L(ft)/A$ (in.$^2$) $= 1$ for all members.

The roller support has a possible displacement $r_1$, and the top joint has possible displacement components $r_2$ and $r_3$, as indicated in Fig. 16-2(b). The nodal forces $R_1$, $R_2$, and $R_3$ correspond to the nodal displacements. Note that $R_1 = R_2 = 0$ and $R_3 = -8$ kips in this problem.

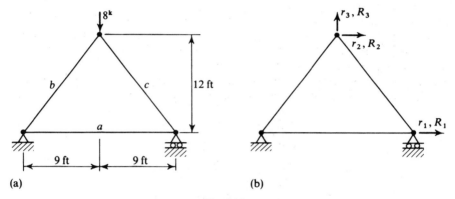

**Fig. 16-2**

Using the results of the example in Sec. 16-2, the displacement transformation matrix is

$$a = \begin{bmatrix} 1 & 0 & 0 \\ 0 & 0.6 & 0.8 \\ 0.6 & -0.6 & 0.8 \end{bmatrix}$$

The individual member stiffness matrix is determined by Eq. 16-12,

$$\left[\frac{AE}{L}\right]$$

Since $A/L = 1$ for all members,

$$k^a = k^b = k^c = E$$

from which we form the diagonal matrix

$$k = E \begin{bmatrix} 1 & & \\ & 1 & \\ & & 1 \end{bmatrix}$$

Thus, the total stiffness matrix is

$$
K = a^T ka = \begin{bmatrix} 1 & 0 & 0.6 \\ 0 & 0.6 & -0.6 \\ 0 & 0.8 & 0.8 \end{bmatrix} (E) \begin{bmatrix} 1 & 0 & 0 \\ 0 & 0.6 & 0.8 \\ 0.6 & -0.6 & 0.8 \end{bmatrix}
$$

$$
= E \begin{bmatrix} 1.36 & -0.36 & 0.48 \\ -0.36 & 0.72 & 0 \\ 0.48 & 0 & 1.28 \end{bmatrix}
$$

$$
K^{-1} = \frac{1}{E} \begin{bmatrix} 1 & 0.5 & -0.375 \\ 0.5 & 1.639 & -0.188 \\ -0.375 & -0.188 & 0.922 \end{bmatrix}
$$

The nodal displacements are then determined by $r = K^{-1} R$:

$$
\begin{Bmatrix} r_1 \\ r_2 \\ r_3 \end{Bmatrix} = \frac{1}{E} \begin{bmatrix} 1 & 0.5 & -0.375 \\ 0.5 & 1.639 & -0.188 \\ -0.375 & -0.188 & 0.922 \end{bmatrix} \begin{Bmatrix} 0 \\ 0 \\ -8 \end{Bmatrix} = \frac{1}{E} \begin{Bmatrix} 3 \\ 1.5 \\ -7.38 \end{Bmatrix}
$$

Using $E = 30,000$ kips/in.$^2$, we obtain

$$
\begin{Bmatrix} r_1 \\ r_2 \\ r_3 \end{Bmatrix} = \begin{Bmatrix} 0.0001 \\ 0.00005 \\ -0.00027 \end{Bmatrix} \text{ft}
$$

Finally, the bar forces are solved by $Q = kar$:

$$
\begin{Bmatrix} Q^a \\ Q^b \\ Q^c \end{Bmatrix} = (E) \begin{bmatrix} 1 & 0 & 0 \\ 0 & 0.6 & 0.8 \\ 0.6 & -0.6 & 0.8 \end{bmatrix} \left(\frac{1}{E}\right) \begin{Bmatrix} 3 \\ 1.5 \\ -7.38 \end{Bmatrix} = \begin{Bmatrix} 3 \\ -5 \\ -5 \end{Bmatrix} \text{ kips}
$$

**Example 16-2.** Solve the bar forces for the truss in Fig. 16-3(a). Assume constant $E$ and $L(\text{ft})/A(\text{in.}^2) = 1$ for all members.

The assigned nodal displacements and the corresponding nodal forces are shown in Fig. 16-3(b). In this problem $R_1 = R_2 = R_3 = 10$ kips, $R_4 = 0$.

From compatibility,

$$
r_1 = 1 \quad r_2 = 1 \quad r_3 = 1 \quad r_4 = 1
$$

$$
\begin{Bmatrix} q^a \\ q^b \\ q^c \\ q^d \\ q^e \end{Bmatrix} = \begin{bmatrix} 0 & 1 & 0 & 0 \\ -1 & 0 & 1 & 0 \\ 0 & 0 & 0 & 1 \\ 0 & 0 & 0.8 & 0.6 \\ -0.8 & 0.6 & 0 & 0 \end{bmatrix} \begin{Bmatrix} r_1 \\ r_2 \\ r_3 \\ r_4 \end{Bmatrix}
$$

$$a$$

Since $L/A = 1$ for all members, the individual member stiffness matrices are

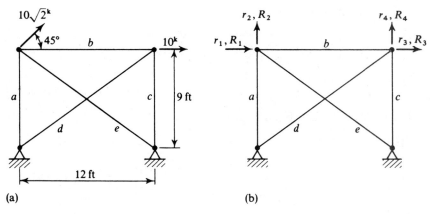

Fig. 16-3

found to be

$$k^a = k^b = k^c = k^d = k^e = E$$

from which

$$k = E \begin{bmatrix} 1 & & & & \\ & 1 & & & \\ & & 1 & & \\ & & & 1 & \\ & & & & 1 \end{bmatrix}$$

Thus, we have the total stiffness matrix:

$$K = a^T k a$$

$$= \begin{bmatrix} 0 & -1 & 0 & 0 & -0.8 \\ 1 & 0 & 0 & 0 & 0.6 \\ 0 & 1 & 0 & 0.8 & 0 \\ 0 & 0 & 1 & 0.6 & 0 \end{bmatrix} (E) \begin{bmatrix} 0 & 1 & 0 & 0 \\ -1 & 0 & 1 & 0 \\ 0 & 0 & 0 & 1 \\ 0 & 0 & 0.8 & 0.6 \\ -0.8 & 0.6 & 0 & 0 \end{bmatrix}$$

$$= E \begin{bmatrix} 1.64 & -0.48 & -1 & 0 \\ -0.48 & 1.36 & 0 & 0 \\ -1 & 0 & 1.64 & 0.48 \\ 0 & 0 & 0.48 & 1.36 \end{bmatrix}$$

$$K^{-1} = \frac{1}{2.150E} \begin{bmatrix} 2.721 & 0.960 & 1.850 & -0.653 \\ 0.960 & 1.920 & 0.653 & -0.230 \\ 1.850 & 0.653 & 2.721 & -0.960 \\ -0.653 & -0.230 & -0.960 & 1.296 \end{bmatrix}$$

The nodal displacements are given by $r = K^{-1} R$:

$$\begin{Bmatrix} r_1 \\ r_2 \\ r_3 \\ r_4 \end{Bmatrix} = \frac{1}{2.150E} \begin{bmatrix} 2.721 & 0.960 & 1.850 & -0.653 \\ 0.960 & 1.920 & 0.653 & -0.230 \\ 1.850 & 0.653 & 2.721 & -0.960 \\ -0.653 & -0.230 & -0.960 & 1.296 \end{bmatrix} \begin{Bmatrix} 10 \\ 10 \\ 10 \\ 0 \end{Bmatrix} = \frac{1}{E} \begin{Bmatrix} 25.7 \\ 16.4 \\ 24.3 \\ -8.57 \end{Bmatrix}$$

The bar forces are then obtained from $Q = kar$.

$$\begin{Bmatrix} Q^a \\ Q^b \\ Q^c \\ Q^d \\ Q^e \end{Bmatrix} = (E) \begin{bmatrix} 0 & 1 & 0 & 0 \\ -1 & 0 & 1 & 0 \\ 0 & 0 & 0 & 1 \\ 0 & 0 & 0.8 & 0.6 \\ -0.8 & 0.6 & 0 & 0 \end{bmatrix} \left(\frac{1}{E}\right) \begin{Bmatrix} 25.7 \\ 16.4 \\ 24.3 \\ -8.57 \end{Bmatrix} = \begin{Bmatrix} 16.40 \\ -1.40 \\ -8.57 \\ 14.30 \\ -10.72 \end{Bmatrix} \text{ kips}$$

**Example 16-3.** Find all the end moments for the frame shown in Fig. 16-4(a). Assume constant $EI$.

We assume unknown nodal displacements $r_1$, $r_2$, and $r_3$ and their corresponding forces $R_1$, $R_2$, and $R_3$, as shown in Fig. 16-4(b). Note that $r_1$ represents the sidesway of the frame and that $r_2$ and $r_3$ represent the rotations at the joints. In the present case $R_1 = 10$ kips, $R_2 = R_3 = 0$. The dashed lines in Fig. 16-4(b) indicate the assumed directions for the member moments, $Q_i^a$, $Q_j^a$, . . . , and their corresponding member rotations, $q_i^a$, $q_j^a$, . . . . From compatibility

$$r_1 = 1 \quad r_2 = 1 \quad r_3 = 1$$

$$\begin{Bmatrix} q_i^a \\ q_j^a \\ q_i^b \\ q_j^b \\ q_i^c \\ q_j^c \end{Bmatrix} = \begin{bmatrix} -\dfrac{1}{10} & 0 & 0 \\ -\dfrac{1}{10} & 1 & 0 \\ 0 & 1 & 0 \\ 0 & 0 & 1 \\ -\dfrac{1}{10} & 0 & 1 \\ -\dfrac{1}{10} & 0 & 0 \end{bmatrix} \begin{Bmatrix} r_1 \\ r_2 \\ r_3 \end{Bmatrix}$$

$a$

Since the members are identical, the individual member stiffness matrices are the same:

$$k^a = k^b = k^c = \frac{EI}{L} \begin{bmatrix} 4 & 2 \\ 2 & 4 \end{bmatrix}$$

See Eq. 16-11.

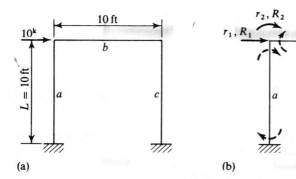

Fig. 16-4

Using Eq. 16-9a, we have

$$k = \frac{EI}{L} \begin{bmatrix} 4 & 2 & & & & \\ 2 & 4 & & & & \\ & & 4 & 2 & & \\ & & 2 & 4 & & \\ & & & & 4 & 2 \\ & & & & 2 & 4 \end{bmatrix}$$

Thus, the total stiffness matrix is

$$K = a^T k a$$

$$= \begin{bmatrix} -\dfrac{1}{10} & -\dfrac{1}{10} & 0 & 0 & -\dfrac{1}{10} & -\dfrac{1}{10} \\ 0 & 1 & 1 & 0 & 0 & 0 \\ 0 & 0 & 0 & 1 & 1 & 0 \end{bmatrix}$$

$$\cdot \left(\frac{EI}{L}\right) \begin{bmatrix} 4 & 2 & & & & \\ 2 & 4 & & & & \\ & & 4 & 2 & & \\ & & 2 & 4 & & \\ & & & & 4 & 2 \\ & & & & 2 & 4 \end{bmatrix} \begin{bmatrix} -\dfrac{1}{10} & 0 & 0 \\ -\dfrac{1}{10} & 1 & 0 \\ 0 & 1 & 0 \\ 0 & 0 & 1 \\ -\dfrac{1}{10} & 0 & 1 \\ -\dfrac{1}{10} & 0 & 0 \end{bmatrix}$$

$$= \frac{EI}{L} \begin{bmatrix} 0.24 & -0.6 & -0.6 \\ -0.6 & 8 & 2 \\ -0.6 & 2 & 8 \end{bmatrix}$$

$$K^{-1} = \frac{L}{EI}\left(\frac{1}{10.08}\right) \begin{bmatrix} 60 & 3.6 & 3.6 \\ 3.6 & 1.56 & -0.12 \\ 3.6 & -0.12 & 1.56 \end{bmatrix}$$

363

The nodal displacements are then obtained from $r = K^{-1}R$:

$$\begin{Bmatrix} r_1 \\ r_2 \\ r_3 \end{Bmatrix} = \frac{L}{EI}\left(\frac{1}{10.08}\right)\begin{bmatrix} 60 & 3.6 & 3.6 \\ 3.6 & 1.56 & -0.12 \\ 3.6 & -0.12 & 1.56 \end{bmatrix}\begin{Bmatrix} 10 \\ 0 \\ 0 \end{Bmatrix} = \frac{L}{EI}\left(\frac{1}{10.08}\right)\begin{Bmatrix} 600 \\ 36 \\ 36 \end{Bmatrix}$$

Finally, the end moments are determined by $Q = kar$. Using the values of $k$, $a$, and $r$ previously obtained gives

$$\begin{Bmatrix} Q_i^a \\ Q_j^a \\ Q_i^b \\ Q_j^b \\ Q_i^c \\ Q_j^c \end{Bmatrix} = \begin{Bmatrix} -28.6 \\ -21.4 \\ 21.4 \\ 21.4 \\ -21.4 \\ -28.6 \end{Bmatrix} \text{ ft-kips}$$

**Example 16-4.** The end moments of the rigid frame shown in Fig. 16-5 were solved by the force method in Example 15-5 and will now be re-solved by the displacement method.

Fig. 16-5

Neglecting the effect of axial deformations, the frame prepared for nodal-force analysis is shown in Fig. 16-6 where $r_1$ and $r_2$ denote the joint rotations and $R_1$ and $R_2$ the corresponding moments. In this problem $R_1 = -R_2 = 10$ ft-kips. The dashed lines indicate assumed directions for member end moments, $Q_i^a, Q_j^a, \ldots$, and their corresponding member end rotations, $q_i^a$, $q_j^a, \ldots$. The latter can be expressed in terms of $r_1$ and $r_2$ as

Fig. 16-6

$$r_1 = 1 \quad r_2 = 1$$

$$\begin{Bmatrix} q_i^a \\ q_j^a \\ q_i^b \\ q_j^b \\ q_i^c \\ q_j^c \end{Bmatrix} = \begin{bmatrix} 0 & 0 \\ 1 & 0 \\ 1 & 0 \\ 0 & 1 \\ 0 & 1 \\ 0 & 0 \end{bmatrix} \begin{Bmatrix} r_1 \\ r_2 \end{Bmatrix}$$

$a$

The diagonal matrix $k$ is the same as that obtained in the preceding example. The stiffness matrix of structure is obtained by

$$K = a^T k a$$

$$= \begin{bmatrix} 0 & 1 & 1 & 0 & 0 & 0 \\ 0 & 0 & 0 & 1 & 1 & 0 \end{bmatrix} \left(\frac{EI}{L}\right) \begin{bmatrix} 4 & 2 & & & & \\ 2 & 4 & & & & \\ & & 4 & 2 & & \\ & & 2 & 4 & & \\ & & & & 4 & 2 \\ & & & & 2 & 4 \end{bmatrix} \begin{bmatrix} 0 & 0 \\ 1 & 0 \\ 1 & 0 \\ 0 & 1 \\ 0 & 1 \\ 0 & 0 \end{bmatrix}$$

$$= \frac{EI}{L} \begin{bmatrix} 8 & 2 \\ 2 & 8 \end{bmatrix}$$

Thus,

$$K^{-1} = \frac{L}{EI} \begin{bmatrix} 8 & -2 \\ -2 & 8 \end{bmatrix}^T \frac{1}{60} = \frac{L}{30EI} \begin{bmatrix} 4 & -1 \\ -1 & 4 \end{bmatrix}$$

and $\qquad r = K^{-1}R$

$$\begin{Bmatrix} r_1 \\ r_2 \end{Bmatrix} = \frac{L}{30EI} \begin{bmatrix} 4 & -1 \\ -1 & 4 \end{bmatrix} \begin{Bmatrix} 10 \\ -10 \end{Bmatrix} = \frac{L}{30EI} \begin{Bmatrix} 50 \\ -50 \end{Bmatrix}$$

Then the member end moments based on the nodal-force analysis (Fig. 16-6) are determined by $Q = kar$. Using the values of $k, a,$ and $r$ previously found gives

$$\begin{Bmatrix} Q_i^a \\ Q_j^a \\ Q_i^b \\ Q_j^b \\ Q_i^c \\ Q_j^c \end{Bmatrix} = \begin{Bmatrix} 3.33 \\ 6.67 \\ 3.33 \\ -3.33 \\ -6.67 \\ -3.33 \end{Bmatrix} \text{ ft-kips}$$

The end moments of member $b$, $Q_i^b$ and $Q_j^b$, must be corrected by adding the fixed-end moments shown in the upper part of Fig. 16-5(b). The final result is

$$\begin{Bmatrix} Q_i^a \\ Q_j^a \\ Q_i^b \\ Q_j^b \\ Q_i^c \\ Q_j^c \end{Bmatrix} = \begin{Bmatrix} 3.33 \\ 6.67 \\ -6.67 \\ 6.67 \\ -6.67 \\ -3.33 \end{Bmatrix} \text{ ft-kips}$$

**Example 16-5.** Obtain the end moments for the frame shown in Fig. 16-7(a). Use the equivalent form in Fig. 16-7(b) for the analysis so that the size of the matrices will be reduced. Assume constant $EI$.

If the effect of axial deformations is neglected, the frame prepared for nodal-force analysis is shown in Fig. 16-8. The frame is subjected to the joint rotations $r_1$ and $r_2$ and the corresponding nodal moments $R_1$ and $R_2$. In this case, $R_1 = 0$ and $R_2 = 10$ ft-kips.

As before, the dashed lines indicate the assumed directions for the member end moments $Q_i^a$, $Q_j^a$, $Q_i^b$, and $Q_j^b$ and their corresponding end rotations $q_i^a$, $q_j^a$, $q_i^b$, and $q_j^b$. The latter can be expressed in terms of $r_1$ and $r_2$ from compatibility as

$$r_1 = 1 \quad r_2 = 1$$

$$\begin{Bmatrix} q_i^a \\ q_j^a \\ q_i^b \\ q_j^b \end{Bmatrix} = \begin{bmatrix} 1 & 0 \\ 0 & 1 \\ 0 & 1 \\ 0 & 0 \end{bmatrix} \begin{Bmatrix} r_1 \\ r_2 \end{Bmatrix}$$

$$a$$

(a)                                      (b)

**Fig. 16-7**

**Fig. 16-8**

Since the members are identical,

$$k = \frac{EI}{L}\begin{bmatrix} 4 & 2 & & \\ 2 & 4 & & \\ & & 4 & 2 \\ & & 2 & 4 \end{bmatrix}$$

The total stiffness matrix is then determined.

$$K = a^T k a = \begin{bmatrix} 1 & 0 & 0 & 0 \\ 0 & 1 & 1 & 0 \end{bmatrix}\left(\frac{EI}{L}\right)\begin{bmatrix} 4 & 2 & & \\ 2 & 4 & & \\ & & 4 & 2 \\ & & 2 & 4 \end{bmatrix}\begin{bmatrix} 1 & 0 \\ 0 & 1 \\ 0 & 1 \\ 0 & 0 \end{bmatrix} = \frac{EI}{L}\begin{bmatrix} 4 & 2 \\ 2 & 8 \end{bmatrix}$$

367

from which

$$K^{-1} = \frac{L}{EI} \frac{\begin{bmatrix} 8 & -2 \\ -2 & 4 \end{bmatrix}^{T}}{28} = \frac{L}{14EI} \begin{bmatrix} 4 & -1 \\ -1 & 2 \end{bmatrix}$$

The nodal displacements are expressed by

$$\begin{Bmatrix} r_1 \\ r_2 \end{Bmatrix} = K^{-1}R = \frac{L}{14EI} \begin{bmatrix} 4 & -1 \\ -1 & 2 \end{bmatrix} \begin{Bmatrix} 0 \\ 10 \end{Bmatrix} = \frac{L}{14EI} \begin{Bmatrix} -10 \\ 20 \end{Bmatrix}$$

Thus, the end moments from nodal-force analysis, based on Fig. 16-8, are found to be

$$\begin{Bmatrix} Q_i^a \\ Q_j^a \\ Q_i^b \\ Q_j^b \end{Bmatrix} = kar = \frac{EI}{L} \begin{bmatrix} 4 & 2 & & \\ 2 & 4 & & \\ & & 4 & 2 \\ & & 2 & 4 \end{bmatrix} \begin{bmatrix} 1 & 0 \\ 0 & 1 \\ 0 & 1 \\ 0 & 0 \end{bmatrix} \left(\frac{L}{14EI}\right) \begin{Bmatrix} -10 \\ 20 \end{Bmatrix} = \begin{Bmatrix} 0 \\ 4.28 \\ 5.72 \\ 2.86 \end{Bmatrix} \text{ ft-kips}$$

After adding the fixed-end moments [see the upper part of Fig. 16-7(b)] to $Q_i^b$ and $Q_j^b$, we obtain the final solution as

$$\begin{Bmatrix} Q_i^a \\ Q_j^a \\ Q_i^b \\ Q_j^b \end{Bmatrix} = \begin{Bmatrix} 0 \\ 4.28 \\ -4.28 \\ 12.86 \end{Bmatrix} \text{ ft-kips}$$

**Example 16-6.** The deflection and slope at $B$ for the system shown in Fig. 16-9(a) were solved by the force method in Example 15-6, and will be resolved by the displacement method. Note that the spring stiffness is $k_s$ or $1/f_s$.

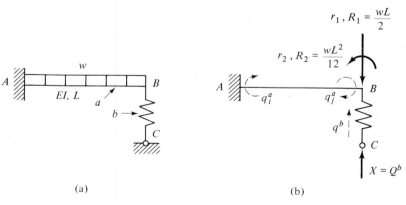

(a)                                                        (b)

**Fig. 16-9**

As in Example 15-6, to obtain the nodal displacements at $B$, it is only necessary to analyze the system under nodal loads shown in Fig. 16-9(b). Also shown in Fig. 16-9(b) by dashed lines are the member deformations $q_i^a$, $q_j^a$, and $q^b$ (spring contraction).

We first relate $q_i^a$, $q_j^a$, and $q^b$ to the external nodal displacements $r_1$ and $r_2$ as

$$\begin{Bmatrix} q_i^a \\ q_j^a \\ q^b \end{Bmatrix} = \begin{bmatrix} -\dfrac{1}{L} & 0 \\ -\dfrac{1}{L} & -1 \\ 1 & 0 \end{bmatrix} \begin{Bmatrix} r_1 \\ r_2 \end{Bmatrix}$$

$$a$$

From the beam stiffness

$$k^a = \begin{bmatrix} \dfrac{4EI}{L} & \dfrac{2EI}{L} \\ \dfrac{2EI}{L} & \dfrac{4EI}{L} \end{bmatrix}$$

and the spring stiffness $k^b = k_s$, we form

$$k = \begin{bmatrix} \dfrac{4EI}{L} & \dfrac{2EI}{L} & \\ \dfrac{2EI}{L} & \dfrac{4EI}{L} & \\ & & k_s \end{bmatrix}$$

With matrices $a$ and $k$ determined, we obtain

$$K = a^T k a = \begin{bmatrix} \dfrac{12EI}{L^3} + K_s & \dfrac{6EI}{L^2} \\ \dfrac{6EI}{L^2} & \dfrac{4EI}{L} \end{bmatrix}$$

$$K^{-1} = \dfrac{\begin{bmatrix} \dfrac{4EI}{L} & -\dfrac{6EI}{L^2} \\ -\dfrac{6EI}{L^2} & \dfrac{12EI}{L^3} + k_s \end{bmatrix}}{\dfrac{4EI}{L}\left(\dfrac{3EI}{L^3} + k_s\right)}$$

Using

$$\begin{Bmatrix} r_1 \\ r_2 \end{Bmatrix} = [K]^{-1} \begin{Bmatrix} R_1 = \dfrac{wL}{2} \\ R_2 = \dfrac{wL^2}{12} \end{Bmatrix}$$

we solve for

$$\begin{Bmatrix} r_1 \\ r_2 \end{Bmatrix} = \frac{1}{\dfrac{3EI}{L^3} + k_s} \begin{Bmatrix} \dfrac{3}{8}wL \\ -\dfrac{1}{2}w + \dfrac{k_s wL^3}{48EI} \end{Bmatrix}$$

which can be checked with the result found in Example 15-6.

## 16-6. USE OF THE MODIFIED MEMBER STIFFNESS MATRIX

Referring to Eq. 16-11, we find the stiffness matrix for a uniform member in frame analysis is given by

$$k^a = \frac{EI}{L}\begin{bmatrix} 4 & 2 \\ 2 & 4 \end{bmatrix}$$

Note that the elements in the first column of the matrix are the member end moments obtained by producing in this end ($i$ end) a unit rotation, the other end ($j$ end) being fixed. The procedure described for obtaining these values is exactly the same as that for finding the stiffness of the $i$ end and its carryover value to the $j$ end in the method of moment distribution, as illustrated in Fig. 16-10. Recall, in the moment-distribution procedures, that if the actual condition of the other end is known, then the computation can be simplified by using the modified stiffness of this end and omitting the presentation of the other portion of structure in the analysis. This technique can also be applied to the displacement method by introducing the *modified member stiffness matrix*.

Stiffness $Q_i^a = \dfrac{4EI}{L}$    $\dfrac{1/2}{\longrightarrow}$    Carry-over moment $Q_j^a = \dfrac{2EI}{L}$

Member "$a$"

$q_j^a = 0$

$q_i^a = 1$

$L$

$EI$ constant

**Fig. 16-10**

With reference to Sec. 13-7 (modified stiffness in moment distribution), we note the following three special cases:

1. When the other end is simply supported, then the moment needed to produce a unit rotation in this end is $3EI/L$; that is, $Q_i^a = 3EI/L$ for $q_i^a = 1$ as indicated in Fig. 16-11.

The modified member stiffness matrix, if we disregard the factors of $Q_j^a$ and $q_j^a$, is then given by

Modified stiffness

$$Q_i^a = \frac{3EI}{L}$$

$q_i^a = 1$

Member "a"

L

EI constant

**Fig. 16-11**

$$(k^a)' = \left[\frac{3EI}{L}\right] \tag{16-21}$$

where $(k^a)'$ denotes the modified member stiffness matrix for member $a$.

2. When the other end rotates an equal but opposite angle to that of this end (the case of symmetry),

$$(k^a)' = \left[\frac{2EI}{L}\right] \tag{16-22}$$

3. When the other end rotates the same angle as that of this end (the case of antisymmetry),

$$(k^a)' = \left[\frac{6EI}{L}\right] \tag{16-23}$$

Application of the procedures described above is illustrated by re-solving Examples 16-3 to 16-5 of Sec. 16-5 as follows:

**Example 16-7.** Re-solve Example 16-3 by using the modified stiffness matrix.

The frame in Fig. 16-4(a) may be put in the form of Fig. 16-12(a). The structure and its loading represent a case of antisymmetry for which we may

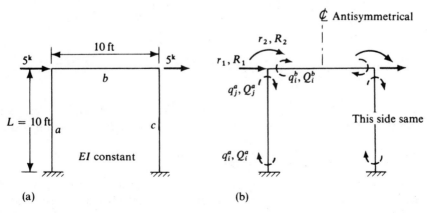

(a)

(b)

**Fig. 16-12**

371

assume the nodal displacements and their corresponding nodal forces as shown in Fig. 16-12(b). Observe that $R_1 = 5$ kips and $R_2 = 0$.

The analysis can be simplified by working with only half the structure if the modified stiffness matrix for member $b$ is used:

$$(k^b)' = \left[\frac{6EI}{L}\right]$$

Hence,

$$k = \begin{bmatrix} k^a & \\ & (k^b)' \end{bmatrix} = \frac{EI}{L}\begin{bmatrix} 4 & 2 & 0 \\ 2 & 4 & 0 \\ 0 & 0 & 6 \end{bmatrix}$$

From compatibility

$$r_1 = 1 \quad r_2 = 1$$

$$\begin{Bmatrix} q_i^a \\ q_j^a \\ q_i^b \end{Bmatrix} = \begin{bmatrix} -\frac{1}{10} & 0 \\ -\frac{1}{10} & 1 \\ 0 & 1 \end{bmatrix}\begin{Bmatrix} r_1 \\ r_2 \end{Bmatrix}$$

$$a$$

Using the values already found for $a$ and $k$, we have

$$K = a^T k a = \frac{EI}{L}\begin{bmatrix} 0.12 & -0.6 \\ -0.6 & 10 \end{bmatrix}$$

$$K^{-1} = \frac{L}{0.84EI}\begin{bmatrix} 10 & 0.6 \\ 0.6 & 0.12 \end{bmatrix}$$

$$r = K^{-1}R = \frac{L}{0.84EI}\begin{bmatrix} 10 & 0.6 \\ 0.6 & 0.12 \end{bmatrix}\begin{Bmatrix} 5 \\ 0 \end{Bmatrix} = \frac{L}{0.84EI}\begin{Bmatrix} 50 \\ 3 \end{Bmatrix}$$

Thus,

$$\begin{Bmatrix} Q_i^a \\ Q_j^a \\ Q_i^b \end{Bmatrix} = kar = \frac{EI}{L}\begin{bmatrix} 4 & 2 & 0 \\ 2 & 4 & 0 \\ 0 & 0 & 6 \end{bmatrix}\begin{bmatrix} -\frac{1}{10} & 0 \\ -\frac{1}{10} & 1 \\ 0 & 1 \end{bmatrix}\left(\frac{L}{0.84EI}\right)\begin{Bmatrix} 50 \\ 3 \end{Bmatrix}$$

$$= \begin{Bmatrix} -28.6 \\ -21.4 \\ 21.4 \end{Bmatrix} \text{ ft-kips}$$

The results, $Q_i^a = Q_j^c = -28.6$ ft-kips, $Q_j^a = Q_i^c = -21.4$ ft-kips, and $Q_i^b = Q_j^b = 21.4$ ft-kips, are the same as previously found in Example 16-3.

**Example 16-8.** Re-solve Example 16-4 by using the modified stiffness matrix.

The portal frame shown in Fig. 16-5(a) is symmetrical about the center line of the beam. The frame assumed for nodal-force analysis may be given as in Fig. 16-13. Referring to Fig. 16-5(b), we note that $R_1 = 10$ ft-kips. The analysis

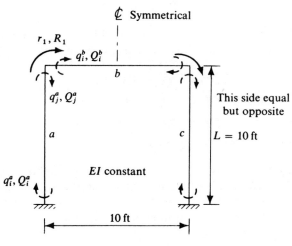

**Fig. 16-13**

can be simplified by working with only half the frame and using the modified stiffness matrix for member $b$:

$$(k^b)' = \left[\frac{2EI}{L}\right]$$

Thus,

$$k = \begin{bmatrix} k^a & \\ & (k^b)' \end{bmatrix} = \frac{EI}{L}\begin{bmatrix} 4 & 2 & 0 \\ 2 & 4 & 0 \\ 0 & 0 & 2 \end{bmatrix}$$

From compatibility,

$$\begin{Bmatrix} q_i^a \\ q_j^a \\ q_i^b \end{Bmatrix} = \begin{Bmatrix} 0 \\ 1 \\ 1 \end{Bmatrix}[r_1]$$

$$a$$

With the values of $a$ and $k$ found, we obtain

$$K = a^T k a = \left[\frac{6EI}{L}\right]$$

It follows that

$$K^{-1} = \left[\frac{L}{6EI}\right]$$

$$r_1 = K^{-1}R_1 = \left[\frac{L}{6EI}\right][10] = \left[\frac{10L}{6EI}\right]$$

With the values of $a$, $k$, and $r_1$ found, we obtain the end moments from nodal-force analysis as

$$\begin{Bmatrix} Q_i^a \\ Q_j^a \\ Q_i^b \end{Bmatrix} = kar_1 = \begin{Bmatrix} 3.33 \\ 6.67 \\ 3.33 \end{Bmatrix} \quad \text{ft-kips}$$

Adding the fixed-end moment of $-10$ ft-kips [see the upper part of Fig. 16-5(b)] to $Q_i^b$ in the result above and using symmetry, we obtain the final solution as

$$Q_i^a = -Q_j^c = 3.33 \text{ ft-kips}$$
$$Q_j^a = -Q_i^c = 6.67 \text{ ft-kips}$$
$$Q_i^b = -Q_j^b = -6.67 \text{ ft-kips}$$

These are the same as previously obtained in Example 16-4.

**Example 16-9.**   Re-solve Example 16-5 by using modified member stiffness.

Refer to Fig. 16-7. Since the moment at the hinged end is zero, we may simplify the computation by using the modified stiffness matrix $[3EI/L]$ for member $a$ and by assuming the frame for nodal-force analysis as in Fig. 16-14. Note that in present case $R_1 = 10$ ft-kips.

$$k = \begin{bmatrix} (k^a)' & \\ & k^b \end{bmatrix} = \frac{EI}{L}\begin{bmatrix} 3 & 0 & 0 \\ 0 & 4 & 2 \\ 0 & 2 & 4 \end{bmatrix}$$

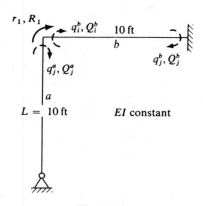

**Fig. 16-14**

From compatibility

$$\begin{Bmatrix} q_j^a \\ q_i^b \\ q_j^b \end{Bmatrix} = \begin{Bmatrix} 1 \\ 1 \\ 0 \end{Bmatrix}[r_1]$$

$$a$$

With $k$ and $a$ determined, we have

$$K = a^T k a = \frac{EI}{L}[7]$$

$$r_1 = K^{-1}R_1 = \frac{L}{EI}\begin{bmatrix} 10 \\ 7 \end{bmatrix}$$

Thus, the end moments from nodal-force analysis are given by

$$
\begin{Bmatrix} Q_j^a \\ Q_i^b \\ Q_j^b \end{Bmatrix} = kar_1 = \frac{EI}{L} \begin{bmatrix} 3 & 0 & 0 \\ 0 & 4 & 2 \\ 0 & 2 & 4 \end{bmatrix} \begin{Bmatrix} 1 \\ 1 \\ 0 \end{Bmatrix} \left(\frac{L}{EI}\right) \begin{bmatrix} 10 \\ 7 \end{bmatrix} = \begin{Bmatrix} 4.28 \\ 5.72 \\ 2.86 \end{Bmatrix} \quad \text{ft-kips}
$$

After the values of $Q_i^b$ and $Q_j^b$ are corrected by adding the fixed-end moment [see the upper part of Fig. 16-7(b)], we obtain the final solution as previously found in Example 16-5:

$$
\begin{Bmatrix} Q_j^a \\ Q_i^b \\ Q_j^b \end{Bmatrix} = \begin{Bmatrix} 4.28 \\ -4.28 \\ 12.86 \end{Bmatrix} \quad \text{ft-kips}
$$

## 16-7. THE GENERAL FORMULATION OF DISPLACEMENT METHOD

It may be interesting to point out that in a frame structure, there exist generally two kinds of node: those with unknown displacements associated with known forces (unprescribed node or free node), and those with known displacements associated with unknown forces (prescribed node). Throughout the previous sections of this chapter, we deal only with the former kind of node, for which we use $r$ to denote the unknown displacements, and $R$ the corresponding known loads. In a more general case, however, the formulation of displacement procedure should include also the unknown forces $X$ (e.g., the support reactions) and the corresponding known displacements $r_X$ at the prescribed nodes. As will be seen, the matrix displacement formulation in this manner leads to a direct analogy with the matrix force presentation as developed in Sec. 15-5. The procedure can be summarized as follows:

1. Separate the nodal displacements into two categories: the unprescribed displacements $r$ and the prescribed displacements $r_X$.

2. Establish the displacement transformation matrix to relate the internal element displacements $q$ with the external nodal displacements $r$ and $r_X$:

$$
q = \underbrace{[a_R \mid a_X]}_{a} \begin{Bmatrix} r \\ \hline r_X \end{Bmatrix} \tag{16-24}
$$

where the transformation matrix $a$ is partitioned into $a_R$ and $a_X$, indicating the separate influences from $r$ (associated with $R$) and $r_X$ (associated with $X$).

3. Assemble the element stiffness matrices $k^a, k^b, \ldots$ to obtain

$$
k = \begin{bmatrix} k^a & & & \\ & k^b & & \\ & & \cdot & \\ & & & \cdot \end{bmatrix} \tag{16-25}
$$

4. Construct the total stiffness matrix by the process

$$K = a^T k a$$

$$= \left\{ \begin{matrix} a_R^T \\ --- \\ a_X^T \end{matrix} \right\} [k][a_R \mid a_X] \qquad (16\text{-}26)$$

$$= \left[ \begin{array}{c|c} a_R^T k a_R & a_R^T k a_X \\ \hline a_X^T k a_R & a_X^T k a_X \end{array} \right]$$

5. The overall stiffness expressed in Eq. 16-26 relates the total nodal forces $R$ and $X$ to the corresponding displacements $r$ and $r_X$. The general formulation can be written as

$$\left\{ \begin{matrix} R \\ --- \\ X \end{matrix} \right\} = \left[ \begin{array}{c|c} K_{RR} & K_{RX} \\ \hline K_{XR} & K_{XX} \end{array} \right] \left\{ \begin{matrix} r \\ --- \\ r_X \end{matrix} \right\} \qquad (16\text{-}27)$$

where
$$\begin{array}{ll} K_{RR} = a_R^T k a_R & K_{RX} = a_R^T k a_X \\ K_{XR} = a_X^T k a_R & K_{XX} = a_X^T k a_X \end{array} \qquad (16\text{-}28)$$

The frequently encountered case is that the prescribed displacements are null (i.e., $r_X = 0$). Then Eq. 16-28 reduces to

$$\left\{ \begin{matrix} R \\ --- \\ X \end{matrix} \right\} = \left[ \begin{array}{c|c} K_{RR} & K_{RX} \\ \hline K_{XR} & K_{XX} \end{array} \right] \left\{ \begin{matrix} r \\ --- \\ 0 \end{matrix} \right\} \qquad (16\text{-}29)$$

**Example 16-10.** Consider the hanger made up by a two-section bar as shown in Fig. 16-15(a). The bar is subjected to load of $P$ at the free end, and load of $2P$ at the point where the change of section takes place. Find the elongations measured at the loaded points, and the reaction at the fixed support [see Fig. 16-15(b)]. Assume that the bar weight is negligible.

The bar elongations $q^a$ and $q^b$ are related to the nodal displacements $r_1$, $r_2$, and $r_3$ ($r_3 = 0$) as

$$\left\{ \begin{matrix} q^a \\ q^b \end{matrix} \right\} = \left[ \begin{array}{cc|c} 1 & -1 & 0 \\ 0 & 1 & -1 \end{array} \right] \left\{ \begin{matrix} r_1 \\ r_2 \\ --- \\ r_3 \end{matrix} \right\}$$

$$\underbrace{\phantom{\left[ \begin{array}{cc} 1 & -1 \end{array} \right]}}_{a_R} \underbrace{\phantom{\left[ 0 \right]}}_{a_X}$$

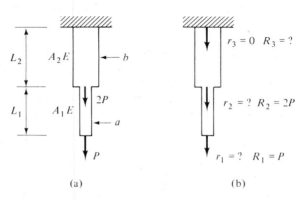

(a)                                            (b)

**Fig. 16-15**

From the element stiffness matrices $k^a = A_1 E/L_1$ and $k^b = A_2 E/L_2$, we obtain

$$k = \begin{bmatrix} \dfrac{A_1 E}{L_1} & \\ & \dfrac{A_2 E}{L_2} \end{bmatrix}$$

With matrices $a_R$, $a_X$, and $k$ determined, the partitioned matrices for the total stiffness matrix are then calculated:

$$K_{RR} = a_R^T k a_R = \begin{bmatrix} \dfrac{A_1 E}{L_1} & -\dfrac{A_1 E}{L_1} \\ -\dfrac{A_1 E}{L_1} & \dfrac{A_1 E}{L_1} + \dfrac{A_2 E}{L_2} \end{bmatrix}$$

$$K_{RX} = a_R^T k a_X = \begin{bmatrix} 0 \\ -\dfrac{A_2 E}{L_2} \end{bmatrix}$$

$$K_{XR} = a_X^T k a_R = \begin{bmatrix} 0 & -\dfrac{A_2 E}{L_2} \end{bmatrix}$$

$$K_{XX} = a_X^T k a_X = \begin{bmatrix} \dfrac{A_2 E}{L_2} \end{bmatrix}$$

Hence, the force–displacement relationship is

$$\begin{Bmatrix} R_1 = P \\ R_2 = 2P \\ \hline R_3 = ? \end{Bmatrix} = \begin{bmatrix} \dfrac{A_1 E}{L_1} & -\dfrac{A_1 E}{L_1} & 0 \\ -\dfrac{A_1 E}{L_1} & \dfrac{A_1 E}{L_1} + \dfrac{A_2 E}{L_2} & -\dfrac{A_2 E}{L_2} \\ \hline 0 & -\dfrac{A_2 E}{L_2} & \dfrac{A_2 E}{L_2} \end{bmatrix} \begin{Bmatrix} r_1 = ? \\ r_2 = ? \\ \hline r_3 = 0 \end{Bmatrix}$$

We solve the unknown displacements $r_1$ and $r_2$ by

$$\begin{Bmatrix} r_1 \\ r_2 \end{Bmatrix} = [K_{RR}]^{-1} \begin{Bmatrix} R_1 \\ R_2 \end{Bmatrix}$$

$$= \begin{bmatrix} \dfrac{A_1 E}{L_1} & -\dfrac{A_1 E}{L_1} \\ -\dfrac{A_1 E}{L_1} & \dfrac{A_1 E}{L_1} + \dfrac{A_2 E}{L_2} \end{bmatrix}^{-1} \begin{Bmatrix} P \\ 2P \end{Bmatrix}$$

$$= \begin{bmatrix} \dfrac{L_1}{A_1 E} + \dfrac{L_2}{A_2 E} & \dfrac{L_2}{A_2 E} \\ \dfrac{L_2}{A_2 E} & \dfrac{L_2}{A_2 E} \end{bmatrix} \begin{Bmatrix} P \\ 2P \end{Bmatrix}$$

Therefore,

$$r_1 = \frac{PL_1}{A_1 E} + \frac{3PL_2}{A_2 E}$$

$$r_2 = \frac{3PL_2}{A_2 E}$$

Finally, we obtain the end reaction $R_3$ by

$$R_3 = [K_{XR}] \begin{Bmatrix} r_1 \\ r_2 \end{Bmatrix}$$

$$= \begin{bmatrix} 0 & -\dfrac{A_2 E}{L_2} \end{bmatrix} \begin{Bmatrix} \dfrac{PL_1}{A_1 E} + \dfrac{3PL_2}{A_2 E} \\ \dfrac{3PL_2}{A_2 E} \end{Bmatrix} = -3P$$

These results can easily be checked.

## 16-8. THE DIRECT STIFFNESS METHOD

In the previous discussion the establishment of total stiffness matrix was performed by

$$K = a^T k a$$

in which $a$ and $k$ were assembled on the basis of the entire structure. Alternatively, we may treat each member or element as a structure (member-structure) and obtain the *stiffness matrix for the member-structure* by using the smaller matrices $a$, $a^T$, and $k$ of the member. The total stiffness matrix of the entire structure is then established by superimposing the stiffness matrix of the individual structural members. This method, referred to as the *direct stiffness method* is particularly convenient for larger structures, for it eliminates the necessity of establishing all the large matrices except one, the final $K$ matrix of the structure. To illustrate, let us consider the frame in Fig. 16-4.

The total stiffness matrix was found to be

$$\frac{EI}{L} \begin{bmatrix} 0.24 & -0.6 & -0.6 \\ -0.6 & 8 & 2 \\ -0.6 & 2 & 8 \end{bmatrix}$$

in Example 16-3 through the multiplication of $a^T k a$, in which $k$ is of order $(6 \times 6)$, $a$ is of order $(6 \times 3)$, and $a^T$ is of order $(3 \times 6)$. The same result may be achieved by the direct stiffness procedures, which enable the use of smaller size matrices, as contained in the following steps:

1. Treat member $a$ as a structure. To find the total stiffness matrix of the member, we note that if only the flexural deformation is considered,

$$\begin{Bmatrix} Q_i^a \\ Q_j^a \end{Bmatrix} = \frac{EI}{L} \begin{bmatrix} 4 & 2 \\ 2 & 4 \end{bmatrix} \begin{Bmatrix} q_i^a \\ q_j^a \end{Bmatrix}$$

and from compatibility

$$\begin{Bmatrix} q_i^a \\ q_j^a \end{Bmatrix} = \begin{bmatrix} -\frac{1}{10} & 0 \\ -\frac{1}{10} & 1 \end{bmatrix} \begin{Bmatrix} r_1 \\ r_2 \end{Bmatrix}$$

since the end rotations of member $a$ are not affected by $r_3$ under the condition that all other nodes are fixed.

Thus, we have

$$a = \begin{bmatrix} -\frac{1}{10} & 0 \\ -\frac{1}{10} & 1 \end{bmatrix} \qquad a^T = \begin{bmatrix} -\frac{1}{10} & -\frac{1}{10} \\ 0 & 1 \end{bmatrix} \qquad k = \frac{EI}{L}\begin{bmatrix} 4 & 2 \\ 2 & 4 \end{bmatrix}$$

for the member-structure, all of order $(2 \times 2)$. If we use $K^a$ to denote the total stiffness matrix of member $a$, we have

$$K^a = a^T k a = \begin{bmatrix} -\frac{1}{10} & -\frac{1}{10} \\ 0 & 1 \end{bmatrix}\left(\frac{EI}{L}\right)\begin{bmatrix} 4 & 2 \\ 2 & 4 \end{bmatrix}\begin{bmatrix} -\frac{1}{10} & 0 \\ -\frac{1}{10} & 1 \end{bmatrix} = \frac{EI}{L}\begin{array}{c}\overset{r_1 \qquad\quad r_2}{\begin{bmatrix} 0.12 & -0.6 \\ -0.6 & 4 \end{bmatrix}}\end{array}$$

2. In the same way, we treat member $b$ as a structure. We find that

$$\begin{Bmatrix} Q_i^b \\ Q_j^b \end{Bmatrix} = \frac{EI}{L}\begin{bmatrix} 4 & 2 \\ 2 & 4 \end{bmatrix}\begin{Bmatrix} q_i^b \\ q_j^b \end{Bmatrix}$$

and

$$\begin{Bmatrix} q_i^b \\ q_j^b \end{Bmatrix} = \begin{bmatrix} 1 & 0 \\ 0 & 1 \end{bmatrix}\begin{Bmatrix} r_2 \\ r_3 \end{Bmatrix}$$

Thus,

$$a = a^T = I \qquad k = \frac{EI}{L}\begin{bmatrix} 4 & 2 \\ 2 & 4 \end{bmatrix}$$

from which

$$K^b = a^T k a = \frac{EI}{L}\begin{array}{c}\overset{r_2 \quad r_3}{\begin{bmatrix} 4 & 2 \\ 2 & 4 \end{bmatrix}}\end{array}$$

3. Member $c$ is in a condition similar to that of member $a$ except that it is subjected to the influences of $r_1$ and $r_3$. Thus,

$$K^c = \frac{EI}{L}\begin{array}{c}\overset{r_1 \qquad\quad r_3}{\begin{bmatrix} 0.12 & -0.6 \\ -0.6 & 4 \end{bmatrix}}\end{array}$$

4. The matrices $K^a$, $K^b$, and $K^c$ above are not immediately subjected to superposition since the columns in the three matrices are under different influences of $r$. To make them compatible for addition, we expand each of the three to the order of $(3 \times 3)$ by adding a column and a row of zeros for the nodal displacement, which is irrelevant for the member in question. In this manner, we obtain

$$K^a = \frac{EI}{L}\begin{array}{c}\overset{r_1 \quad\;\; r_2 \quad\; r_3}{\begin{bmatrix} 0.12 & -0.6 & 0 \\ -0.6 & 4 & 0 \\ 0 & 0 & 0 \end{bmatrix}}\end{array} \qquad K^b = \frac{EI}{L}\begin{array}{c}\overset{r_1 \; r_2 \; r_3}{\begin{bmatrix} 0 & 0 & 0 \\ 0 & 4 & 2 \\ 0 & 2 & 4 \end{bmatrix}}\end{array}$$

$$K^c = \frac{EI}{L} \begin{matrix} r_1 & r_2 & r_3 \\ \begin{bmatrix} 0.12 & 0 & -0.6 \\ 0 & 0 & 0 \\ -0.6 & 0 & 4 \end{bmatrix} \end{matrix}$$

The order of the total stiffness matrix for each member is now the same as the order of $K$ for the entire structure. By directly adding $K^a$, $K^b$, and $K^c$, we obtain the total structure stiffness matrix $K$ as

$$K = K^a + K^b + K^c = \frac{EI}{L} \begin{bmatrix} 0.24 & -0.6 & -0.6 \\ -0.6 & 8 & 2 \\ -0.6 & 2 & 8 \end{bmatrix}$$

After $K$ is found, the rest of analysis is the same as previously discussed for the displacement method. Thus, the direct stiffness method differs from the displacement method only in the manner by which the total stiffness matrix of structure is constructed. For complex structures this difference is important, since the direct stiffness method is a direct and easier way to establish the overall stiffness matrix by the superposition of total stiffness matrices of individual elements.

## 16-9. COMPARISON OF THE FORCE METHOD AND THE DISPLACEMENT METHOD

The force method and the displacement method represent two parallel ways of analyzing structures. The basic procedures for the two methods may be briefly recapitulated as in Table 16-1. The duality between the two methods is apparent. The choice of the methods mainly lies in the accuracy of the solution and the ease of computation which in turn would depend upon the idealization of the structure, the rounding off of error, and the type of and formulation of the problem. Generally, except for structures that involve many joint displacements but few force redundants, the displacement method is often preferred. Some of the reasons are as follows:

TABLE 18-1

| Force Method | Displacement Method |
|---|---|
| (1) Select member forces as basic unknowns. | (1) Select nodal displacements as basic unknowns. |
| (2) Establish the force transformation matrix | (2) Establish the displacement transformation matrix. |
| (3) Evaluate member flexibility matrices | (3) Evaluate member stiffness matrices. |
| (4) Obtain the total flexibility matrix. | (4) Obtain the total stiffness matrix. |
| (5) Express the nodal displacement in terms of the nodal forces | (5) Express the nodal forces in terms of the nodal displacements |

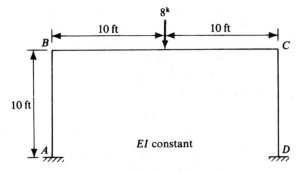

Fig. 16-21

**16-9.** Obtain, by the displacement method, all the end moments for the frame of Fig. 16-22, using the modified member stiffness matrix for the center beam due to antisymmetry.

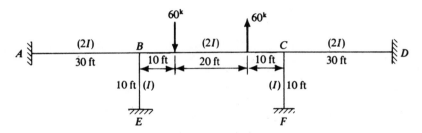

Fig. 16-22

**16-10.** For the composite system in Fig. 16-23, construct the displacement transformation matrix consistent with the nodal displacements shown.

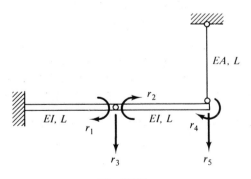

Fig. 16-23

**16-11.** Use the displacement method to obtain the end moments for the frame shown in Fig. 16-24. Assume constant *EI*.

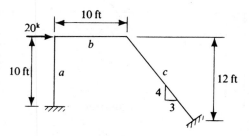

**Fig. 16-24**

**16-12.** Use the direct stiffness method to obtain the total stiffness matrix for the frame of Fig. 16-24.

# 17

## THE TREATMENT OF
## NONPRISMATIC MEMBERS

### 17-1. GENERAL

In the previous discussion of framed-structure analysis, we have concerned only with structures composed of uniform elements. In many instances, however, the members forming a structure are nonprismatic, or for more general case, have variable rigidities such as those shown in Fig. 17-1. The fundamental concepts of analysis remain the same as if the structure were built up by prismatic members. However, the expression for member constants, including the fixed-end actions, the flexibility and stiffness coefficients, and the stiffness and carry-over factor necessary for a moment distribution, derived specifically for prismatic members, are no longer valid for nonprismatic members. These constants applied for nonprismatic members must be first determined so that the analysis of structure either by slope-deflection equations, or by moment distribution, or by matrix procedure can be carried out in the usual manner.

In this chapter we develop various integral formulas expressing these constants and demonstrate how to employ a numerical approach to approximate an integration.

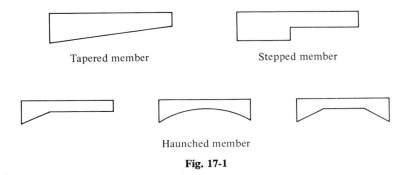

Tapered member        Stepped member

Haunched member

**Fig. 17-1**

## 17-2. FIXED-END ACTIONS

Consider a member of varying fluxural rigidity with both ends fixed and subjected to the bending action caused by member loads as shown in Fig. 17-2. The general expression for the fixed-end moments may be found by the method of least work, using Eqs. 10-5 and 10-6:

**Fig. 17-2**

$$\int_0^l \frac{M\,dx}{EI} = 0 \qquad (17\text{-}1)$$

$$\int_0^l \frac{Mx\,dx}{EI} = 0 \qquad (17\text{-}2)$$

If the member is made of the same material, we can assume that $E$ is a constant. The expressions above thus become

$$\int_0^l \frac{M\,dx}{I} = 0 \qquad (17\text{-}3)$$

$$\int_0^l \frac{Mx\,dx}{I} = 0 \qquad (17\text{-}4)$$

where both $M$ and $I$ are functions of $x$.

As an illustration, let us find $M_1$ and $V_1$ for the fixed beam shown in Fig. 17-3 due to a uniform load over the entire span.

The moment at any section distance $x$ from the left end is

$$M = M_1 + V_1 x - \frac{wx^2}{2}$$

**Fig. 17-3**

Substituting $M$ in Eqs. 17-3 and 17-4 gives

$$M_1 \int_0^l \frac{dx}{I} + V_1 \int_0^l \frac{x\,dx}{I} - \frac{w}{2} \int_0^l \frac{x^2\,dx}{I} = 0 \qquad (17\text{-}5)$$

$$M_1 \int_0^l \frac{x\,dx}{I} + V_1 \int_0^l \frac{x^2\,dx}{I} - \frac{w}{2} \int_0^l \frac{x^3\,dx}{I} = 0 \qquad (17\text{-}6)$$

In matrix form,

$$\begin{bmatrix} \displaystyle\int_0^l \frac{dx}{I} & \displaystyle\int_0^l \frac{x\,dx}{I} \\[2mm] \displaystyle\int_0^l \frac{x\,dx}{I} & \displaystyle\int_0^l \frac{x^2\,dx}{I} \end{bmatrix} \begin{Bmatrix} M_1 \\[2mm] V_1 \end{Bmatrix} = \frac{w}{2} \begin{Bmatrix} \displaystyle\int_0^l \frac{x^2\,dx}{I} \\[2mm] \displaystyle\int_0^l \frac{x^3\,dx}{I} \end{Bmatrix} \qquad (17\text{-}7)$$

Eliminating $V_1$ from Eqs. 17-5 and 17-6 yields the fixed-end moment at 1:

$$M_1 = \left(\frac{w}{2}\right) \frac{\left(\int_0^l (x^2\,dx)/I\right)^2 - \int_0^l (x\,dx)/I \int_0^l (x^3\,dx)/I}{\int_0^l dx/I \int_0^l (x^2\,dx)/I - \left(\int_0^l (x\,dx)/I\right)^2} \qquad (17\text{-}8)$$

The fixed-end moment at 2 can also be obtained from Eq. 17-8, by taking the integral origin at 2 and using reverse sign. For a member of varying $I$, $M_1$ and $M_2$ are not equal except in a symmetrical system. For a member of uniform section, Eq. 17-8 reduces to

$$M_1 = -M_2 = \left(\frac{w}{2}\right) \frac{\left(\int_0^l x^2\,dx\right)^2 - \int_0^l x\,dx \int_0^l x^3\,dx}{\int_0^l dx \int_0^l x^2\,dx - \left(\int_0^l x\,dx\right)^2} = -\frac{wl^2}{12} \qquad (17\text{-}9)$$

With $M_1$ and $M_2$ obtained, we can find $V_1$ and $V_2$ by simple statics.

## 17-3. THE ROTATIONAL FLEXIBILITY MATRIX
## OF A BEAM ELEMENT

Consider the beam element with a variable cross section subjected to end moments $R_1$ and $R_2$ with the corresponding rotations $r_1$ and $r_2$ as in Fig. 17-4(a). In order to find flexibility coefficients, we must have the element properly supported. If we regard the element as a simple beam as in Fig. 17-4(b), then the bending moment at any section distance $x$ from the left end is

$$M = R_1 - \frac{(R_1 + R_2)x}{l} \qquad (17\text{-}10)$$

By the method of least work, we obtain the end rotations $r_1$ and $r_2$ as

$$r_1 = \frac{\partial W}{\partial R_1} = \int_0^l \frac{M(\partial M/\partial R_1)\,dx}{EI}$$

$$= \int_0^l \frac{[R_1 - (R_1 + R_2)(x/l)][1 - (x/l)]\,dx}{EI} \qquad (17\text{-}11)$$

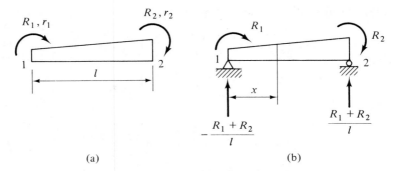

(a)                                                    (b)

**Fig. 17-4**

$$r_2 = \frac{\partial W}{\partial R_2} = \int_0^l \frac{M(\partial M/\partial R_2)\, dx}{EI}$$

$$= \int_0^l \frac{[R_1 - (R_1 + R_2)(x/l)][-x/l]\, dx}{EI} \tag{17-12}$$

Setting $R_1 = 1$ and $R_2 = 0$ in Eqs. 17-11 and 17-12, respectively, leads to the rotational flexibility coefficients $f_{11}$ and $f_{21}$:

$$f_{11} = \int_0^l \frac{(l - x)^2\, dx}{EIl^2} \tag{17-13}$$

$$f_{21} = -\int_0^l \frac{(l - x)(x)\, dx}{EIl^2} \tag{17-14}$$

Similarly, setting $R_1 = 0$ and $R_2 = 1$ in Eqs. 17-11 and 17-12, respectively, yields

$$f_{12} = -\int_0^l \frac{(l - x)(x)\, dx}{EIl^2} \tag{17-15}$$

$$f_{22} = \int_0^l \frac{x^2\, dx}{EIl^2} \tag{17-16}$$

Collecting Eqs. 17-13 to 17-16 and assuming a constant $E$, we form the rotational flexibility matrix:

$$[f] = \begin{bmatrix} f_{11} & f_{12} \\ f_{21} & f_{22} \end{bmatrix}$$

$$= \frac{1}{EI^2} \begin{bmatrix} \displaystyle\int_0^l \frac{(l - x)^2\, dx}{I} & -\displaystyle\int_0^l \frac{(l - x)(x)\, dx}{I} \\ -\displaystyle\int_0^l \frac{(l - x)(x)\, dx}{I} & \displaystyle\int_0^l \frac{x^2\, dx}{I} \end{bmatrix} \tag{17-17}$$

For a member of uniform cross section, the expression above reduces to

$$[f] = \frac{l}{6EI}\begin{bmatrix} 2 & -1 \\ -1 & 2 \end{bmatrix} \tag{17-18}$$

## 17-4. THE ROTATIONAL STIFFNESS MATRIX OF A BEAM ELEMENT

Consider the same beam element in Fig. 17-4(a). By definition, the rotational stiffness coefficients $s_{11}$ and $s_{21}$ are the respective moments at coordinates 1 and 2 due to a unit rotation at coordinate 1 ($r_1 = 1$). Similarly, the rotational stiffness coefficients $s_{12}$ and $s_{22}$ are the respective moments at coordinates 1 and 2 due to a unit rotation at coordinate 2 ($r_2 = 1$). They are illustrated in Fig. 17-5(a) and (b).

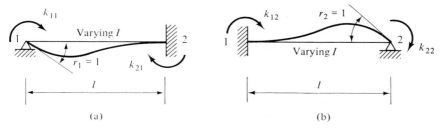

(a)                              (b)

**Fig. 17-5**

To find the integral expression for the rotational stiffness matrix, we need only to perform the inversion for the rotational flexibility matrix given by Eq. 17-17. Thus,

$$[k] = \begin{bmatrix} k_{11} & k_{12} \\ k_{21} & k_{22} \end{bmatrix} = \begin{bmatrix} f_{11} & f_{12} \\ f_{21} & f_{22} \end{bmatrix}^{-1}$$

$$= \begin{bmatrix} \dfrac{1}{EI^2} \displaystyle\int_0^l \dfrac{(l-x)^2\,dx}{I} & -\dfrac{1}{EI^2} \displaystyle\int_0^l \dfrac{(l-x)(x)\,dx}{I} \\ -\dfrac{1}{EI^2} \displaystyle\int_0^l \dfrac{(l-x)(x)\,dx}{I} & \dfrac{1}{EI^2} \displaystyle\int_0^l \dfrac{x^2\,dx}{I} \end{bmatrix}^{-1}$$

$$= \frac{E}{\displaystyle\int_0^l \frac{dx}{I} \int_0^l \frac{x^2\,dx}{I} - \left(\int_0^l \frac{x\,dx}{I}\right)^2} \begin{bmatrix} \displaystyle\int_0^l \dfrac{x^2\,dx}{I} & \displaystyle\int_0^l \dfrac{(l-x)(x)\,dx}{I} \\ \displaystyle\int_0^l \dfrac{(l-x)(x)\,dx}{I} & \displaystyle\int_0^l \dfrac{(l-x)^2\,dx}{I} \end{bmatrix}$$

$$\tag{17-19}$$

Equation 17-19 shows that the stiffness matrix is symmetrical (i. e., $k_{12} = k_{21}$).

For a member of uniform cross section, Eq. 17-19 reduces to

$$[k] = \frac{2EI}{l}\begin{bmatrix} 2 & 1 \\ 1 & 2 \end{bmatrix} \tag{17-20}$$

## 17-5. THE GENERALIZED SLOPE-DEFLECTION EQUATIONS

The slope-deflection equations derived for a typical uniform member 1–2 (see Sec. 12-2) are

$$M_{12} = 2E\frac{I}{l}\left[2\theta_1 + \theta_2 - 3\left(\frac{\Delta}{l}\right)\right] + M_{12}^F \tag{17-21}$$

$$M_{21} = 2E\frac{I}{l}\left[2\theta_2 + \theta_1 - 3\left(\frac{\Delta}{l}\right)\right] + M_{21}^F \tag{17-22}$$

In cases involving a variable moment of inertia, these equations are not valid and some generalized slope deflection equations must be formed. To do this, we recall that the basic slope-deflection equations are derived from the sum of four separate effects:

1. The rigid body translation $\Delta$ between two ends. This is equivalent to a rigid body rotation $\Delta/l$ or $R$ of the whole member.
2. The rotation $(\theta_1 - \Delta/l)$ at end 1 only.
3. The rotation $(\theta_2 - \Delta/l)$ at end 2 only.
4. The application of member loads with end displacements prevented.

These steps are recaptured as in Fig. 17-6 and are also applicable to the instances of nonuniform member.

Now the rigid body motion of step 1 causes no moments. Step 2 produces

$$M_{12} = k_{11}\left(\theta_1 - \frac{\Delta}{l}\right) \tag{17-23}$$

$$M_{21} = k_{21}\left(\theta_1 - \frac{\Delta}{l}\right) \tag{17-24}$$

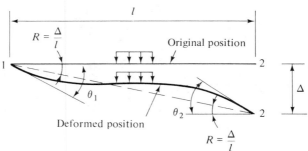

**Fig. 17-6**

Similarly, step 3 will give

$$M_{12} = k_{12}\left(\theta_2 - \frac{\Delta}{l}\right) \tag{17-25}$$

$$M_{21} = k_{22}\left(\theta_2 - \frac{\Delta}{l}\right) \tag{17-26}$$

Note that $k_{11}$, $k_{12}$, $k_{21}$, and $k_{22}$ are the rotational stiffness coefficients defined in Eq. 17-19.

The end moments corresponding to step 4 are the fixed end moments that have been discussed in Sec. 17-2. They are usually denoted by $M_{12}^F$ and $M_{21}^F$ in slope-deflection equations.

Collecting these effects, we arrive at the generalized slope-deflection equations applicable to cases involving variable moment of inertia:

$$M_{12} = k_{11}\theta_1 + k_{12}\theta_2 - (k_{11} + k_{12})\frac{\Delta}{l} + M_{12}^F \tag{17-27}$$

$$M_{21} = k_{21}\theta_1 + k_{22}\theta_2 - (k_{21} + k_{22})\frac{\Delta}{l} + M_{21}^F \tag{17-28}$$

Assembling Eqs. 17-27 and 17-28 in a matrix and using $R = \Delta/l$, we have

$$\begin{Bmatrix} M_{12} - M_{12}^F \\ M_{21} - M_{21}^F \end{Bmatrix} = \begin{bmatrix} k_{11} & k_{12} \\ k_{21} & k_{22} \end{bmatrix} \begin{Bmatrix} \theta_1 - R \\ \theta_2 - R \end{Bmatrix} \tag{17-29}$$

which gives the relationship between member end forces and displacements. For a member of uniform section,

$$\begin{bmatrix} k_{11} & k_{12} \\ k_{21} & k_{22} \end{bmatrix} = \frac{2EI}{l}\begin{bmatrix} 2 & 1 \\ 1 & 2 \end{bmatrix} \tag{17-30}$$

and Eqs. 17-27 and 17-28 reduce to Eqs. 17-21 and 17-22.

With the fixed-end moments and rotational stiffness coefficients for each member determined, application of generalized slope-deflection equations 17-27 and 17-28 in analyzing frames composed of nonuniform members can be carried out by the procedure given in Chapter 12.

## 17-6. THE STIFFNESS AND CARRY-OVER FACTOR
### FOR MOMENT DISTRIBUTION

To develop an expression for stiffness and carry-over factor for a member of varying $I$ in moment distribution, let us recall the definition of rotational stiffness for an end of a member as the end moment required to produce a unit rotation at this end (simple end) while the other end is fixed; and the definition of carry-over factor as the ratio of induced moment at the other end (fixed) to the applied moment at this end. See Fig. 17-7, in which we use the conventional notation $S_{12}$ in moment distribution as the rotational stiffness of end 1 (this end) of member 1–2; $C_{12}$ as the carry-over factor from end 1 to end 2. By defini-

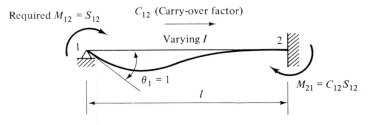

**Fig. 17-7**

tion, $S_{12}$ is the same stiffness coefficient $k_{11}$ given in Eq. 17-19. Thus,

$$S_{12} = \frac{E \int_0^l (x^2\, dx)/I}{\int_0^l dx/I \int_0^l (x^2\, dx)/I - \left(\int_0^l (x\, dx)/I\right)^2} \tag{17-31}$$

The rotational stiffness of end 2 of member 1–2, denoted by $S_{21}$, can similarly be explained and is equal to $k_{22}$ given in Eq. 17-19. Thus,

$$S_{21} = \frac{E \int_0^l (l - x)^2\, dx/I}{\int_0^l dx/I \int_0^l (x^2\, dx)/I - \left(\int_0^l (x\, dx)/I\right)^2} \tag{17-32}$$

Note that for a member of nonuniform cross section, $S_{21}$ is usually not equal to $S_{12}$. For a member of constant cross section,

$$S_{12} = S_{21} = \frac{4EI}{l} \tag{17-33}$$

Again, refer to the setup in Fig. 17-7. The induced moment $M_{21}$ at end 2 is the carry-over moment from end 1 and is therefore equal to $C_{12}S_{12}$ or $C_{12}k_{11}$. But it can also be interpreted as the moment restraint at end 2 due to a unit rotation only at end 1, and therefore is equal to the stiffness coefficient $k_{21}$ in Eq. 17-19. Thus,

$$C_{12} = \frac{k_{21}}{k_{11}} \tag{17-34}$$

Using the expressions of $k_{11}$ and $k_{21}$ in Eq. 17-19, we obtain

$$C_{12} = \frac{\int_0^l (l - x)x\, dx/I}{\int_0^l x^2\, dx/I} \tag{17-35}$$

By the same reasoning,

$$C_{21} = \frac{k_{12}}{k_{22}} \tag{17-36}$$

Using the expressions of $k_{12}$ and $k_{22}$ in Eq. 17-19 gives

$$C_{21} = \frac{\int_0^l (l - x)x \, dx/I}{\int_0^l (l - x)^2 \, dx/I} \tag{17-37}$$

Note that for a member of nonuniform cross section, $C_{12}$ and $C_{21}$ are usually not equal. For a member of constant cross section,

$$C_{12} = C_{21} = \tfrac{1}{2} \tag{17-38}$$

Since $C_{12}S_{12} = k_{21}$, $C_{21}S_{21} = k_{12}$, and $k_{12} = k_{21}$, we reach

$$C_{12}S_{12} = C_{21}S_{21} \tag{17-39}$$

This relationship provides a check on separately computed value of the stiffness and carry-over factors.

The determination of fixed-end moments, stiffnesses, and carry-over factors prerequisite to a moment distribution procedure usually involves a large amount of computation. Fortunately, the values of considerable number of these factors for the more common types of nonprismatic members have been published for the convenience of structural engineers. One such source is the "Handbook of Frame Constants" published by the Portland Cement Association. The member stiffness matrix and therefore flexibility matrix required for a matrix analysis of frames can be deduced from these data by using the relationships

$$\begin{bmatrix} k_{11} & k_{12} \\ k_{21} & k_{22} \end{bmatrix} = \begin{bmatrix} S_{12} & C_{21}S_{21} \\ C_{12}S_{12} & S_{21} \end{bmatrix} = \begin{bmatrix} S_{12} & C_{12}S_{12} \\ C_{12}S_{12} & S_{21} \end{bmatrix} \tag{17-40}$$

## 17-7. FIXED-END MOMENT DUE TO JOINT TRANSLATION

When frame sideway is involved, it is necessary to determine the fixed-end moments due to joint translation for the relevant members before we can carry out a moment distribution. We recall that the fixed-end moment developed at either end of a prismatic member because of relative end displacement equals $-6EI\Delta/l^2$. This is no longer valid for a member of nonuniform cross section.

Consider member 1-2 having varying $I$ subjected to a pure relative end translation $\Delta$ as shown in Fig. 17-8. The moment restraints at end 1 and 2 can

$M_{12} = ?$

$\theta_1 = \theta_2 = 0$　　$M_{21} = ?$

**Fig. 17-8**

readily be obtained by setting $\theta_1 = \theta_2 = M^F_{12} = M^F_{21} = 0$ in the generalized slope-deflection equations 17-27 and 17-28. Thus,

$$M_{12} = -(k_{11} + k_{12})\frac{\Delta}{l} \tag{17-41}$$

$$M_{21} = -(k_{21} + k_{22})\frac{\Delta}{l} \tag{17-42}$$

in which $k_{11}, k_{12}, k_{21}$, and $k_{22}$ are stiffness coefficients defined by Eq. 17-19. Equations 17-41 and 17-42 in terms of the conventional notation of moment distribution are given as

$$M_{12} = -(S_{12} + C_{21}S_{21})\frac{\Delta}{l}$$
$$= -S_{12}(1 + C_{12})\frac{\Delta}{l} \tag{17-43}$$

$$M_{21} = -(S_{21} + C_{12}S_{12})\frac{\Delta}{l}$$
$$= -S_{21}(1 + C_{21})\frac{\Delta}{l} \tag{17-44}$$

by using the relationship $C_{12}S_{12} = C_{21}S_{21}$. For member of uniform cross section, $S_{12} = S_{21} = 4EI/l$, $C_{12} = C_{21} = \frac{1}{2}$, and the expressions above reduce to

$$M_{12} = M_{21} = -\frac{6EI\Delta}{l^2} \tag{17-45}$$

If end 2 is hinged, the modified fixed-end moment at 1, called $M'_{12}$ resulting from the relative end translation $\Delta$, can be found by first assuming both ends fixed and subsequently restoring end 2 to its original hinged condition. Thus,

$$M'_{12} = M_{12} - C_{21}M_{21}$$

Using Eqs. 17-43 and 17-44, we obtain

$$M'_{12} = -S_{12}(1 - C_{12}C_{21})\frac{\Delta}{l} \tag{17-46}$$

For a member of uniform cross section, the expression above reduces to

$$M'_{12} = -\frac{3EI\Delta}{l^2} \tag{17-47}$$

as previously found.

### 17-8. MODIFIED STIFFNESS FOR MOMENT DISTRIBUTION

The modified stiffness of this end of a member may be defined as the end moment required to produce this end (simple end) a unit rotation while the other end remains in the actual condition other than being fixed.

*Case* 1 The modified stiffness for end 1 of member 1–2, if end 2 is simply supported, is given by

$$S'_{12} = S_{12}(1 - C_{12}C_{21})   \tag{17-48}$$

in which $S'_{12}$ denotes the modified stiffness and $S_{12}$ is the stiffness found by the usual manner. The equation above can be proved as follows.

By definition, the setup in Fig. 17-9(a) gives the configuration for finding the modified stiffness for end 1 of member 1–2, the other end being simply supported. To accomplish this, we may break member 1–2, down into two separate steps, as shown in Fig. 17-9(b) and (c). In Fig. 17-9(b) we temporarily lock end 2 against rotation ($\theta_2 = 0$). A moment $S_{12}$ applied at end 1 will produce a unit rotation at 1 and induce a carry-over moment of $C_{12}S_{12}$ at 2. In Fig. 17-9(c) we release end 2 to its actual condition of zero moment and at the same time lock end 1 against further rotation. A moment of $-C_{12}S_{12}$ must be developed at end 2 and, consequently, $-C_{21}C_{12}S_{12}$ will be carried over to end 1. The sum of foregoing two steps for end 1 gives

$$S'_{12} = S_{12} - C_{21}C_{12}S_{12}$$
$$= S_{12}(1 - C_{12}C_{21})$$

as asserted.

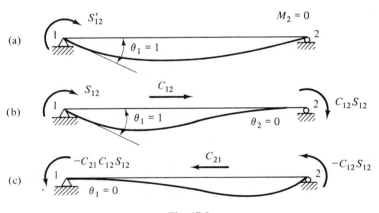

**Fig. 17-9**

For a prismatic member $C_{12} = C_{21} = \frac{1}{2}$; therefore,

$$S'_{12} = \tfrac{3}{4}S_{12}   \tag{17-49}$$

*Case* 2 The modified stiffness for end 1 of member 1–2, if end 2 rotates an equal but opposite angle to that of end 1 as in the case of symmetry, is given by

$$S'_{12} = S_{12}(1 - C_{12})   \tag{17-50}$$

To prove this, we refer to Fig. 17-10(a), which is postulated according to the definition of modified stiffness for the present case.

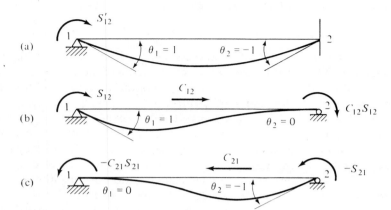

**Fig. 17-10**

As before, we break this into two separate steps as shown in Fig. 17-10(b) and (c). Figure 17-10(b) shows the usual way of determining $S_{12}$. In Fig. 17-10(c) a moment of $-S_{21}$ is applied at end 2 necessary to bring it back to its actual position. Consequently, a moment of $-C_{21}S_{21}$ is carried over to end 1. The sum of the results of these two steps for end 1 gives

$$S'_{12} = S_{12} - C_{21}S_{21}$$

On substituting $C_{12}S_{12}$ for $C_{21}S_{21}$ in the expression above, we obtain

$$S'_{12} = S_{12}(1 - C_{12})$$

as asserted.

For a prismatic member $C_{12} = \frac{1}{2}$; therefore,

$$S'_{12} = \tfrac{1}{2}S_{12} \tag{17-51}$$

*Case* 3 The modified stiffness of end 1 of member 1–2, if end 2 rotates an angle equal to that of end 1 as in the case of antisymmetry, is given by

$$S'_{12} = S_{12}(1 + C_{12}) \tag{17-52}$$

To prove this, we refer to Fig. 17-11(a) for the setup of $S'_{12}$. As before, this may be considered as the superposition of two separate cases, as shown in Fig. 17-11(b) and (c). Consequently,

$$S'_{12} = S_{12} + C_{21}S_{21}$$

On substituting $C_{12}S_{12}$ for $C_{21}S_{21}$, we obtain

$$S'_{12} = S_{12}(1 + C_{12})$$

as asserted.

For a prismatic member $C_{12} = \frac{1}{2}$; therefore,

$$S'_{12} = \tfrac{3}{2}S_{12} \tag{17-53}$$

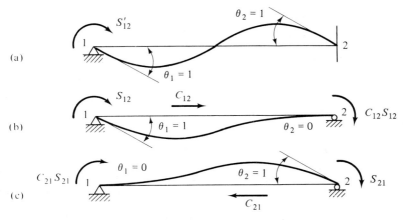

**Fig. 17-11**

## 17-9. A NUMERICAL SOLUTION

One of the most frequently used methods of approximate integration is the Simpson's one-third rule. Consider an integral $\int f(x)\,dx$ between the limit $a$ and $b$. If the integral from $x = a$ to $x = b$ is divided into $n$ equal parts, where $n$ is an even number, and if $y_0, y_1 \ldots, y_{n-1}, y_n$ are the ordinates of the curve $y = f(x)$ at these points of subdivision (Fig. 17-12), then according to Simpson's one-third rule,

$$\int_a^b f(x)\,dx \approx \frac{\Delta}{3}(y_0 + 4y_1 + 2y_2 + 4y_3 + \cdots + 2y_{n-2} + 4y_{n-1} + y_n)$$

$$(17\text{-}54)$$

where $\Delta = (b - a)/n$. Simpson's rule gives accurate result if $f(x)$ is a linear, quadratic, or cubic function.

To illustrate, let us solve the rotational flexibility matrix of a beam element

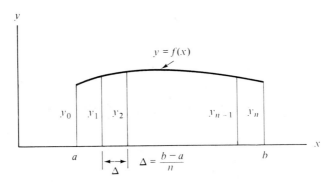

**Fig. 17-12**

expressed in Eq. 17-17. Consider the coefficient $f_{22}$,

$$\frac{1}{EI^2} \int_0^l \frac{x^2 \, dx}{I} \qquad [I = I(x)]$$

We divide the length $l$ into 10 equal segments, each with $0.1l$. An application of Eq. 17-54 yields

$$f_{22} = \frac{1}{EI^2} \int_0^l \frac{x^2 \, dx}{I}$$

$$= \frac{0.1l}{3E} \left[ \frac{(1)(0)^2}{I_0} + \frac{(4)(0.1)^2}{I_1} + \frac{(2)(0.2)^2}{I_2} + \frac{(4)(0.3)^2}{I_3} + \frac{(2)(0.4)^2}{I_4} + \frac{(4)(0.5)^2}{I_5} \right.$$

$$\left. + \frac{(2)(0.6)^2}{I_6} + \frac{(4)(0.7)^2}{I_7} + \frac{(2)(0.8)^2}{I_8} + \frac{(4)(0.9)^2}{I_9} + \frac{(1)(1)^2}{I_{10}} \right] \qquad (17\text{-}55)$$

with $I_0 = I(0)$, $I_1 = I(0.1l)$, $I_2 = I(0.2l)$, and so on. In matrix form

$$f_{22} = \frac{0.1l}{3E}[I_0^{-1} \quad I_1^{-1} \quad I_2^{-1} \quad I_3^{-1} \quad I_4^{-1} \quad I_5^{-1} \quad I_6^{-1} \quad I_7^{-1} \quad I_8^{-1} \quad I_9^{-1} \quad I_{10}^{-1}] \begin{Bmatrix} 0 \\ 0.04 \\ 0.08 \\ 0.36 \\ 0.32 \\ 1 \\ 0.72 \\ 1.96 \\ 1.28 \\ 3.24 \\ 1 \end{Bmatrix}$$

$$(17\text{-}56)$$

In the same manner, we obtain

$$f_{11} = \frac{1}{EI^2} \int_0^l \frac{(l - x)^2 \, dx}{I}$$

$$= \frac{0.1l}{3E} \left[ \frac{(1)(1)^2}{I_0} + \frac{(4)(0.9)^2}{I_1} + \frac{(2)(0.8)^2}{I_2} + \frac{(4)(0.7)^2}{I_3} + \frac{(2)(0.6)^2}{I_4} + \frac{(4)(0.5)^2}{I_5} \right.$$

$$\left. + \frac{(2)(0.4)^2}{I_6} + \frac{(4)(0.3)^2}{I_7} + \frac{(2)(0.2)^2}{I_8} + \frac{(4)(0.1)^2}{I_9} + \frac{(1)(0)^2}{I_{10}} \right]$$

$$= \frac{0.1l}{3E}[I_0^{-1} \quad I_1^{-1} \quad I_2^{-1} \quad I_3^{-1} \quad I_4^{-1} \quad I_5^{-1} \quad I_6^{-1} \quad I_7^{-1} \quad I_8^{-1} \quad I_9^{-1} \quad I_{10}^{-1}] \begin{Bmatrix} 1 \\ 3.24 \\ 1.28 \\ 1.96 \\ 0.72 \\ 1 \\ 0.32 \\ 0.36 \\ 0.08 \\ 0.04 \\ 0 \end{Bmatrix}$$

$$(17\text{-}57)$$

$$f_{12} = f_{21} = -\frac{1}{EI^2} \int_0^l \frac{(l-x)(x)\,dx}{I}$$

$$= -\frac{0.1l}{3E}\left[ \frac{(1)(1)(0)}{I_0} + \frac{(4)(0.9)(0.1)}{I_1} + \frac{(2)(0.8)(0.2)}{I_2} + \frac{(4)(0.7)(0.3)}{I_3} \right.$$

$$+ \frac{(2)(0.6)(0.4)}{I_4} + \frac{(4)(0.5)^2}{I_5} + \frac{(2)(0.4)(0.6)}{I_6} + \frac{(4)(0.3)(0.7)}{I_7}$$

$$\left. + \frac{(2)(0.2)(0.8)}{I_8} + \frac{(4)(0.1)(0.9)}{I_9} + \frac{(1)(0)(1)}{I_{10}} \right]$$

$$= \frac{0.1l}{3E}[I_0^{-1} \quad I_1^{-1} \quad I_2^{-1} \quad I_3^{-1} \quad I_4^{-1} \quad I_5^{-1} \quad I_6^{-1} \quad I_7^{-1} \quad I_8^{-1} \quad I_9^{-1} \quad I_{10}^{-1}]$$

$$\cdot \begin{Bmatrix} 0 \\ -0.36 \\ -0.32 \\ -0.84 \\ -0.48 \\ -1 \\ -0.48 \\ -0.84 \\ -0.32 \\ -0.36 \\ 0 \end{Bmatrix}$$

$$(17\text{-}58)$$

If we let $c_{11} = f_{11}$, $c_{12} = f_{12}$, $c_{21} = f_{22}$, and $c_{22} = f_{21}$, we have

$$
\begin{bmatrix} c_{11} & c_{12} \\ c_{21} & c_{22} \end{bmatrix}
$$

$$
= \frac{0.1l}{3E} \begin{bmatrix} I_0^{-1} & I_1^{-1} & I_2^{-1} & I_3^{-1} & I_4^{-1} & I_5^{-1} & I_6^{-1} & I_7^{-1} & I_8^{-1} & I_9^{-1} & I_{10}^{-1} \\ I_{10}^{-1} & I_9^{-1} & I_8^{-1} & I_7^{-1} & I_6^{-1} & I_5^{-1} & I_4^{-1} & I_3^{-1} & I_2^{-1} & I_1^{-1} & I_0^{-1} \end{bmatrix} \cdot \begin{bmatrix} 1 & 0 \\ 3.24 & -0.36 \\ 1.28 & -0.32 \\ 1.96 & -0.84 \\ 0.72 & -0.48 \\ 1 & -1 \\ 0.32 & -0.48 \\ 0.36 & -0.84 \\ 0.08 & -0.32 \\ 0.04 & -0.36 \\ 0 & 0 \end{bmatrix}
$$

or simply

$$[C] = [A][B] \tag{17-59}$$

For a prismatic member, the $I$-value for each segment is constant. Therefore,

$$
\begin{bmatrix} c_{11} & c_{12} \\ c_{21} & c_{22} \end{bmatrix} = \frac{l}{6EI} \begin{bmatrix} 2 & -1 \\ 2 & -1 \end{bmatrix} \tag{17-60}
$$

That is, the rotational flexibility matrix becomes

$$
\begin{bmatrix} f_{11} & f_{12} \\ f_{21} & f_{22} \end{bmatrix} = \frac{l}{6EI} \begin{bmatrix} 2 & -1 \\ -1 & 2 \end{bmatrix}
$$

as expected.

With element rotational flexibility coefficients determined, the element rotational stiffness coefficients as well as other constants can readily be derived.

# PROBLEMS

**17-1.** Figure 17-13 shows a beam of varying cross section with both ends fixed and subjected to the bending action of a concentrated load $P$. Find the integral expressions for the fixed-end moments.

**Fig. 17-13**

**17-2.** For the beam element of constant width shown in Fig. 17-14, find, by the numerical method, the rotational flexibility matrix. Assume that the element is simply supported.

**Fig. 17-14**

**17-3.** Use the result of Prob. 17-2 to find the rotational stiffness matrix for the same element.

**17-4.** For the haunched member of constant width shown in Fig. 17-15, find the stiffnesses and carry-over factors by the numerical method or from the table of frame constants if available.

**Fig. 17-15**

**17-5.** Using the result of Prob. 17-4, give the rotational stiffness and flexibility matrices for the element shown in Fig. 17-15.

**17-6.** Use the numerical method or table of frame constants to determine the fixed-end moments for the haunched beam of Fig. 17-15 due to a uniform load of 5 kN/m over the entire span.

**17-7.** Use the relevant calculation from Probs. 17-4 to 17-6 to find, by moment distribution, the end moments for the beam in Fig. 17-16 subjected to a uniform load of 5 kN/m over the entire span.

**Fig. 17-16**

**17-8.** Repeat Prob. 17-7 by the method of slope deflection.

**17-9.** Repeat Prob. 17-7 by the matrix force method.

**17-10.** Repeat Prob. 17-7 by the matrix displacement method.

# ANSWERS TO SELECTED PROBLEMS

**CHAPTER 2**

2-1. (a) Stable and indeterminate to the fifth degree
(b) Unstable
(c) Stable and determinate
(d) Stable and indeterminate to the second degree

2-2. (b) Stable and indeterminate to the third degree
(c) Unstable externally
(d) Unstable internally

2-3. (b) Stable and indeterminate to the fifth degree
(e) Stable and indeterminate to the fourth degree
(f) Unstable externally
(g) Stable and indeterminate to the 102nd degree

**CHAPTER 3**

3-2. $R_A$ = 8.33 kips (down)    $R_B$ = 23.33 kips (up)
$R_D$ = 5 kips (up)         $M_B$ = −80 ft-kips
$V_C$ = 5 kips

3-3. $R_A$ = 7.5 kN(up)     $M_A$ = 2.5 kN·M
$R_E$ = 22.5 kN(up)    $V_B$ = 7.5 kN
$M_B$ = 17.5 kN·m

**CHAPTER 4**

4-1. (e) Simple    (g) Simple    (h) Complex
The rest are compound trusses.

4-2. (b) $S_{ab}$ = $S_{bc}$ = +14.6 kN     $S_{aB}$ = +9.1 kN
         $S_{Bb}$ = +17.5 kN              $S_{Bc}$ = -18.6 kN

(d) $S_{bc}$ = $S_{cd}$ = $S_{de}$ = $S_{ef}$ = 0
    $S_{ag}$ = $-S_{bg}$ = 26.9 kN

(e) $S_{ab}$ = $S_{bc}$ = 0     $S_{aB}$ = -11.3 kN
    $S_{Bc}$ = -14.4 kN

4-3. (a) $S_a$ = -32 kips     $V_b$ = +12 kips     $S_c$ = +24 kips
     (b) $H_a$ = +18.75 kips     $V_b$ = -15 kips     $S_c$ = -6.25 kips
     (c) $V_a$ = +11.6 kips     $H_b$ = +166.7 kips     $S_c$ = -20 kips
     (d) $S_a$ = +20 kips     $V_b$ = +100 kips     $V_c$ = -45 kips
     (e) $S_a$ = -50 kips

4-4. $V_a$ = -20 kips     $V_b$ = +10 kips     $V_c$ = -20 kips     $S_d$ = 0

4-5. $S_{AB}$ = $S_{BC}$ = $S_{CD}$ = $S_{BF}$ = $S_{CF}$ = 0     $S_{AD}$ = +15.4 kips
     $S_{AE}$ = +11.2 kips     $S_{DE}$ = -13.6 kips     $S_{EF}$ = -10 kips

4-5. $S_{ad}$ = +7.85 kN


# CHAPTER 5

5-1. (d) $M_c$ = +108 ft-kips
     (e) $M_b$ = +31.25 ft-kips
     (f) $M_b$ = -144 ft-kips

5-2. (a) $M_c$ = +33.3 kN·m
     (b) $M_b$ = +71.4 kN·m

5-4. $S_{de}$ = 32 kN

5-6. $S_{ch}$ = $S_{hi}$ = $S_{ei}$ = 0     $S_{cd}$ = -15 kips
     $S_{de}$ = +5 kips     $V_{bd}$ = $-V_{df}$ = 8.33 kips


# CHAPTER 6

6-1. $V$ = 49.33 kips     $M$ = 458.67 ft-kips

6-4. $R_A$ = 45 kN     $R_C$ = 175 kN     $R_E$ = 30 kN
     $V_B$ = -55 kN     $M_B$ = -90 kN·m     $M_C$ = -180 kN·m


# CHAPTER 7

7-1. (a) $R$ = 22.78 kips
     (b) $V$ = 14.44 kips     $M$ = 216.67 ft-kips

7-2. $V_{1-2}$ = 27.5 kips    $M_2$ = 690 ft-kips

7-3. (a) $V$ = 83.75 kips
   (b) $V$ = 57.5 kips
   (c) $M$ = 975 ft-kips
   (d) $M$ = 978.5 ft-kips

7-4. $V_a$ = +29.68 kips (maximum tension)
   $V_a$ = -4.4 kips (maximum compression)
   $S_b$ = +52.75 kips

7-5. $R_C$ = 156 kN    $M_D$ = 111 kN·m

7-6. Max. $M$ = 208.1 kN·m

7-7. $M_E$ = 135 kN·m    $R_E$ = 55 kN

# CHAPTER 8

8-1. $\Delta_c$ = $5wl^4/384EI$ (down)    $\theta_a$ = $wl^3/24EI$ (clockwise)

8-2. $\Delta_p$ = $0.0147Pl^3/EI$ (down)    $\Delta$(at midspan) = $0.0236Pl^3/EI$ (down)

8-3. $\Delta$ = 0.54 in. (down)    $\theta$ = 0.006 rad (counterclockwise)

8-4. $\Delta_h$ = 0.43 ft (right)    $\Delta_v$ = 0.696 in. (down)
   $\Delta_r$ = 0.006 rad (counterclockwise)

8-5. $\theta_b$ = 0.0023 radian (clockwise)    $\Delta_b$ = 0.375 cm (down)
   $\theta_c$ = 0.0038 radian (clockwise)    $\Delta_c$ = 1.35 cm (down)

8-6. $\Delta_1$ = 1.15 cm (right)    $\Delta_2$ = 0.59 cm (down)
   $\Delta_3$ = 0.0039 radian (counterclockwise)

8-7. $\Delta_1$ = 0.1 cm (right)    $\Delta_2$ = 0.12 cm (down)

8-8. (a) $\Delta_B$ = 0.00746 ft (down)
   (b) $\Delta_C$ = 0.00278 ft (right)
   (c) $\Delta_{bC}$ = 0.0002 ft (toward each other)
   (d) $\theta$ = 0.000433 rad (clockwise)

8-20. $\theta_c$ = 0.003 radian (counterclockwise, at the left side)
   $\theta_c$ = 0.0017 radian (clockwise, at the right side)
   $\Delta_b$ = 0.267 cm (down)

# CHAPTER 9

9-1. (a) $R_b$ = 50 kips (up)
   (b) $M_b$ = -100 ft-kips

9-2.   $R_b$ = 10 kips (up)

9-3.   $R_b$ = 7.28 kips (up)

9-4.   $R_e$ = 0.916 kN (up)

9-5.   (b) $M_a$ = wm $(m^4 + 4mn^3 + 3n^4)/8(m^3 + n^3)$

9-6.   $H_c$ = 15.24 kN (left)

9-7.   $H_a$ = 1.25 kips (right)    $V_a$ = 5 kips (up)
        $M_a$ = 4.16 ft-kips (clockwise)

9-9.   $S_{bc}$ = 5.25 kN

9-10.  (a) $S_{bC}$ = -0.75 kip
      (b) $H_d$ = 30 kips (left)
      (c) $S_{bC}$ = -5.4 kips    $H_d$ = 31.1 kips (left)

9-11.  $S_{Bc}$ = $S_{bC}$ = +8.75 kips


# CHAPTER 10

10-10.  $S_{ac}$ = +4.25 kips    $M_a$ = -3.25 ft-kips
        The effect of axial force in beam is neglected.

10-11.  21.75 kips, 17.4 kips, and 13.05 kips

10-12.  $S_{ab}$ = 13.5 kN    $S_{ad}$ = -15 kN    $S_{cd}$ = 0


# CHAPTER 11

11-1.   The influence ordinates are 1.3, 1, 0.704, 0.432, 0.208, 0.056, 0

11-2.   The influence ordinates are 0, 2.09, 2.67, 2.25, 1.33, 0.417, 0

11-3.   (a) The influence ordinates are 1, 0.578, 0.222, 0, -0.05, 0
      (b) The influence ordinates are 0, -0.532, -0.668, 0, -0.30, 0
      (c) The influence ordinates are 0, -0.422, 0.222, 0, -0.05, 0

11-5.   (a) The influence ordinates for $S_{cd}$ are 0, 0.320, 0.666, 0.293, 0,
         -0.039, -0.038, 0
      (b) $S_{BC}$ = $-S_{cd}$


# CHAPTER 12

12-1.  $M_{ab}$ = -22 ft-kips    $M_{ba}$ = $-M_{bc}$ = 28 ft-kips
      $M_{cb}$ = 31 ft-kips

12-2.  $M_{ab}$ = -23.55 ft-kips    $M_{ba}$ = 16.89 ft-kips
      $M_{bd}$ = -8.89 ft-kips    $M_{db}$ = -4.44 ft-kips

12-3.   $M_{ab}$ = -22.1 ft-kips     $M_{ba}$ = -$M_{bc}$ = 68.3 ft-kips
$M_{cb}$ = 20 ft-kips

12-5.   $M_{ab}$ = -44.8 ft-kips     $M_{ba}$ = -$M_{bc}$ = -34.4 ft-kips

12-6.              $M_{ab}$ = 4.9 ft-kips     $M_{ba}$ = -$M_{bc}$ = -0.9 ft-kip
$M_{cb}$ = - $M_{cd}$ = 2.24 ft-kips          $M_{dc}$ = -1.76 ft-kips

12-7.   Spring force = 3.58 kN (compression)

12-8.   $R_c$   = 7.35 kN     $M_{ab}$ = -3.26 kN·m

12-9.   $M_{ab}$ = -0.63 kN·m     $M_{ba}$ = -$M_{bc}$ = 2.49 kN·m
$M_{dc}$ = -1.87 kN·m

12-10.  $R_d$   = 3.03 kN

12-11.  (a)            $M_{ab}$ = -14.9 ft-kips     $M_{ba}$ = - $M_{bc}$ = 14.9 ft-kips
$M_{cb}$ = 39.4 ft-kips          $M_{ce}$ = -29.8 ft-kips

(b)            $M_{ab}$ = -120.8 ft-kips     $M_{ba}$ = -$M_{bc}$ = -112.6 ft-kips
$M_{cb}$ = -$M_{cd}$ = 104 ft-kips          $M_{dc}$ = -103.6 ft-kips

(c)            $M_{ab}$ = 8.52 ft-kips     $M_{ba}$ = - $M_{bc}$ = 9.58 ft-kips
$M_{cb}$ = -$M_{cd}$ = -1 ft-kip          $M_{dc}$ = 26.74 ft-kips

(d) $M_{ba}$ = $M_{ef}$ = -150 ft-kips     $M_{be}$ = $M_{eb}$ = 156.25 ft-kips
$M_{bc}$ = $M_{ed}$ = -6.25 ft-kips     $M_{cb}$ = $M_{de}$ = -43.75 ft-kips
$M_{cd}$ = $M_{dc}$ = 43.75 ft-kips

(e)            $M_{ab}$ = -118.1 ft-kips     $M_{ba}$ = - $M_{bc}$ = -82 ft-kips
$M_{cb}$ = -$M_{cd}$ = 11.5 ft-kips     $M_{dc}$ = -$M_{de}$ = 25.8 ft-kips
$M_{ed}$ = -74.1 ft-kips

12-12.  (a) $M_{ac}$ = 18.5 kN·m     $M_{bc}$ = 14.2 kN·m
(b) $M_{ac}$ = -$M_{bc}$ = -24 kN·m

12-15.  $M_{ab}$ = -41.25 kN·m     $M_{ba}$ = -$M_{bc}$ = -6.75 kN·m

## CHAPTER 13

13-5.   End moment = 14.6 ft-kips

13-6.   Moments at interior supports are 36.5 ft-kips and 27.3 ft-kips

13-7.   $M_{ba}$ = -$M_{bc}$ = 11.4 ft-kips     $M_{cb}$ = -$M_{cd}$ = 0.6 ft-kip

13-8.   $R_c$ = 57.27 kN

13-9.   $K_{ab}$ = $3EIl_1{}^2/(l_1{}^3 + l_2{}^3)$

## CHAPTER 14

14-1.  $M_{ba}$ = -$M_{bc}$ = $\dfrac{57wl^2}{816}$ - $\dfrac{288EI\Delta}{17l^2}$

14-2.  (a) $M_{bc} = 61.2$ kN·m

(b) $M_{bc} = -33.8$ kN·m

(c) $M_{bc} = -26.7$ kN·m

(d) $M_{bc} = 33.9$ kN·m

14-4.  End moment $= 22.10$ ft-kips

14-6.      $M_{ad} = 50.9$ ft-kips        $M_{bc} = -12.2$ ft-kips

$M_{cg} = -35.6$ ft-kips    $M_{da} = -M_{de} = 85.8$ ft-kips

$M_{ed} = 119.8$ ft-kips        $M_{eb} = -48.5$ ft-kips

$M_{ef} = -71.3$ ft-kips    $M_{fe} = -M_{fg} = 108.8$ ft-kips

$M_{gf} = -M_{gc} = 78.1$ ft-kips

14-7.  The influence ordinates for the left half of structure are 0, 0.268, 0.428, 0.375, 0, −0.697, −1.286, −1.232, 0

## CHAPTER 15

15-1.   $r_p = 5PL^3/6EI$ (down)

15-3.   $r$(under 10 kips load) $= 0.01608$ ft (down)

$r$(under 20 kips load) $= 0.00548$ ft (down)

15-4.   $r_1 = (4L^3R_1 - L^2R_2)/6EI$      $r_2 = (-L^2R_1 + 2LR_2)/6EI$

15-5.   Deflection $= 5PL^3/6EI$      Slope $= PL^2/EI$

15-6.   
$$\begin{Bmatrix} Q^a \\ Q^b \\ Q^c \end{Bmatrix} = \begin{Bmatrix} 3.36 \\ 8.28 \\ 2.76 \end{Bmatrix} \text{ kips} \qquad \begin{Bmatrix} r_1 \\ r_2 \end{Bmatrix} = \frac{1}{AE} \begin{Bmatrix} 83.2 \\ -15.72 \end{Bmatrix} \text{ ft}$$

15-7.   $S = -6.725$ kN

15-9.   $Q_i^a = -44$ kN·m      $Q_j^a = -10$ kN·m      $Q_j^c = -23.3$ kN·m

15-10.  slope $= \dfrac{1}{96EI} (wL^3 + \dfrac{51\, wL^3}{1 + L^3/3EIf_s})$

deflection $= \dfrac{17\, wL^4}{48EI(1 + L^3/3EIf_s)}$

15-11.  $Q_i^a = 0.080LR_1 - 0.236LR_2 - 0.274R_3$

$Q_j^d = -0.094LR_1 - 0.392LR_2 + 0.030R_3$

## CHAPTER 16

16-5.   The moment at the fixed support $= -11.25$ kN·m

16-7.   $M_B = wL^2/10$

16-8.    $M_{AB}$ = 8 ft-kips    $M_{BA}$ = 16 ft-kips

16-9.    $M_{AB}$ = 31 ft-kips    $M_{BA}$ = 62 ft-kips
$M_{BE}$ = 93.2 ft-kips    $M_{EB}$ = 46.6 ft-kips
$M_{BC}$ = -155.2 ft-kips

16-10.

$$a = \begin{bmatrix} \dfrac{2EI}{L} & 0 & -\dfrac{6EI}{L^2} & 0 & 0 \\[2mm] \dfrac{4EI}{L} & 0 & -\dfrac{6EI}{L^2} & 0 & 0 \\[2mm] 0 & \dfrac{4EI}{L} & \dfrac{6EI}{L^2} & \dfrac{2EI}{L} & -\dfrac{6EI}{L^2} \\[2mm] 0 & \dfrac{2EI}{L} & \dfrac{6EI}{L^2} & \dfrac{4EI}{L} & -\dfrac{6EI}{L^2} \\[2mm] 0 & 0 & 0 & 0 & \dfrac{AE}{L} \end{bmatrix}$$

16-11.    $Q_i^a$ = -46.48 ft-kips    $Q_j^a = -Q_i^b$ = -42.25 ft-kips
$Q_j^b = -Q_i^c$ = 33.80 ft-kips    $Q_j^c$ = -30.98 ft-kips

16-12.

$$K = \frac{EI}{L} \begin{bmatrix} 0.243 & -0.150 & 0.117 \\ -0.150 & 8 & 2 \\ 0.117 & 2 & 6.667 \end{bmatrix}$$

## CHAPTER 17

17-2.  $f = \dfrac{l}{3EI} \begin{bmatrix} 0.73 & -0.27 \\ -0.27 & 0.40 \end{bmatrix}$

17-3.  $k = \dfrac{EI}{l} \begin{bmatrix} 5.48 & 3.70 \\ 3.70 & 10 \end{bmatrix}$

17-4.  $C_{ab}$ = 0.743    $C_{ba}$ = 0.462
$S_{ab}$ = 4.62 $EI_a/l$    $S_{ba}$ = 7.43 $EI_a/l$

17-5.  $k = \dfrac{EI_a}{l} \begin{bmatrix} 4.62 & 3.43 \\ 3.43 & 7.43 \end{bmatrix}$

$f = \dfrac{l}{EI_a} \begin{bmatrix} 0.329 & -0.152 \\ -0.152 & 0.205 \end{bmatrix}$

17-6.  $M_{ab}^F$ = -7.5 kN·m    $M_{ba}^F$ = 14.4 kN·m

17-7.  $M_{ba}$ = 20.3 kN·m    $M_{cb}$ = 22.3 kN·m

# INDEX